Symmetry and Collective Fluctuations in Evolutionary Games

Symmetry and Collective Fluctuations in Evolutionary Games

Eric Smith
Santa Fe Institute, New Mexico, USA and George Mason University, Virginia, USA

Supriya Krishnamurthy
Stockholm University, Stockholm, Sweden

IOP Publishing, Bristol, UK

ISBN 978-0-7503-1137-3 (ebook)
ISBN 978-0-7503-1138-0 (print)
ISBN 978-0-7503-1139-7 (mobi)

DOI 10.1088/978-0-7503-1137-3

Version: 20150101

IOP Expanding Physics
ISSN 2053-2563 (online)
ISSN 2054-7315 (print)

British Library Cataloguing-in-Publication Data: A catalogue record for this book is available from the British Library.

Published by IOP Publishing, wholly owned by The Institute of Physics, London

IOP Publishing, Temple Circus, Temple Way, Bristol, BS1 6HG, UK

US Office: IOP Publishing, Inc., 190 North Independence Mall West, Suite 601, Philadelphia, PA 19106, USA

Contents

Preface

Today evolutionary population models are being applied to a wide variety of problems, reaching from traditional organism competition and reproduction upward to include demes and species, and downward into cellular and molecular dynamics and to mechanisms responsible for development and regulation. Evolutionary modeling has expanded laterally into domains of social and cultural dynamics, economics, and theories of learning and optimal inference. The expanding scope of evolutionary explanations and metaphors has led to renewed interest in the nature, and the origin in real systems, of abstractions such as individuality, which are the foundation for Darwinian dynamics. Improved understanding of the role of partial autonomy and competition in creating developmental programs, and of group-level coordinated action in ecosystems, has led to efforts to incorporate modern understanding of development and ecology more integrally within the population-based framework that formalizes evolutionary dynamics.

The statistical sophistication of evolutionary modeling has also increased in the past two decades. Topics of interest include multiple forms and levels of individuality, dynamics at many scales of time or of aggregation, strong selection and feedbacks through population states, interactions among multiple genes or multiple criteria of selection and a more statistical approach to the gene concept itself. The wish to apply evolutionary models as more than proofs of concept—as falsifiable quantitative theories of the causes of order—has also led to efforts toward less parametric and hence less biased model selection, and to inference from incompletely specified models.

These developments create collaborative opportunities for physics, as biologists, economists and others are independently reformulating basic concepts of entity, agency and interaction in statistical terms, along lines similar to those that condensed matter and quantum mechanics followed for particles and forces in the latter half of the 20th century, as well as studying new kinds of dynamical order in populations that have no direct analogs in equilibrium. In parallel, the past decade especially has seen significant advances in the large-deviations theory of non-equilibrium stochastic processes, by both mathematicians and physicists. While technical difficulties abound, it is becoming possible to speak of a coherent non-equilibrium thermodynamics based on sound first principles of path ensembles and path entropies, and in the process to understand equilibrium statistical mechanics in terms that are less dependent on mechanics and more plainly rooted in inference.

In this monograph we bring together a conceptual treatment of evolutionary dynamics and a path-ensemble approach to non-equilibrium stochastic processes. Our framework is evolutionary game theory, in which the map from individual types and their interactions to the fitness that determines their evolutionary success is modeled as a game played among agents in the population. Our approach, however, is not anchored either in analogy to play or in motivations to interpret particular interactions as games. Rather, we argue that games are a flexible and reasonably generic framework to capture, classify and analyze the processes in development and some forms of inter-agent interaction that lie behind arbitrary frequency-dependent fitness models. Games are generic in evolutionary dynamics in the same way as abstractions such as the individual or the gene are; the scientific problem is to decompose their structure and understand what dynamics different structures imply.

Readers familiar with the conceptual shifts in condensed matter physics and quantum field theory in the 20th century will anticipate that symmetry and collective fluctuations will be central to determining which distinctions among games matter qualitatively, and their consequences for dynamics. We explain why symmetry plays this role and illustrate with several examples that are in some ways similar to cases in equilibrium thermodynamics and in other ways conceptually new and intrinsically dynamical.

The presentation is meant to introduce quantitative methods while emphasizing the concepts they capture rather than mere computational technique. The mathematical development is self-contained and all results can be reproduced by the reader from the inputs provided, mostly with elementary methods. We do not, however, provide pedagogical introductions to basic ideas of evolutionary dynamics, game theory, or statistical mechanics. Readers who have had an introductory exposure to each of these topics will find the material here more familiar and intuitive than it will be to those for whom this is a first exposure.

Our work grew from more than a decade of stimulating exchange with colleagues and friends at the Santa Fe Institute, including Cosma Shalizi, David Krakauer, Steve Frank, Martin Shubik, Jessica Flack, Walter Fontana, Doug Erwin, Martin Nowak, Duncan Foley, Jeremy van Cleve and Sam Bowles. Each of them, through years of patience, guidance to literature and shared work, explained to us aspects of evolutionary theory, stochastic processes, game theory, or the scientific problems that were most central in applying these ideas to a variety of disciplines.

Early stages of work by ES were carried out under the generous hospitality of Martin Nowak and the Harvard Program for Evolutionary Dynamics, and the last year was hosted by George Mason University. ES also acknowledges two month-long visits to Stockholm in 2012 (hosted by the Theoretical Computer Science department of the Royal Institute of Technology (KTH)) and in 2014 (hosted by the KTH ACCESS Linnaeus Center, School of Electrical Engineering, Royal Institute of Technology (KTH)). Finally, ES is grateful for financial support from Insight Venture Partners and from William Melton. SK acknowledges funding from the Swedish Research Council.

Author biography

Eric Smith

Eric Smith received his BSc in Physics and Mathematics from the California Institute of Technology in 1987 and his PhD in Physics from The University of Texas at Austin in 1993, with a dissertation on problems in string theory and high-temperature superconductivity.

He has worked as a post-doctoral researcher at the Los Alamos National Laboratory, the Applied Research Labs at the University of Texas at Austin, and the Santa Fe Institute. Since 2006 he has been a faculty member of the Santa Fe Institute and since 2012 has been a Research Professor at George Mason University.

His work is broadly concerned with the origin of robust order in physical, chemical, biological and social systems. Specific areas of interest include the origin and role of institutions in economic dynamics and the transition from the geochemistry of the early earth to the first levels of biological organization.
(Photo credit Nerissa Escanlar.)

Supriya Krishnamurthy

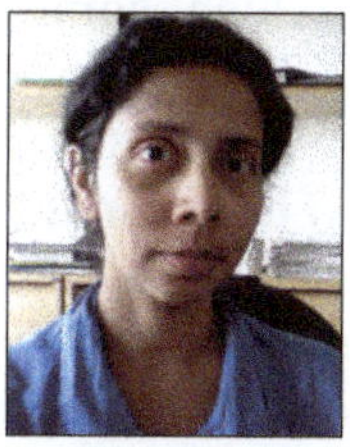

Supriya Krishnamurthy received her BSc and MSc degrees in Physics from Hansraj College, Delhi University and the Indian Institute of Technology Kanpur, respectively, and her PhD, in non-equilibrium statistical mechanics, from the Tata Institute of Fundamental Research in Bombay in 1998. After her PhD she worked as a post-doctoral researcher in Paris at the Ecole Supérieure de Physique et de Chimie Industrielles, at the Theoretical Physics Department of the University of Oxford and at the Santa Fe Institute. She has worked at the Swedish Institute of Computer Science (SICS), Stockholm, as a senior researcher (2004–2009) and as an assistant professor at the Royal Institute of Technology at Stockholm (2006–2008). Currently, she is an associate professor at the Department of Physics at the University of Stockholm (2008–). She was also associated with the Santa Fe Institute during 2005–2011 as external faculty.
Her research interests include understanding both fundamental as well as inter-disciplinary applications of non-equilibrium statistical mechanics.

IOP Publishing

Symmetry and Collective Fluctuations in Evolutionary Games

Eric Smith and Supriya Krishnamurthy

Chapter 1

Introduction: bringing together Darwinian evolution and games

Evolutionary game theory describes a class of population models in which the individuals subject to Darwinian selection are also the agents who play games to determine their fitnesses. The merger of the two domains has been a fertile source of models which can be given many interpretations. Our subject is statistical estimation and stochastic dynamics of evolutionary systems, in which games provide a taxonomy for major families of structure and behavior. Here we introduce the central concepts of symmetry, collective fluctuation, robustness and scale dependence that will be the themes of the following chapters. The technical problems of computing robust forms of scale-dependent stochastic dynamics will lead us to reconsider the basic abstractions that unify population models and games. We argue in favor of a statistical formulation of concepts such as individuality and agency, similar to the reformulation of the concepts of particles and forces in 20th century statistical physics, and a good match to modern efforts to incorporate principles of developmental biology integrally within our understanding of evolution.

1.1 The content and interpretation of evolutionary games

Evolutionary game theory [1–12] refers to a class of population models that bring together the formalization of evolutionary dynamics from population genetics and the structured models of interaction from game theory. At a minimum, the connection between population genetics and games is made at two points: the elementary entities in the evolving population (when described at an appropriate resolution) are the agents who play the games and the payoffs that result from play define their levels of fitnesses in the evolutionary dynamic. Depending on the model employed and the question of interest, many other points of contact may also be formed, drawing from the diverse inventory of game structures and the many kinds

doi:10.1088/978-0-7503-1137-3ch1

of possible interaction among individuals in populations. We will discuss some of these connections in later chapters.

The move to combine evolutionary dynamics with games can be approached as an extension or refinement from within either evolutionary theory or game theory. Within population-genetic formalizations of evolution, fitness models are effectively 'black boxes', chosen arbitrarily along with other models for heredity and for the mechanisms that generate variation. Games provide a structured way to unpack those black boxes, assigning meaning to the way interactions among individuals generate fitness from an individual's type in a population context. They may also offer ways to link the interactions that create fitness with mechanisms that generate variation, thus embedding knowledge about developmental programs or ecological interactions more integrally within the formalization of evolutionary dynamics.

Within game theory, many different solution concepts[1] may be applied to the same structured interaction; which solution concept is used determines which strategies or collections of strategies will be favored. Evolutionary updating provides an alternative solution concept to rational-choice solutions, which has many desirable properties statistically and, for some applications, empirically.

Evolutionary game theory potentially offers a very rich synthesis of concepts and tools. It can draw on all the methods to treat assortation, replication, transmission and selection formalized within modern population genetics [15–21], representing much of what is understood about the multilevel structure of interactions that converts the general mechanism of selection into the panoply of distinct evolutionary situations [22, 23]. It may also employ the full range of descriptions of structured individual and group interactions from game theory [14, 24, 25]. These include the extensive form [26], which maps out the dynamics and sub-structure during the course of a particular interaction, and cooperative solution concepts [13], which abstract some forms of institutional agreement or other group-level constraints on joint actions.

1.2 The approach to evolutionary games in this monograph

1.2.1 The foundation in prior work

Compared to either of its parent fields, mathematical population genetics and rational-choice game theory, evolutionary game theory is still a relatively young field. At present, a growing list of example models has been worked out (reviewed in [4, 7, 8]); some classification has been carried out based on symmetry, particularly from a dynamical-systems perspective; and suites of analytic methods now exist, drawing from the established fixed-point analysis of strategic games [25], from non-linear dynamics [4, 7, 12], and to a limited extent from stochastic process theory [3, 6].

[1] A *solution concept* is any formal procedure that uses the specification of the game and the assignment of payoffs to select a strategy or a distribution over strategies [13]. Solution concepts may employ the normal (also called 'strategic') form, the extensive form, or the coalitional form representation of the game. For a sense of the diversity of solution concepts that may be defined within any one of these representations, see [14].

A large amount of analysis has been carried out for games in the strategic form[2] (also called 'normal' form) and some work has been done using the extensive form [26]. Since the extensive form is a refinement of the strategic form in which the *structure* of play is made explicit [13], the relatively limited treatment it has received represents only a very early stage of exploration of a potentially rich and important topic.

The majority of the existing literature on evolutionary games grows out of the study of either fixed points or dynamical systems and has been based on the classical *replicator equation* [4], which is a mean-field equation[3]. The concept of *evolutionary stability* of equilibria, introduced by Maynard Smith and Price [27], is defined in terms of infinitesimal perturbations about the solution given by the replicator equation. Some research has been carried out on non-infinitesimal population fluctuations (inevitable in finite-sized populations and, as we will show, sometimes important even in infinite-population limits). Work on fluctuations divides into studies concerned with refining equilibrium selection to exclude ambiguity and studies that treat ongoing dynamics as an empirical consequence of ambiguity that models should not seek to exclude.

In general, games will admit multiple Nash equilibria or evolutionary stable states [28–31]. A tradition in economics has been to seek *equilibrium refinements* [14], which reduce this multiplicity by placing further restrictions to rule out sub-sets of equilibria in different contexts. In the presence of finite fluctuations, the long-run probabilities for a population to be found within basins of attraction of different equilibria will generally differ and the ratios of these probabilities can generally be made to diverge with large population size or small fluctuation strengths[4]. Therefore, by a process analogous to annealing, stochasticity may be used to reduce the number of equilibria that are populated with non-zero measure in the long run [32], providing an evolutionary argument for refinement. Adopting an alternative emphasis, a few studies have been performed in which stochastic dynamics in the presence of multiple equilibria was the primary focus [33–37]. As in the study of the extensive form in evolutionary dynamics, the topic of stochastic aggregate dynamics is an exceedingly broad and important area into which only introductory forays have been made. It will be our main area of emphasis.

1.2.2 Using symmetries to classify and understand the robust forms of stochastic dynamics

We will study evolutionary game theory in its stochastic form. Stochasticity arises from the population-level events which are already recognized in population

[2] The *strategic form* of a game is the most widely seen representation in terms of 'payoff matrices', in which entire strategies simply appear as indices to the rows and columns (and further indices, for k-player games), without reference to the structure of play that the strategy represents.

[3] To the extent that one views all evolutionary game phenomena as fundamentally stochastic—a point of view that we will strongly advocate—the replicator equation is more importantly a form of mean-field *approximation*.

[4] An important exception to this generalization arises when multiple equilibria reflect the presence of an underlying *symmetry* and this will be one of the reasons symmetry is important in our treatment that follows.

genetics: sampling of individuals who will interact (here, by playing a game together), culling (death) and replacement (replication). In some of the examples we will also consider stochasticity within the course of play of a single game. In games for which strategies consist of moves that could be shuffled as part of the reproductive process, the place at which crossover occurs is an additional source of randomness. Beyond the mean-field analysis of the replicator equation, new phenomena arise that are not possible in deterministic systems. We are also led to ask which outcomes predicted by the replicator dynamic are robust in the presence of fluctuations, a seemingly innocent question that in statistical mechanics and field theory has led to a radical reconceptualization of the nature of objects and interactions [38–40].

In suitable weak-fluctuation limits, we will recover standard results from the replicator analysis concerning bifurcations to multiple equilibria or limit cycles (other attractors could be included but are not pursued here). These phenomena are an important source of multilevel dynamics in ontogeny[5] and evolution. They cause individual dynamics, through mutually reinforcing cooperative effects, to become entrained by population states which then take on dynamics of their own. We classify bifurcations according to symmetry, as is done in the dynamical-systems approach [4]. However, making a slightly different emphasis than the typical one from dynamical systems, we view bifurcations not as fundamental changes of symmetry groups, but as changes in the *representation* of symmetries by dynamical states. The presence of underlying symmetries that are merely hidden, in the stochastic domain, is the basis for proofs that multilevel dynamics is a *robust* property against all orders of fluctuation corrections, even if we cannot compute or efficiently simulate them. The existence of hidden symmetries causes multiple equilibria to escape the filters of equilibrium refinement *inherently*, allowing us to use stochastic approaches such as annealing [32], not with the goal of singling out a unique static equilibrium, but to identify sources of long-run dynamics that are not sensitive to fine modeling assumptions. We will recover important symmetry-derived theorems of condensed matter physics and field theory, such as *Goldstone's theorem*, and show the forms that they take in evolutionary dynamics, particularly as these result from new roles of time in irreversible stochastic processes.

We then consider effects that cannot be produced at all in the deterministic approximation, including fluctuation-controlled dynamical regimes that persist in infinite-population limits, creep and forms of symmetry breaking that resemble glass phases, where the number of ordered macrostates and the complexity can be 'open-ended'. The potential for open-ended complexity is an often cited property of evolving systems [41] that simple bifurcations do not possess, so it is important to have examples in evolutionary game theory where at least the rudiments of an unlimited diversity of macrostates can be exhibited.

[5] *Ontogeny* refers to the entire sequence of events in the life of an organism, which occur between its origin through replication and its death or fissioning into offspring.

1.3 Empirical bases for the abstractions behind models and model selection

To a large extent evolutionary game theory has been used as a cornucopia of 'toy models'—proofs of concept demonstrating certain types of dynamics that might occur—but only in rare cases [42] have these been required to serve as empirically calibrated analytic frameworks to show that a certain causal explanation *must* hold. Most conventional use of evolutionary game theory thus stands in contrast to other areas of quantitative evolutionary modeling, such as quantitative genetics [43], which aspire to less mechanistically rich descriptions of processes than games (they employ merely linear regressions on identified alleles), but which seek to show that any correct account must be equivalent to, or a refinement of, a statistically defended regression model.

An exploration of toy models is an essential part of developing the phenomenology of a domain as rich as the merger of population genetics and game theory, but on its own it leaves a literature that is to some extent a collection of ad hoc cases[6], rather than the application of a set of overarching principles. The motivation for game models is often drawn heavily from the scientific narrative for each particular case [8, 27], obscuring the role that games *as a system* play in modeling evolutionary dynamics. Toy modeling fills the level of ∃ (there exists) in propositional logic, whereas an empirical defense of causality must fill the role of ∀ (for all). Proofs of concept therefore do not expose the basic abstractions of a theory to certain tests of robustness or generality that more quantitative methods require. Although it may not be apparent upon first consideration, 'stress-testing' the choice and interpretation of models, especially against the pervasive effects of stochastic perturbations, can lead to a reconceptualization of the basic abstractions underlying a theory, ultimately making it better able to incorporate advances in understanding in other areas.

1.3.1 Introducing games from a starting point in regression modeling of population processes

We will be interested in common mathematical elements of evolutionary game theory as a system for studying evolutionary dynamics, with an eye toward empirical applications. Therefore, we will bypass the appeals to scientific narratives that often play a large role in the motivation of game models for particular cases. We focus instead on the consequences of stochasticity that must affect the choice and interpretation of *all* game models from empirical observations. Any attempt to use evolutionary games to understand natural phenomena will inevitably include sampling fluctuations from observations and hence uncertainty in model identification[7], as well as stochasticity in model dynamics that will affect analysis and prediction.

---

[6] Here the term ad hoc—literally 'for this'—does not carry a pejorative connotation. In final applications, all models must be justified by appeal to the details of the particular case.

[7] In the most general case, model identification includes not only parameter estimation, but specification of the basic concepts of agency, move and interaction sequence that define the structure of a model and its connection to observations [42].

De-emphasizing narrative forces us to ask what can be justified statistically in the choice and interpretation of a game model, in the presence of uncertainty, noise and error. By defining games statistically, we obtain a clearer abstraction of the role of games as a general framework in evolutionary dynamics: we will argue that the proper abstraction for the role of games is as models of development, which complement the models of information dynamics formalized in population genetics. We will introduce the problem of identifying a game model as a problem of non-linear regression following the empirically motivated methods of Fisher's theorem [44] and the more general Price equation [45–47]. Regression estimates begin with the lowest-order (generally linear) models of fitness and recursively *construct* the dynamics of the game through the addition of higher-order interaction terms as required. We will thus embed evolutionary game theory within the larger suite of formal evolutionary methods, so that approaches such as quantitative genetics coincide (tautologically) with the lowest-order estimators for games. Since any finite sample supports the inference of only a limited number and precision of model coefficients, we will be forced to address the problem of justifying models that formally involve infinite hierarchies of coefficients (even when these are all set to zero, they nonetheless exist as modeling choices[8]), and the related problem of determining which predictions from a statistically estimated model are robust.

1.3.2 Symmetry and scale

An immediate consequence of incorporating stochasticity in all elements of model selection, analysis and interpretation is that regression coefficients generically come to depend on *scale*. Relevant scales may be the population size, or the time interval or number of interaction events over which samples are drawn. The scale dependence of regression coefficients that is readily demonstrated in models reminds us that scale is also an inherent property of empirical observations. Therefore the concept of calibrating the coefficients in a game model to describe a natural system is one that inherently invokes the scale at which the model is to be estimated and analyzed. We will go further to propose that the *meaning* of the fundamental abstractions of evolutionary game theory, such as individuals, genes, or strategies, should come to be understood as scale-relative concepts, for the same reason that elementary particles and forces are now understood to be scale-relative concepts in physics.

In a modeling framework where the detailed model description becomes scale-dependent, symmetries take on elevated importance because they are the invariant properties that identify systems across all scales. The changes of symmetry representation by population states, as population size or interaction strength are changed, then define the robust dynamical regimes or *phases* in which variation, interaction, replication and selection act. This view is very compatible

[8] This is the fundamental insight behind *effective field theory* [40]. Although first appreciated in the contexts of condensed matter and elementary particle physics, the statistical arguments that force the effective-theory interpretation apply equally to population processes. Current work in other areas of population genetics [15, 16, 48–52] is already re-deriving similar results, in the course of defining statistically valid methods to treat multilocus interactions, strong selection and other phenomena.

with the modern understanding of multilevel selection [47, 53, 54] and the hierarchical role played by evolutionary dynamics in the evolution of developmental complexity [22, 23]. We believe it leads to a conceptualization of the fundamental abstractions of evolutionary game theory in keeping with the best understanding in modern evolutionary theory. In addition to demonstrating some qualitative categories of scale-dependent description, and showing why they matter, we derive methods to quantitatively compare model descriptions that differ at multiple scales because they incorporate different degrees of correlation and we show that these methods match well against simulation results, even in some cases of large fluctuations.

1.4 A summary of the key ideas and the topics to be developed

The following is a brief summary of the major conceptual commitments that we believe underlie a systematic and principled understanding of evolutionary games, and which will guide our presentation of a general framework and a few illustrative examples in the following chapters. We provide an overview here without attempting full explanations, so that readers from different backgrounds will see the relation of some topics that they will know as familiar foundations with others that may be new to them. Complex topics for which we can only provide a brief summary description in this list are developed in detail in later chapters. The list below is not exhaustive of concepts that could be developed in this area, but we believe it provides a reliable foundation that can be elaborated without needing to be overturned.

1.4.1 The Price equation: accounting identities, fitness and closures in population genetics

We construct evolutionary game theory as a general framework to classify and interpret fitness models—their quantitative dependence on population state and, if desired, explicit representations of the interaction sequences that determine fitness—within the axiomatic structure of population genetics. We begin in chapter 2 with the *Price equation*, an accounting identity for any process satisfying the assumptions of population genetics, in which *fitness* universally appears as a summary statistic [47].[9]

Fitness is defined in terms of the number of offspring relative to the number of their parents, grouped by the parents' type. It is a descriptive statistic, which can be computed for any given realization of an evolutionary process. If the purpose of an evolutionary account is not only description or historical reconstruction of a particular instance, but also estimation of a process model for change, then fitness (as well as other parameters) must be given a model in terms of properties of individuals and populations. The model estimation problem is to determine what

[9] Economists will appreciate the importance of accounting identities as non-trivial constraints, despite their 'tautological' nature: by construction they apply to all well-formed models in the domain for which they are derived; and thus they identify that domain.

structure and what coefficients can be justified from empirical observation. Once a population process model is chosen, one must often also define *closures* for it[10], which are approximations that permit calculations from finite orders of terms or parameters. The empirical calibration of fitness models leads to the approach of Fisher [44] and Price [45], who replace fitness (the summary statistic) with models that are meant to match regression coefficients of fitness on individual and population states, obtained from statistical samples. Since regressions can be performed on interaction terms of arbitrarily high order, in principle this approach provides a full basis for the identification of those aspects of a process that affect fitness; one can then ask in a principled way how much detail is supported by empirical evidence and attempt to systematically construct least-committal models [55, 56] for undetermined parameters.

1.4.2 The incorporation of information and development are the two complements that govern evolutionary dynamics

The mechanism of heredity in any population process determines which consequences of events affecting parents persist as features of the population state of the offspring. Since the filter of natural selection—the part of the population process represented explicitly in population-genetic models—acts to narrow the distribution of properties of offspring, population genetics comprises the information incorporating aspects of evolutionary dynamics[11]. The formal equivalence between the replicator equation and Bayes's theorem for updating probability distributions [58] provides a way to quantify this concept of information and also to show that selection is a statistically *optimal* method for incorporating information within a population about its environment.

Not all properties of organisms are directly preserved by mechanisms of heredity and the difference between what is preserved and what is generated and acted on by selection is the difference between genotype and phenotype. The complement to the information transmitted via a distribution of genotypes is the collection of all other aspects of phenotype, which are constructed through non-heritable interactions within generations. We will refer to these as *development*, broadly construed. From the perspective of classical population genetics, development consists simply of a genotype/environment → phenotype/fitness map, but we are concerned with the actual generating processes responsible for that map.

The complementarity between the information incorporating function of selection and heredity, and the constructive role of development, for us *defines* the respective roles of population genetics and games within evolutionary game theory.

[10] 'Closure' is used as a general term in economics; in population genetics it normally refers to the more specific problem of *moment closure*, which we will show can be handled in a variety of ways.

[11] Here we are referring specifically to those aspects of information incorporated and preserved within the Darwinian paradigm. Other modes of propagation of ordered states—in particular modification of an environment that persists through mechanisms different from replication by populations of individuals undergoing Darwinian competition—are also relevant to evolutionary dynamics but they are a different topic [57] which we leave out of the scope of this discussion.

1.4.3 The emergence of games as a framework to systematically model development

Starting from the Price equation and a need for closures, a general polynomial expansion of frequency-dependent regression models for fitness is equivalent, at order k, to treating development as a k-player normal-form game with uniform matching of individuals. This equivalence is a tautology, meaning both that the normal-form game interpretation is always available and at the same time that it is highly ambiguous about the mechanism that constitutes the 'game' and about its interpretation. If, beyond the aggregate statistics of fitness, we are given more information about the frequencies with which individuals are sampled to interact in the population, we may resolve the normal form into contributions from assortative matching and a set of payoffs which differ from the mere coefficients in the fitness function. If we know more about the internal structure of interactions—which may be temporal sequence, signaling or imitation, or even just linkage—then we may refine the normal form to a particular extensive-form game [26]. Further elaborations, to include constraints on joint actions by multiple individuals, could be developed to make contact with the coalitional-form representation from cooperative game theory, but we do not pursue those systematically in this monograph. The sequence of one-to-many mappings, from the normal-form to the extensive-form and the coalitional-form solution concepts, constitutes a well-understood approach to refining the definition and interpretation of games in classical game theory [13] and we think it provides a useful level of discipline also for the interpretation of evolutionary games.

In this way games emerge as a highly general, if not all encompassing, framework to model development.

1.4.4 Symmetry and collective fluctuations in evolutionary games

A central theme in our approach to the topic of evolutionary games is that interactions among individuals in single events may produce population behaviors that, in aggregate, are describable with games of a similar form but with coefficients that may differ from those that the individuals directly experience. Most obviously, individual traits may polarize population states or lead to distributions of fluctuations that feed back so that higher-order correlations become part of the best average estimates for individual fitness.

In the domain of toy models, we often have the option to regard such differences as artifacts of the adoption of coarse-grained descriptions, but we think that if games are to become a serious tool for the analysis and interpretation of empirical phenomena, it is better to start to think of such scale dependence of parameters as an essential feature of the *definition* of such concepts as individuality, agency and interaction. Such a revision in the notion of what constitutes an elementary particle have been fundamental to a radical reformulation of the conceptualization of objects and interactions in physics [38–40]. Modern writing on evolution makes a serious effort to understand the way structured interactions, replication and selection interact at many scales, to produce multiple novel levels of individuality both in development [22] and as levels of selection [47, 53, 54]. We believe that a statistical notion of individuality and interaction is already inherent in modern biology and we make that integral to the way we present game models.

When the use of evolutionary games is altered from toy modeling of a hypothesized 'fundamental' interaction and its scale-dependent approximations, to an attempt to represent data in which all estimation and prediction problems involve uncertainties that depend on scale, a different approach is required to specifying what constitutes a model of a particular actual phenomenon, *at all scales of interest*. We therefore introduce fitness models and their classification in terms of *symmetry groups and the representation of symmetries by population states*. Symmetries are scale-invariant properties of systems and changes in their representations (known as *symmetry breaking* [59–61]) imply robust predictions for multiscale dynamics.

Stochasticity and correlation are the causes of parameter change in models of the same system at different scales and the stochastic effects that are robust within symmetry classes (and that lead to symmetry breaking) are *collective fluctuations* [62]. We introduce in chapter 2, and develop in detail in chapters 5–8, a set of examples of major classes of symmetry groups and categories of symmetry breaking and show how each implies a distinctive form of scale dependence in fitness or dynamics. These include the emergence of new units of selection or of coalitional behavior from interactions that are non-cooperative at the scale of individual interactions.

1.4.5 Large-deviations theory as the central organizing concept for calculations

The preceding four points provide a set of abstractions of evolutionary games that frees the fundamentals of the theory from arbitrary narratives invoked to justify particular cases and also acts as a guard against over-interpretation. However, these points are not useful in practice without ways to identify the relevant classes of collective fluctuations or their consequences for the expression of symmetry and for parameter changes across scales.

Evolutionary population processes are extended-time, irreversible Markov processes. In general, their distributions over collections of events could be too complicated to permit *any* robust characterizations. However, a feature of even moderately large populations or times that can make such processes tractable and can make games a stable and useful class of models is the tendency for probability distributions to converge toward a small number of exponential families. Within these families, the combinatorics of large numbers of agents or events may produce leading-log probabilities of fluctuations with a scaling relation known as the *large-deviations property*, in which the dependence on system scale separates from the dependence on the structure of the fluctuation [63].

Chapter 4 is devoted to a derivation of the large-deviations theory of discrete population processes tailored to the structure of evolutionary game models introduced in chapter 2. We show how the large-deviations limit singles out classes of collective fluctuations and how their properties can be computed to connect game descriptions across scales.

Good treatments of the consequences of large numbers and aggregation exist for evolutionary dynamics [3, 6] and for population processes more generally [64, 65].

Many of these draw from the probability literature and are concerned with convergence and laws of large numbers. Our approach will be one more familiar to physicists and will emphasize the extraction of terms most directly responsible for multiscale dynamics and multilevel selection.

These first five points have addressed general conceptual foundations of evolutionary game theory. The next four points concern particular applications, which are nonetheless of wide interest within either population biology or game theory.

1.4.6 The event structure of single interactions and uses of the extensive form

For many applications of evolutionary games it is not necessary (or not empirically warranted) to go beyond the normal form and the assumption of random matching. For others, though, the sub-structure of play when a collection of agents is brought together in a single interaction is central to the question being asked. The standard way to represent event sequences and information conditions in game theory [13, 25] is to refine the normal form to the *extensive-form* representation, most familiar as a 'game tree'.

Therefore, after introducing the general role of games in population processes in chapter 2, we provide a non-exhaustive but systematic introduction to extensive forms in chapter 3. A dedicated treatment of the extensive form is given in [26] and many more applications still could be developed. Here we will mostly be concerned with the two concepts of *neutrality* and *repetition*.

Neutrality [66–71] arises in evolutionary dynamics when distinct genotypes have the same fitness in populations that they themselves produce. It is an important form of symmetry under interchange of agent types, and a property that complex developmental programs can be expected to produce very frequently. A summary of some combinatorial counts of game trees enables us to provide explicit examples.

Repetition is a property of game interactions studied extensively in the economics literature [72–75] to model the relation between long-term and short-term incentives. We consider it because it is a source of some very well-known models of neutral evolution, for which we can demonstrate new consequences of collective fluctuations. A widely used framework to study repetition in rational-choice game theory is the *repeated game*, a particular kind of extensive form built up by recursive attachment of a normal-form *stage game* to build up the game tree. Repeating the play of a stage game is the simplest and most generic way to produce complexity and with it the problems of understanding error and limitations on strategic capacity.

The next three topics consider other, more specific, uses of the extensive form and of repetition.

1.4.7 The statistical gene in relation to modularity in development

Mendelian heredity—the property that the sources of traits are replicated as discrete units which are shuffled like cards rather than being mixed like paint—was one pillar of the modern evolutionary synthesis [54, 76] and is incorporated as an essential assumption within standard population genetics. The convention of formalizing evolution in terms of invariant hereditary 'particles' has presented one of two important challenges to the use of evolution to describe social, behavioral, or institutional dynamics [36, 77].[12]

Given the importance of particulate heredity to current evolutionary thinking, we briefly consider its origin in biological systems. We argue that the standard invariant hereditary unit in biology—the *gene* understood as a non-recombining region of DNA—results not from the properties of DNA as a system for storing information, but from modularity in development which is either reinforced by selection or recapitulated in transmission mechanisms. The extensive form of a game inherently produces a kind of modularity in the sequence of play and offers a natural set of elementary units (the moves) to vary through mutation or to shuffle via crossover. We show, first in chapter 7 and in a different way in chapter 8, how the 'gene' description that arises out of the modularity of game trees leads in the presence of crossover to a multilevel evolutionary dynamic that must select both genes and their covariance within genotypes. In this way we connect the use of the extensive-form game as a general framework to the statistical and developmental origins of modularity that justify the gene concept in biology.

1.4.8 Repetition in evolutionary and rational-choice game theory: re-directing the forces of selection

The repeated games, with payoffs accumulated from the moves in each stage of play, have become a standard framework, in both evolutionary [8, 11, 79] and rational-choice game theory [25, 72, 74], to study the relations between short-term and long-term rewards for nominally equivalent move profiles. Repeated games are most often invoked as a framework to study the 'paradox of cooperation'.

In the play of a repeated game, correlations between moves in different stages can re-direct the force of selection (in the evolutionary context), or the non-cooperative equilibrium condition (in the rational-choice framing) to favor outcomes very different from those that would be favored by the stage game in isolation. The approaches to repeated games in evolutionary and rational-choice game theory have diverged sharply in the study of how repetition produces these differences. The divergence of approaches is at first surprising because for finite normal-form games, a result known as the *fundamental theorem of evolutionary game theory* [7][13] asserts that the evolutionary stable strategies are the same fixed points as Nash equilibria. The forward looking self-consistency of rational choice and the recursive filtering of natural selection might thus be expected to yield the same solutions.

Rational-choice repeated game theory, however, concerns cases with a very large or even infinite degeneracy of Nash equilibria, or with ambiguities in the way

[12] The other difficulty, perhaps more fundamental, is the possibility that a given offspring may draw inherently from multiple parents, obviating the existence of fitness as a universal summary statistic in the Price equation, which we take to be a defining feature of the classical population-genetic formalization of evolutionary dynamics. This difficulty is particularly evident in attempts to characterize technological innovation with evolutionary terms [78].

[13] Sometimes this is alternatively referred to as the *'folk' theorem of evolutionary game theory.*

equilibrium should be formalized, typically arising from an infinite or indefinite number of stage repetitions and associated normal forms that are not finite. Ambiguities in equilibrium selection are treated with a poorly formalized notion of 'prior agreement', which requires only conditions of feasibility and individual rationality of outcomes[14]. A set of results known as *(the) folk theorems* [75] of repeated game theory show that any outcome in the large set of feasible and individually rational move profiles can be supported as an equilibrium.

In evolutionary models with repeated games, the notion of prior agreement is replaced by direct specification of the set of strategies that may exist within evolving populations and the ways they are generated. Depending on how strategies are restricted, or on the dynamics of type changes, limited sets of marginally stable or even unique stable equilibria may be selected.

In chapter 9 we construct the mapping between the evolutionary and rational-choice approaches to strategy selection in repeated games. We show how different forms of coordination in the repeated game, which appear as features of development in the evolutionary setting, may also be cast as signaling about strategy types, or even as kinds of public information that have been introduced in rational-choice game theory to expand the notion of Nash (*aka* non-cooperative) equilibrium to include correlated equilibria. Our interest is to use the explicitly dynamical and constructive approach to strategy generation in the evolutionary approach to define systematic approximations to the concepts of indefinite repetition and prior agreement in the rational-choice setting and to understand how these approximations imply bounds on strategic complexity.

1.4.9 Evolutionary mechanics and thermodynamics

The classical *replicator equation* [4] often used to study evolutionary games is very restrictive: by omitting explicit fluctuations, it is effectively a single-scale description. It also uses a particular set of closure assumptions, known as *mean-field assumptions*, which can sometimes be invalid even in infinite-population limits. For us, the stochastic population process is fundamental at all levels because stochasticity is the basis for criteria of robustness.

It is now well understood that the large-deviations scaling limit is the essential property behind the concept of entropy and thermodynamic limits [62, 63, 80]. A game model at the scale of individual interactions is the mechanical description, from which a (possibly different) game model of the aggregate population behavior is derived as the corresponding thermodynamic limit. The difference between individual and aggregate models results from *entropy corrections*, equivalent to those in any thermodynamic theory, but derived for extended-time irreversible Markov processes. A low-dimensional example in which the control over dynamics is shifted from

[14] Compared to the emphasis of every other aspect of game theory on the precise formalization of models of interaction [11], the notion of prior agreement is so permissive as to be almost unformalized. It is something like a solution concept with a potentially infinite set of distinct choices that the players are not forced but are simply declared to 'agree upon' before play. The way the choices are made then selects one rather than another equilibrium.

deterministic parameters (mean fitness and mutation) to fluctuation entropies is solved in detail in chapter 7. We expect that the most important and common use of evolutionary entropy will come not from low-dimensional systems with imposed symmetries, but from the high-dimensional neutrality produced by complex developmental trajectories, sketched in chapter 3.

Bibliography

[1] Smith J M 1982 *Evolution and the Theory of Games* (London: Cambridge University Press)

[2] Foster D and Young P 1990 Stochastic evolutionary game dynamics *Theor. Population Biol.* **38** 219–32

[3] Weibull J W 1997 *Evolutionary Game Theory* (Cambridge, MA: MIT Press)

[4] Hofbauer J and Sigmund K 1998 *Evolutionary Games and Population Dynamics* (New York: Cambridge University Press)

[5] Samuelson L 1998 *Evolutionary Games and Equilibrium Selection* (Cambridge, MA: MIT Press)

[6] Benaim M and Weibull J W 2003 Deterministic approximation of stochastic evolution in games *Econometrica* **71** 873–903

[7] Hofbauer J and Sigmund K 2003 Evolutionary game dynamics *Bull. Am. Math. Soc.* **40** 479–519

[8] Nowak M A 2006 *Evolutionary Dynamics: Exploring the Equations of Life* (New York: Belknap)

[9] Reichenbach T, Mobilia M and Frey E 2006 Coexistence versus extinction in the stochastic cyclic lotka–volterra model *Phys. Rev.* E **74** 051907

[10] Reichenbach T, Mobilia M and Frey E 2007 Mobility promotes and jeopardizes biodiversity in rock-paper-scissors games *Nature* **448** 1046–9

[11] Gintis H 2009 *Game Theory Evolving: a Problem-Centered Introduction to Modeling Strategic Interaction* 2nd edn (Princeton, NJ: Princeton University Press)

[12] Frey E 2009 Evolutionary game theory: non-equilibrium and non-linear dynamics of interacting particle systems *Nonequilibrium Statistical Mechanics: Fundamental Problems and Applications (Boulder School for Condensed Matter and Materials Physics*, July 6–July 24, 2009) pp 1–39 http://boulder.research.yale.edu/Boulder-2009/ReadingMaterial-2009/Frey/frey_lecture_notes_games.pdf

[13] Shubik M 1984 *Game Theory in the Social Sciences: Concepts and Solutions* (Cambridge, MA: MIT Press)

[14] Harsanyi J C and Selten R 1988 *A General Theory of Equilibrium Selection in Games* (Cambridge, MA: MIT Press)

[15] Barton N H and Turelli M 1991 Natural and sexual selection on many loci *Genetics* **127** 229–55

[16] Kirkpatrick M, Johnson T and Barton N 2002 General models of multilocus evolution *Genetics* **161** 1727–50

[17] Gillespie J H 2004 *Population Genetics: a Concise Introduction* (Baltimore, MD: Johns Hopkins University Press)

[18] Lehmann L and Keller L 2006 The evolution of cooperation and altruism—a general framework and a classification of models *J. Evol. Biol.* **19** 1365–76

[19] Lehmann L, Keller L, West S and Roze D 2007 Group selection and kin selection: two concepts but one process *Proc. Natl Acad. Sci. USA* **104** 6736–9
[20] Roze D and Rousset F 2008 Multilocus models in the infinite island model of population structure *Theor. Population Biol.* **73** 529–42
[21] Lehmann L and Rousset F 2009 Perturbation expansions of multilocus fixation probabilities for frequency-dependent selection with applications to the Hill–Robertson effect and to the joint evolution of helping and punishment *Theor. Population Biol.* **76** 35–51
[22] Buss L W 2007 *The Evolution of Individuality* (Princeton, NJ: Princeton University Press)
[23] Gerhart J and Kirschner M 1997 *Cells, Embryos, and Evolution* (New York: Wiley)
[24] Fudenberg D and Levine D K 1998 *The Theory of Learning in Games* (Cambridge, MA: MIT Press)
[25] Fudenberg D and Tirole J 1991 *Game Theory* (Cambridge, MA: MIT Press)
[26] Cressman R 2003 *Evolutionary Dynamics and Extensive Form Games* (Cambridge, MA: MIT Press)
[27] Maynard Smith J and Price G R 1973 The logic of animal conflict *Nature* **246** 15–18
[28] Negishi T 1960 Welfare economics and the existence of an equilibrium for a competitive economy *Metroeconomica* **12** 92–97
[29] Hahn F and Negishi T 1962 A theorem on non-tâtonnement stability *Econometrica* **30** 463–9
[30] Debreu G 1987 *Theory of Value* (New Haven, CT: Yale University Press)
[31] Arrow K J and Hahn F H (ed) 1971 *General Competitive Analysis* (New York: Elsevier)
[32] Kandori M, Mailath G J and Rob R 1993 Learning, mutation, and long run equilibria in games *Econometrica* **61** 29–56
[33] Young H P 1993 The evolution of conventions *Econometrica* **61** 57–84
[34] Young H P 1996 The economics of conventions *J. Econ. Perspectives* **10** 105–22
[35] Young H P 1998 Conventional contracts *Rev. Econ. Stud.* **65** 776–92
[36] Young P H 1998 *Individual Strategy and Social structure: An Evolutionary Theory of Institutions* (Princeton, NJ: Princeton University Press)
[37] Young H P and Burke M A 2001 Competition and custom in economic contracts: a case study of illinois agriculture *Am. Econ. Rev.* **91** 559–73
[38] Wilson K G and Kogut J 1974 The renormalization group and the ε expansion *Phys. Rep.* **12** 75–199
[39] Polchinski J G 1984 Renormalization group and effective lagrangians *Nucl. Phys.* B **231** 269–95
[40] Weinberg S 1995 *The Quantum Theory of Fields* vol 1 (New York: Cambridge)
[41] Vasas V, Szathmary E and Santos M 2010 Lack of evolvability in self-sustaining autocatalytic networks: a constraint on metabolism-first path to the origin of life *Proc. Natl Acad. Sci. USA* **107** 1470–5
[42] deDeo S, Krakauer D C and Flack J 2010 Inductive game theory and the dynamics of animal conflict *PLoS Comp. Biol.* **6** e1000782
[43] Falconer D S and Mackay T F C 1996 *Introduction to Quantitative Genetics* 4th edn (New York: Benjamin Cummings)
[44] Fisher R A 2000 *The Genetical Theory of Natural Selection* (London: Oxford University Press)
[45] Price G R 1972 Fisher's 'fundamental theorem' made clear *Ann. Human Genetics* **36** 129–40
[46] Frank S A 1995 George Price's contributions to evolutionary genetics *J. Theor. Biol.* **175** 373–88

[47] Frank S A 1997 The Price equation, Fisher's fundamental theorem, kin selection, and causal analysis *Evolution* **51** 1712–29
[48] Baake E and Wagner H 2001 Mutation–selection models solved exactly with methods of statistical mechanics *Genetics Res.* **78** 93–117
[49] Barton N H and de Vladar H P 2009 Statistical mechanics and the evolution of polygenic quantitative traits *Genetics* **181** 997–1011
[50] Barton N H and Coe J B 2009 On the application of statistical physics to evolutionary biology *J. Theor. Biol.* **259** 317–24
[51] de Vladar H P and Barton N H 2011 The contribution of statistical physics to evolutionary biology *Trends. Ecol. Evolution* **26** 242–32
[52] de Vladar H P and Barton N H 2011 The statistical mechanics of a polygenic character under stabilizing selection, mutation, and drift *J. R. Soc. Interface* **8** 720–39
[53] Hamilton W D 1970 Selfish and spiteful behavior in an evolutionary model *Nature* **228** 1218–20
[54] Gould S J 2002 *The Structure of Evolutionary Theory* (Cambridge, MA: Harvard University Press)
[55] Jaynes E T 1957 Information theory and statistical mechanics *Phys. Rev.* **106** 620–30
[56] Jaynes E T 1957 Information theory and statistical mechanics: II *Phys. Rev.* **108** 171–90
[57] Odling-Smee F J, Laland K N and Feldman M W 2003 *Niche Construction: The Neglected Process in Evolution* (Princeton, NJ: Princeton University Press)
[58] Shalizi C R 2009 Dynamics of Bayesian updating with dependent data and misspecified models *Electron. J. Stat.* **3** 1039–74
[59] Ma S-K 1976 *Modern Theory of Critical Phenomena* (New York: Perseus)
[60] Lifshitz E M and Pitaevskii L P 1980 *Statistical Physics Part II* (New York: Pergamon)
[61] Goldenfeld N 1992 *Lectures on Phase Transitions and the Renormalization Group* (Boulder, CO: Westview)
[62] Smith E 2011 Large-deviation principles, stochastic effective actions, path entropies, and the structure and meaning of thermodynamic descriptions *Rep. Prog. Phys.* **74** 046601
[63] Touchette H 2009 The large deviation approach to statistical mechanics *Phys. Rep.* **478** 1–69
[64] Kurtz T G 1981 *Approximation of Population Processes*. Society for Industrial and Applied Mathematics, Philadelphia, PA, CMBS-NSF Regional Conference Series in Applied Mathematics vol 36
[65] Feng J and Kurtz T G 2006 *Large Deviations for Stochastic Processes (Mathematical Surveys and Monographs* vol 131) (Providence, RI: American Mathematical Society)
[66] Fontana W, Wagner G and Buss L W 1994 Beyond digital naturalism *Artificial Life* **1** 211–7
[67] Gruener W, Giegerich R, Strothmann D, Reidys C, Weber J, Hofacker I L, Stadler P F and Schuster P 1996 Analysis of RNA sequence structure maps by exhaustive enumeration: I. Neutral networks *Monatsh. Chem.* **127** 355–74
[68] Gruener W, Giegerich R, Strothmann D, Reidys C, Weber J, Hofacker I L, Stadler P F and Schuster P 1996 Analysis of RNA sequence structure maps by exhaustive enumeration: II. Structures of neutral networks and shape space covering *Monatsh. Chem.* **127** 375–89
[69] Reidys C, Stadler P F and Schuster P 1997 Generic properties of combinatory maps: neutral networks of RNA secondary structures *Bull. Math. Biol.* **59** 339–97
[70] Ancel L W and Fontana W 2000 Plasticity, evolvability and modularity in RNA *J. Exp. Zool. (Mol. Dev. Evol.)* **288** 242–83
[71] Fontana W 2002 Modeling 'evo-devo' with RNA *Bioessays* **24** 1164–77

[72] Fudenberg D and Maskin E 1991 The folk theorem in repeated games with discounting or with incomplete information *Econometrica* **54** 533–54
[73] Aumann R J and Shapley L S 1994 Long-term competition—a game-theoretic analysis in *Essays in Game Theory: In Honor of Michael Maschler* ed N Megiddo (Heidelberg: Springer) pp 1–15
[74] Rubinstein A and Wolinsky A 1995 Remarks on infinitely repeated extensive-form games *Games Econ. Behavior* **9** 110–5
[75] Ratliff J 2010 *A folk theorem sampler* pp 1–34 http://www.virtualperfection.com/gametheory/Section5.3.html
[76] Provine W B 2001 *The Origins of Theoretical Population Genetics* (Chicago, IL: University of Chicago Press)
[77] Nelson R R and Winter S G 1985 *An Evolutionary Theory of Economic Change* (Cambridge, MA: Balknap Press)
[78] Arthur W B 2009 *The Nature of Technology* (New York: The Free Press Simon & Schuster)
[79] Nowak M A 2006 Five rules for the evolution of cooperation *Science* **314** 1560–3
[80] Ellis R S 1985 *Entropy, Large Deviations, and Statistical Mechanics* (New York: Springer)

Eric Smith and Supriya Krishnamurthy

Chapter 2

Transmission, development, selection: the Price equation and the role of games

This chapter introduces the two general frameworks that make up evolutionary game theory: population genetics formalizes the relation between Mendelian heredity and Darwinian selection, and game theory captures the general form of maps from genotypes and population states to phenotypes and fitness. Three consequences of the assumptions of population genetics are the existence of a summary statistic that furnishes a well-defined concept of fitness, an accounting identity known as the Price equation, and Fisher's fundamental theorem of natural selection. An application of the Price equation to the numbers of agent types that make up the population leads to the well-known replicator equation. From a perspective that emphasizes stochastic population processes as the defining framework for evolutionary dynamics, we introduce the role of symmetries as the foundation for model classification as well as the robust estimation and subsequent statistical analysis of models, and the role of collective fluctuations as governing multiscale dynamics and multilevel selection.

2.1 Formal evolutionary modeling as a foundation for games

Evolutionary games are first and foremost evolutionary processes. A bulky, complicated and often fractious literature has built up around what counts as evolution [1, 2], but for our purposes the important criteria are those that lead to particulate, forward-branching descent, which can be formalized without ambiguity. They are the assumptions that have become the axioms of classical population genetics [3, 4], which grew out of the modern evolutionary synthesis [2, 5] of Mendelian heredity with Darwinian selection.

The importance of the axioms of population genetics is that they entail the existence of fitness as a summary statistic for population dynamics. This was first consistently formulated by Irving Fisher [6], but only given a thorough exposition

doi:10.1088/978-0-7503-1137-3ch2

which was broadly understood [7] in the work of George Price [8] more than four decades later. We begin this chapter with the definition of fitness and the accounting identity known as the *Price equation* and out of the problems of estimating, modeling and computing the consequences of fitness functions, we introduce the most general role and form of games.

2.1.1 The basic elements of population-genetic models of evolving systems

Evolving systems may have a whole host of forms and levels of organization. The key step in arriving at a population-genetic model is choosing a level of representation at which the elementary units can be modeled with Mendelian heredity[1]. The resulting representation should be understood as an explicit modeling commitment: its abstractions may be only approximations to elements in real systems, or may hold only probabilistically. When it has been understood that this is the case, it becomes clear that for some systems more than one level of description may have a representation in terms of Mendelian descent [2]. The different representations then become distinct levels of selection[2]. Before introducing algebraic representations, we summarize the common elements shared by population-genetic formalizations of evolving systems.

2.1.1.1 Replicating individuals, populations, generations

The part of a system that undergoes Darwinian dynamics consists of a set of elementary units that reproduce by *replication*. We will refer to these as *individuals*, because they are taken to be *indivis*ible, and therefore appear as discrete members in populations[3]. Replication is *defined* by the property that *one parent may leave multiple offspring, but each offspring has a unique parent*, which sets it apart from mechanisms of descent that may involve inheritance from multiple ancestors. The individuals in such Darwinian populations are the elementary 'hereditary particles', the distribution of which incorporates and transmits information over time. The existence of a unique parent for each offspring causes lineages of descent to branch forward in time, but not in reverse.

The events of replication organize the population into *generations*. Reproduction may be synchronous or asynchronous; we are free to treat successive states of a population in which only one individual has replicated or died (asynchronous reproduction) as successive generations in which every other individual is the unique offspring of a parent, which was the same individual in the immediately preceding state.

---

[1] Here we invoke the discrete-particle character of Mendelian heredity, not any particular model of the way these characters may be shuffled during descent, such as the sexual recombination that was Mendel's particular subject of study.

[2] Arguments over this point, which fail to acknowledge the approximate and probabilistic nature of the abstractions in *all* cases, and which fail to recognize that when they have been accounted for, they can be applied at multiple levels of representation, seem to be at the base of ongoing and often acrimonious disputes over the possibility for multilevel selection [9] and its applicability to particular cases [10].

[3] Whether to use 'individual', 'replicator', or some other term, is another subject that enjoys a large literature of *ex cathedra* proclamations and scholastic debate over their authority; see [10] and references therein. Our concern is with the behavior of the elements in discrete stochastic processes and their names are for us matters of convenience.

We have emphasized that evolving systems may be organized at many levels because the entities that satisfy the model of descent by replication need not be organisms. They may be genes, collections of genes, entire chromosomes, haploid organisms (if recombination is ruled out), or even higher-level collectives. Note that in descent by sexual reproduction, which combines chromosomes from two parents, the 'individual' in ontogeny, who develops, experiences the forces of selection and eventually dies, possibly having left progeny, is not a replicating individual, but only a transient association between two lines of chromosomal material. When chromosomal crossing is important—whether due to sexual or bacterial recombination, the insertion of transposable elements, etc—the chromosome may cease to be a replicating individual by this definition and a smaller unit such as a gene may become the elementary unit of selection [9]. All these cases can be handled, and in what follows we suppose that an appropriate level of representation has been chosen for the system and question of interest.

2.1.1.2 Types and type spaces

The differences among individuals that enable adaptation are represented by giving individuals *types*. A well-defined formal evolutionary model must have a sufficiently complete type space to label any individual that can be produced by the mechanisms of descent. The essential parameters in a statistical analysis of an evolving population are a representation of the type space, the total population number, a way to represent the generation structure and the vector whose components are the numbers of individuals of each type in the population at each time.

Often the choice of a type space requires non-trivial coarse-graining so that some fine distinctions are not recognized; in practical applications the goal is to maximize tractability and statistical power in a type space that is still fine enough to capture the necessary distinctions to explain dynamics. Richard Lewontin has termed the latter requirement *dynamical sufficiency* [11]. The descriptions of what is transmitted during descent and how variation is generated are subordinate to the way individuals are grouped into elementary types.

2.1.1.3 Transmission, variation, selection

All remaining specification of the conditions under which the elementary, Darwinian individuals in a parent generation can reproduce, or configurations in which the individuals in an offspring generation are brought into existence, jointly make up the *transmission process* for the system. For example, if the phenomenon of evolutionary interest is a developmental program involving many genes in *cis* interactions[4], statistics and models must reflect the fact that the genes involved do not replicate in isolation, but in the context of assembled chromosomes encoding the complete program. If dominance is of interest, the full complement of chromosomes in a (diploid, tetraploid, hexaploid, aneuploid, etc) cell or organism may be required. Games provide a convenient framework in which to impose such restrictions. Fitness,

[4] *Cis* interactions are those between genes within a chromosome, distinguished from *trans* interactions among genes on different chromosomes.

or even the event of replication, may be made dependent on the outcome of a game that can only be played when a pre-specified complex of players is assembled.

An important aspect of the transmission process is the set of mechanisms that produce variation and the patterns of variation that arise. We will refer to any change in the type of a replicating unit between the parent and offspring generations as *mutation*. The most fundamental insight about development is that a complex constructive algorithm, itself the product of evolution [12], and also sensitive to environmental states, lies between variations in genotype and variations in the resulting phenotype. Most phenotypic variation therefore does not admit a simple decomposition in which independent variations in traits result from independent variations in genes and some variation need not reflect a change in the genotype at all. Games again provide a framework to capture structural relations that fall between changes in genotype and phenotype, for which we will provide a few simple examples.

Paradoxically, selection, the oldest and most widely invoked aspect of evolution, has the least simple correspondence to any single class of elementary events. It refers to the aggregate consequences of many mechanisms that impact longevity and fecundity. Both in descriptive statistics and in formal models, once the transmission process and the mechanisms of variation have specified the dynamics of descent, the remaining aspects of selection are aggregated into fitness and its covariance with other characteristics of the individual or the population. Since our constructions will use well-defined concepts of transmission process and fitness, the complexity behind the concept of selection will not be a source of trouble.

2.2 The Price equation and closures for processes described by population genetics

We now show how formal models instantiate these concepts, reviewing the definition of fitness, a few key relations including the Price equation and Fisher's fundamental theorem, the problem of closures and the systematic elaboration of games.

2.2.1 The accounting identity for one-step processes with a given initial state

Transmission models in population genetics are *Markovian*, meaning both that they are memoryless if a sufficiently rich type space is assumed[5] and that the stochastic process is defined by its time-local behavior in the form of a *generator* acting on the instantaneous population state [13].

Because our interest is the role of games, we will minimize our discussion of the structure of transmission processes. Individuals will be permitted to reproduce independently. In particular, we will suppose that we may pass freely back and forth between the discrete-time and continuous-time formulation of Markov processes.

[5] The type space required for the Markov assumption may be too rich to be tractable in many cases. Although in formal models this is not a problem, in practice the Markov assumption is sometimes considered a strong constraint. How much the Markov assumption is sacrificed by coarse-graining the type space, into a description in terms of genes, is a similar question to how much information is lost by approximating fitness with additive or other low-order models.

Our introduction of the Price equation will allow general branching processes (subject to fixed population), but in examples we will assume for simplicity that the discrete-time process can be modeled as a *Moran process* [13, 14][6] on sufficiently short intervals[7]. The Moran process is a model of asynchronous reproduction and lends itself to a continuous-time representation in master equations. We fix population size N because this parameter controls fluctuation strengths in models whose other parameters are stationary. Constant fluctuation strength simplifies examples, though our methods (primarily developed in chapter 4) admit changing populations with no formal difference.

In the discrete-time formulation, a change model known as the *one-step process* is the generator of the Markov chain. We first describe the accounting identity for one step with a known initial state and then consider the problem of estimating the generator from moments of a distribution over initial states. The moments of the distribution may be more robust summary statistics in small samples, but they are incomplete to identify the one-step process.

2.2.1.1 Asynchronous branching-and-removal processes with fixed population size

A population of N individuals is partitioned into D types indexed $i \in 1, \ldots, D$. The population state at any instant is a vector $\mathrm{n} \equiv (\mathrm{n}_1, \ldots, \mathrm{n}_D)$ giving the number of individuals n_i with type i. We leave open the possibility that contacts among individuals in the population that determine fitness may not all happen with equal probability (a condition known as *assortative interaction*), but we assume that any such interaction probabilities are based on type alone, and that individuals of the same type are interchangeable.

Our types i correspond to 'genotypes', in the sense that they govern descent and are not directly assigned fitness values until a development model has been chosen[8]. Although we retain the term 'genotype' to make use of biological intuition for the potential size and complexity of the type space, we will not assume that the space of types $\{i\}$ is built up from more elementary genes. We instead suppose that the type space is primitive and is empirically defined. The identification of genes will follow as we attempt to capture structure in the type space or in development, on which fitness or transmission depend.

[6] The Moran process is a special case of the *birth–death process*, which conserves total population number.

[7] We wish to emphasize the conceptual distinctions behind model classes. The *inability* to pass freely between discrete and continuous time implies a *synchronization* of generations. Similarly, a continuous-time Markov process aggregated within discrete intervals may approximate a *Wright–Fisher* process [13, 15], which may be more or less well approximated by a Moran process acting over multiple generations. The *inability* to substitute the Wright–Fisher process by a Moran process implies synchronized reproduction of multiple offspring or multiple deaths.

[8] To the extent that development depends on interaction of multiple genotypes, the closest concept to 'phenotype' that appears in this construction will be the vector of payoffs at the outcome of a game. Phenotype is therefore a property of tuples of player genotypes, which in game theory corresponds to *strategy profiles* [16, 17]. The notion of a 'genotype–phenotype' map for the individual who is the unit of transmission is not required in this analysis, but could be defined as a summary statistic from projection of game outcomes onto the focal player's payoff, conditioned on the population state.

Branching and removal (birth and death) occur on discrete time intervals of length Δt. We do not assume that all individuals in the population reproduce in a single time interval (reproduction is asynchronous) and one or more (or zero) offspring may be produced. At each branching event, a set of individuals from the existing population (which may include the parent) are chosen for removal at random to keep the number of individuals, N, fixed. The type of the offspring may preserve or may differ from that of the parent, but we suppose that the offspring types j have the same values $j \in 1, \ldots, D$ (the type space is complete even if not all types are populated in a generation). This set of branching processes is rich enough to describe imitation dynamics or other forms of oblique transmission [18, 19] as well as literal reproduction.

2.2.1.2 *Accounting for population number: defining fitness and parameterizing the transmission process*

Fisher's contribution to clarifying the concept of fitness came from grouping offspring according to the parent's type [6, 20]. As understood by Fisher and subsequently presented systematically by Price [8][9], this way of partitioning populations is the step that ensures that fitness is always well defined for population processes in which each offspring has a unique parent. In the accounting of Fisher and Price, numbers in the offspring generation are conventionally denoted by primes.

We let $\mathrm{n}'_{j|i}$ denote the number of offspring of type j from all parents of type i over a single time step Δt. Fitness is expressed in samples from the one-step branching/removal process as a random variable w_i giving the total number of offspring to parents of type i in that pair of generations,

$$\mathrm{n}_i \mathrm{w}_i \equiv \sum_j \mathrm{n}'_{j|i}. \tag{2.1}$$

The change in the number of individuals of any type j between the parent and offspring generations results from both fitnesses in the parent generation and mechanisms that cause offspring type to differ from parent type. These changes are denoted $\Delta \mathrm{n}_j$, and are given by

$$\sum_i \mathrm{n}'_{j|i} \equiv \mathrm{n}_j + \Delta \mathrm{n}_j. \tag{2.2}$$

If total population number is conserved, we must have $\sum_i (\mathrm{w}_i - 1)\mathrm{n}_i = 0$, so the $\{\mathrm{w}_i\}$ cannot all be independent, and we have the constraint

$$N = \sum_i \mathrm{n}_i = \sum_i \sum_j \mathrm{n}'_{j|i}. \tag{2.3}$$

[9] Fisher's own writing, which heavily emphasized the importance of covariance of traits with fitness, but sometimes neglected to mention other sources of change in population states, led to considerable confusion in the use and interpretation of his results for several decades [20]. The cause for confusion was removed by Price's grouping of other sources of change in offspring type or condition which are conditionally independent of fitness into 'environment' terms, in the Price equation which is a complete accounting identity.

Selection comprises all processes affecting offspring number that depend only on the type of the parent, whatever type the offspring have[10]. All other processes affecting the distribution of offspring types are subsumed under the transmission process[11]. To reflect this division, any distribution of type-j offspring from type-i parents may be separated into a type-preserving term that simply counts offspring, and a remainder whose sum over types j is zero, as

$$\mathrm{n}'_{j|i} = \delta_{ji}\mathrm{n}_i\mathrm{w}_i + \left(\mathrm{n}'_{j|i} - \delta_{ji}\mathrm{n}_i\mathrm{w}_i\right). \tag{2.4}$$

The average over realizations of the one-step update process, starting from a definite composition n, is denoted $\langle\ \rangle_{\mathrm{n}}$. When applied to the reproduced number $\mathrm{n}'_{j|i}$ in (2.4), the conventional notation used is

$$\left\langle \mathrm{n}'_{j|i}\right\rangle_{\mathrm{n}} = \left\{\delta_{ji}\left[1 + \Delta t\left(f_i(\mathrm{n}) - \phi(\mathrm{n})\right)\right] + \Delta t\mu_{ji}(\mathrm{n})\right\}\mathrm{n}_i. \tag{2.5}$$

The scaling of terms to linear order in Δt is the basis for a continuous-time limit if the coefficients of Δt are stable in sample averages as $\Delta t \to 0$. By construction from (2.4), $\sum_j \mu_{ji}(\mathrm{n}) \equiv 0$, so $\mu(\mathrm{n})$ is a stochastic matrix. We will refer to it as the *mutation matrix*[12].

For population processes that preserve total number, the D fitness terms f_i cannot be independently determined from the $D - 1$ independent components of the $\{\mathrm{w}_i\}$. We therefore write fitness f_i offset from a population-averaged fitness $\phi(\mathrm{n}) \equiv (1/N)\sum_i \mathrm{n}_i f_i(\mathrm{n})$.[13] Although (2.5) resembles the classical *replicator equation* [21] in form, it differs in the important respect that it describes actual population numbers n_i. The replicator equation, defined for fractions n_i/N, must *by construction* include the fitness offset ϕ in (2.5), which is introduced by the dynamics of the normalizing factor $1/N$ even when the population size itself is not constant. The incorporation of ϕ in models of actual population size, as we have noted, implies restrictions on the time dependence of fluctuations.

In the examples below, with D types, for the sake of simplicity and in order to isolate other roles of symmetry, we consider only the isotropic mutation process

$$\mu_{ij} = 1 - D\delta_{ij}. \tag{2.6}$$

[10] Note that this dependence can be trivial, if fitness is equal for all parent types.

[11] Transmission effects may include consequences of mutation, selective reproduction such as meiotic drive reflecting competition at other levels not modeled explicitly, feedbacks from the environment, or other factors.

[12] When we return to consider models for f and μ in the next sub-section, we will admit arbitrary polynomial expansions over $\{\mathrm{n}_i\}$ for both terms. Therefore fitness and mutation have some overlap in the terms they can contain. We may identify $\mu_{ji}(\mathrm{n})$ as any collection of terms that form a stochastic matrix and that include a sum over types i different from the focal offspring type j. This contrasts with the fitness terms, which must all contain at least one factor of n_j.

[13] When we derive minimal stochastic process models from their first-moment constraints in chapter 4, the overall magnitude assigned to fitness will appear in the strength of symmetric diffusion, even though it cancels from selection effects.

2.2.1.3 *Consequences for arbitrary observable functions of population state: the Price equation*

For any function $g_i(\mathrm{n})$ defining some attribute for each type i in the context of a population composition n, we will denote the population average by

$$\bar{g}(\mathrm{n}) \equiv \frac{1}{N}\sum_i \mathrm{n}_i g_i(\mathrm{n}). \tag{2.7}$$

The Price equation provides a decomposition of the intergenerational change in such population-averaged observables, into effects attributable to selection and those from environmental change.

To make use of Fisher's definition of fitness, the changes in each term $\mathrm{n}_i g_i(\mathrm{n})$ in the sum (2.7) are written as a sum of three rather than only two effects[14]. The first effect is the change from a distribution $\{\mathrm{n}_i\}$ in the parent generation to a distribution $\{\mathrm{w}_i\mathrm{n}_i\}$ of *offspring*, partitioned not by the offspring's own types but by their parents' types. The second effect comes from the dependence of $g_i(\mathrm{n} + \Delta\mathrm{n})$ on a changed population state and the third effect comes from whatever change occurs between parents' types and their offsprings' types during transmission. Only the first of these three contributions is directly due to selection; the remaining two terms are grouped together as environmental sources of change, written as

$$\Delta g_i(\mathrm{n}) \equiv g_i(\mathrm{n} + \Delta\mathrm{n}) - g_i(\mathrm{n}) + \sum_j g_j(\mathrm{n} + \Delta\mathrm{n})\left(\frac{\mathrm{n}'_{j|i}}{\mathrm{n}_i\mathrm{w}_i} - \delta_{ji}\right). \tag{2.8}$$

The difference between the population average of the $\{g_i\}$ in the parent and offspring generations in a sample may then be decomposed as

$$\begin{aligned}
\sum_j\left(\mathrm{n}_j + \Delta\mathrm{n}_j\right)g_j(\mathrm{n} + \Delta\mathrm{n}) - \sum_i \mathrm{n}_i g_i(\mathrm{n}) &= \sum_i\left\{\sum_j \mathrm{n}'_{j|i} g_j(\mathrm{n} + \Delta\mathrm{n}) - \mathrm{n}_i g_i(\mathrm{n})\right\}\\
&= \sum_i \mathrm{n}_i g_i(\mathrm{n})\left(\mathrm{w}_i - 1\right) + \sum_i \mathrm{n}_i\mathrm{w}_i\Delta g_i(\mathrm{n})\\
&= \sum_i \mathrm{n}_i\left[g_i(\mathrm{n}) - \bar{g}(\mathrm{n})\right]\left(\mathrm{w}_i - 1\right) + \sum_i \mathrm{n}_i\mathrm{w}_i\Delta g_i(\mathrm{n})\\
&= N\mathrm{Cov}_\mathrm{n}(g, \mathrm{w}) + N\mathrm{E}_\mathrm{n}(\mathrm{w}\cdot\Delta g(\mathrm{n})).
\end{aligned} \tag{2.9}$$

In the third line of (2.9), we have used the fact that the averaged fitness satisfies $\sum_i \mathrm{n}_i\mathrm{w}_i = N$ to subtract a (free) factor of $\bar{g}$, leaving the covariance in the last line. Equation (2.9) is the standard form for the *Price equation.*

---

[14] One's first impulse is to simply track the change in total numbers n_i and the functional dependence of $g_i(\mathrm{n})$ on all n_j. The split of the change in n_i into two separate terms is the step due to Fisher and Price that at first is unfamiliar.

The key to understanding the division of terms in the Price equation is recognizing that its averages are performed in the parent distribution, though they are averages of functions computed from offspring properties. It is under the logic of this division that the first (covariance) term defines the contribution of fitness for any population process and any observables $\{g_i\}$ and the remaining 'environment' term includes not only effects from interaction in the offspring generation, but also type changes upon transmission that are treated as falling outside the scope of total fecundity used to define fitness.

It is also essential to remember that the existence of a relation of the form (2.9) for any population-genetic process and any observable g is only ensured because the variables $\{\mathrm{w}_i\}$, $\{g_i\}$, and $\{\mathrm{n}'_{j|i}\}$ are summary statistics computed from each realization of the population process. The Price equation as a relation among summary statistics is not yet a hypothesis about the evolutionary process or a model for any of its parameters. Moreover, the observable $g_i(\mathrm{n})$ is a function of type i and population state n, but the fitness is only computable as a summary statistic of the parents' population state by virtue of the ability to assign a unique parent to each offspring. To go beyond the accounting identity we must propose models for fitness.

2.2.2 From summary statistics to models, bringing in the problem of moment closure

To replace summary statistics $\{f_i\}$ and $\{\mu_{ij}\}$ with models, we must make further assumptions about the ensemble in which $\{f_i\}$ and $\{\mu_{ij}\}$ sample the generating process[15]. Steven Frank has emphasized [20] that the Price equation relates three empirical quantities—initial population state, final population state and process description—and that from any two of these, it may be used to estimate the third[16]. However, summary statistics do not determine the evolution of population states or probability distributions except under additional (often rather strong) assumptions known as *closures*. Therefore most problems of model estimation, phylogenetic reconstruction, or prediction, are undefined prior to the choice of closures.

The standard problem in population genetics is the choice of *moment closures* [22, 23], which we consider in this section. Other problems of closure also arise, however, including the selection of robust parametric models of fitness in the presence of dynamical noise or estimation error. We return to the latter in section 3.3.

---

[15] This ensemble may result from accumulation of generations in a single population, reflecting time-stationarity, but it could instead come from replication of populations whose short-term evolution is tested for reproducibility, even if it is not time-stationary.

[16] Each estimation process reflects one aspect of evolution. Reconstruction of the diachronic past from synchronic variations and a process model (usually chosen to be random over variables that are not specifically constrained) is the object of phylogenetics. Estimation of process models is the attempt to extract evolutionary regularities or laws, and estimation of future states from inferred or assumed laws is the standard problem of prediction.

2.2.2.1 One-step averages over samples with a fixed initial state

The covariance term in (2.9) is linear in the only variable (w_i) that depends on the one-step update. If it is possible to obtain a sample of updates from a common initial state n, a fitness model may be directly assigned, as

$$\left\langle w_i - 1 \right\rangle_{n} = \Delta t\left[f_i(n) - \phi(n)\right]. \tag{2.10}$$

The transmission process is not generally identifiable in terms of linear expectations alone, as it may involve covariance of w with the change Δn. g itself may also be a non-linear function of n. A linear estimator in the $\{n'_{j|i}\}$ that ignores these covariance terms may be used to approximate $\mu(n)$ by

$$\frac{1}{\left\langle w_i \right\rangle_{n}}\left\langle \frac{n'_{j|i}}{n_i} - \delta_{ji} w_i \right\rangle_{n} = \frac{\Delta t \mu_{ji}(n)}{1 + \Delta t\left(f_j(n) - \phi(n)\right)}. \tag{2.11}$$

If we use these sample estimators as models, supposing they are stationary in time, and substitute $f_i(n)$ for the observables $g_i(n)$ in (2.9), then the covariance term in the last line expresses *Fisher's fundamental theorem* [6].

Note that for the expected number n_j, the linear estimator (2.11) is exact. We therefore have, as a definition, the relation for first moments under the one-step process

$$\frac{\left\langle n_j + \Delta n_j \right\rangle_{n} - n_j}{\Delta t} = \left[f_j(n) - \phi(n)\right]n_j + \sum_i \mu_{ji}(n) n_i. \tag{2.12}$$

In many situations, the formula (2.12) may be taken from discrete to continuous time, where its form is the basis for the deterministic differential equation known as the replicator equation.

2.2.2.2 Averages over the initial state and the mean-field approximation for moment closure

Suppose that the robust observables are sample estimators for mean population numbers, where we assume that the true mean values are the first moments of an underlying distribution that we do not know and are trying to infer. We denote the true means by[17]

$$\bar{n}_i \equiv \sum_{n} n_i \rho_n, \tag{2.13}$$

where ρ_n is the (generally time-dependent) underlying distribution appropriate to the system context we assume and the set of questions we wish to ask[18].

[17] We may regard n_i in (2.13) as the expectation in the one-step process of an observable g_j which for any type j is just the indicator function δ_{ji} for some particular i. It is the entire one-step process that is then averaged under the distribution ρ_n.

[18] That is, ρ_n might be a smooth distribution in the context where a stationary population process has acted for a long time, or it might be singular because it is the outcome of a shock, or because it reflects a particular initial sample over types made by the experimenter but outside the model of the evolutionary dynamics.

If (2.13) gives the average in a generation that we treat as parents, the corresponding quantity in the offspring generation is obtained from (2.12) as

$$\begin{aligned}\bar{n}'_j &\equiv \sum_{\mathrm{n}} \rho'_{\mathrm{n}} \mathrm{n}_j \\ &= \sum_{\mathrm{n}+\Delta\mathrm{n}} \rho'_{\mathrm{n}+\Delta\mathrm{n}} \left(\mathrm{n}_j + \Delta\mathrm{n}_j\right) \\ &= \sum_{\mathrm{n}} \rho_{\mathrm{n}} \sum_i \left\langle \mathrm{n}'_{j|i} \right\rangle_{\mathrm{n}} \\ &= \bar{n}_j + \Delta t \sum_{\mathrm{n}} \rho_{\mathrm{n}} \left\{ \left[f_j(\mathrm{n}) - \phi(\mathrm{n}) \right] \mathrm{n}_j + \sum_i \mu_{ji}(\mathrm{n}) \mathrm{n}_i \right\}. \end{aligned} \tag{2.14}$$

The map (2.14) from ρ_{n}, $\{f_i\}$ and $\{\mu_{ij}\}$ to $\{\bar{n}'_j\}$ is exact but non-invertible. In addition to higher-order moments of ρ'_{n} that are not identified by $\{\bar{n}'_j\}$, the estimation of $\{f_i\}$ itself is no longer drawn from a simple mean of w_i in a fixed background n as in (2.10), but now involves an average over n which may incorporate fluctuation corrections from undetermined higher-order moments of ρ_{n}.

A strong *approximate* moment-closure assumption, valid when the distributions ρ_{n} and ρ'_{n} are sharply peaked around a single value, is the *mean-field approximation* (MFA). It replaces (2.14) by a relation of similar form, in which $\{f_i\}$ and $\{\mu_{ij}\}$ are simply evaluated at $\bar{\mathrm{n}}$, which is assumed to change slowly as well as having only small fluctuations. From the assumption of slow change, we also replace $\left(\bar{n}'_j - \bar{n}_j\right)/\Delta t$ with a derivative in continuous time, to write

$$\frac{\mathrm{d}\bar{n}_j}{\mathrm{d}t} = \left[f_j(\bar{n}) - \phi(\bar{n}) \right] \bar{n}_j + \sum_{i=1}^{D} \mu_{ji}(\bar{n}) \bar{n}_i. \tag{2.15}$$

The MFA is a useful approximation but not a proper moment closure, in that it will generally be violated even within stochastic models that it is used to define.

2.2.3 *k*-player normal forms provide a polynomial expansion in frequency-dependent selection

Suppose we restrict fitness models to be approximated by polynomials in the agent numbers $\{\mathrm{n}_i\}$.[19] Then the coefficients of all order k terms, in the expansion of the vector $\left[f_i(\mathrm{n})\mathrm{n}_i\right]$, form a rank k matrix that may be interpreted as the expected-payoff matrix in a k-player normal-form game with random matching.

To order $k \leqslant 2$ this is the affine expansion

$$f_i = \alpha_i + \sum_j a_{ij} \left(\mathrm{n}_j - \delta_{ij}\right), \tag{2.16}$$

[19] These are simply the fitness functions that converge to their Taylor's series expansions.

in which α_i is an inherent fitness of type i, and a_{ij} is the payoff to type i when matched with type j in a two-player normal-form game[20]. By the freedom to define α_i we adopt the convention of subtracting a term δ_{ij}, so that in the game interpretation no agent plays against itself[21]. The two-player normal form is the form in which evolutionary game theory was introduced by Maynard Smith and Price [24][22].

Given sufficiently many independent samples of the set $\{\mathrm{w}_i\}$ in different population contexts n, estimators for the parameters $\{\alpha_i\}$ and $\{a_{ij}\}$ may be determined by linear regression. (In the examples to follow we will further take all $\alpha_i = 0$ as a simplification.) One writes the relation between population-corrected fitness (2.10) and the population state parametrically as

$$\frac{\mathrm{w}_i - 1}{\Delta t} = \sum_j a_{ij}\left(\mathrm{n}_j - \delta_{ij}\right) - \frac{1}{N}\sum_k \mathrm{n}_k a_{kj}\left(\mathrm{n}_j - \delta_{kj}\right) + \epsilon_i \tag{2.17}$$

with a residual error term that we denote ϵ_i, which also absorbs any additional (unmodeled) higher-order frequency-dependent contributions to fitness.

The residual error will have zero mean if the regression coefficients are chosen to minimize the residual sum of squares (RSS)

$$\sum_i \mathrm{n}_i \epsilon_i^2 = \sum_i \mathrm{n}_i \left(\frac{\mathrm{w}_i - 1}{\Delta t} - \sum_j a_{ij}\left(\mathrm{n}_j - \delta_{ij}\right) + \frac{1}{N}\sum_k \mathrm{n}_k a_{kj}\left(\mathrm{n}_j - \delta_{kj}\right)\right)^2. \tag{2.18}$$

The values of $\{a_{ij}\}$ obtained by minimizing the RSS of (2.18) are sample estimators, which fluctuate as a result of sampling noise. They become less reliable as the magnitude of sample fluctuation increases and in non-linear systems they may also be biased. For systems with neutral or nearly neutral directions, in which some sub-set of the $\{\mathrm{w}_i\}$ take very similar values along extended sub-sets of configurations n, fluctuations are large precisely where accurate parameter estimation is most difficult. The impact of sample fluctuations may be mitigated by coarse-graining the state space indices n, or estimating $\{a_{ij}\}$ implicitly from population-level properties such as approximate rest points in long-term time averages.

We will show in a repeated prisoner's dilemma (RPD) example in section 7.1 that time-averaged steady population states may be consistent with a model in the same symmetry class as the microscopic interaction, but that the best estimators for the $\{a_{ij}\}$ from the population-level time averages may give parameters different from

[20] By freedom to vary the magnitude of a_{ij}, we may reference fitness to the case that each agent plays the others, on average, exactly once in a generation.

[21] This convention leads to a simple map from the parameters in the fitness function and those in the stochastic-process model, which we develop in chapter 4.

[22] The place of low-order polynomials and therefore low-order games may be more important than their restricted form suggests. The central limit theorem for many kinds of distributions projects infinite-dimensional manifolds onto low-dimensional sub-spaces under aggregation [25, 26]. In the large-deviations rate functions computed in later chapters, the parameters in low-order polynomial models of fitness will occupy the place of stable distribution parameters in the central limit theorem. Therefore low-order normal form games may be general in population-level regressions even when they do not describe individual interactions.

those actually experienced by agents when they interact in individual events to play the game.

Another reason regression coefficients may not correspond directly to individual-level payoffs is that the default mapping from an affine regression of fitness to a normal form game is one in which agents are matched randomly with uniform probability. Agent matching may be assortative, or non-random with respect to types, but we will not pursue the additional problem of estimating matching statistics in this monograph. Some degree of players' responses to other players' types is captured in game models with repeated play, in which early rounds can be used to 'signal' agents' types to one another. For many purposes this offers a more explicit process model of many of the same interaction biases as can be achieved with assortative matching.

2.3 Symmetry: the unifier of scale-dependent effective theories

The polynomial expansion of fitness is a general framework and as such it leaves an infinite set of parameters to be chosen. Section 2.2.2.2 showed that even the dependence of these on mean population dynamics cannot be determined from samples drawn from an unknown distribution over population states. The mean also is only one moment; an infinite set of observables $g_i(\mathrm{n})$ would be required to identify the full functional forms of fitness from regressions in the Price equation (2.9). Finally, because higher-order moments of a distribution ρ_{n} will generally contribute to sample estimators for robust observables such as the mean, even the question of what a set of parameters stand for as a representation of a system cannot be answered in general without a reference to scale.

We address all three of these problems of identification by grouping games and the phenomena they predict according to symmetries, which are properties associated with a system at all scales. The representations of symmetries by individual- and population-level dynamics, although they may exhibit different regimes for a single generating model, are nonetheless invariant under changes of parameters within open sub-sets of parameter space and are therefore robust aspects of evolutionary dynamics.

2.3.1 Effective theories, universality classes and closures

A polynomial expansion of a game model, in which all orders are implicitly present, and which is the most general form consistent with the assumed symmetries of a process, is called an *effective theory* [27]. We keep all parameters in an effective theory (at least implicitly) because to further restrict them would be to impose more information than is required by the symmetry classification. Such excess information is not a robust feature and generally is not preserved under averaging anyway[23]. The reason admitting an infinite set of parameters—almost all of which we do not know

[23] We may choose to restrict the parameters in the effective theory to fall within the open parameter domains associated with particular representations of symmetries, in cases of symmetry breaking, without losing the robustness of the approach.

and cannot estimate precisely—does not destroy prediction is that under averaging, infinite-dimensional sets of effective theories will generally converge on finite-dimensional stable families, by generalizations of the central limit theorem [28]. The domain of attraction of such a flow to a low-dimensional limiting family of models is called a *universality class*. Although the stable family may still have infinitely many parameters in a polynomial expansion, it is only finite-dimensional because most parameter variations in the microscopic model self-average away, leaving the differences between one microscopic model and another to be expressed in only finitely many dimensions of difference in the aggregate-level theory [26].

Effective theories and universality classification provide an alternative approach to closure, from the use of moment-closure approximations. In the moment-closure approach, we regard a finite-order model as the least-biased description of a stochastic process and then we further compute its consequences using moment-closure approximations. Instead, we may regard the infinite-order but finite-dimensional manifold of theories that are self-consistent under rescaling as the least-biased description of a stochastic process known only to finite precision. The self-consistent manifold is closed *exactly*, although we generally do not have an expression for the infinite set of parameters that define it. Finite-order models used in computation are then *approximate theories*. As long as they are in the same universality class as the correct theory, quantities such as scale-dependent *change* of parameters due to correlation effects, as calculated in the approximate theory, will also approximate the true changes as these would be calculated in the self-consistent theory (if it were known).

Errors in high-order terms may arise because the approximate (finite-order) theory differs from the self-consistent theory, or they may reflect additional information about the real system. In either case, they do not destroy the robustness of predictions because they can be absorbed in small shifts of the leading-order terms, which are calibrated to the robust summary statistics of the empirical phenomenon. In a major conceptual revolution in the 20th century, *all* fundamental theories in condensed matter and elementary particle physics have come to be understood as effective theories and used in this manner [26] and this interpretation is the basis of their strength as theories of causation. The constructions leading to their reconceptualization are mathematically the same as those we will use in the analysis of stochastic evolutionary dynamics.

2.3.2 Symmetry as a fundamental concept for evolutionary games

The last conceptual dimension of a general theory of evolutionary games is therefore an abstraction of the major robust effects defined in terms of symmetry groups and their representations by population states. The most important of these is spontaneous symmetry breaking, which is the principle source of complexity and large-scale dynamics in the hierarchy of matter [27, 29]. Symmetry breaking may occur within either continuous or discrete symmetry groups, leading either to episodic or continuous large-scale dynamics. Some types of symmetry breaking lead to groups of possible ordered population states that offer limited complexity in higher-level structure or dynamics, while other types of symmetry breaking may create potential complexity that grows without bound as the dimensionality of the agent-level state space grows. The fact that symmetry remains a property of the system as a whole can be used to prove the existence of properties such as persistent dynamics, in regimes where direct calculation by approximation methods is infeasible.

2.3.3 Collective fluctuations mediate large-scale dynamics and bridge effective-theory representations at different scales

The elementary events in a Moran process are the death or the birth (possibly accompanied by mutation) of single agents[24]. Single-agent events do not, however, provide an efficient basis in which to represent the causes of macroscopic dynamics, or the correlation effects that feed back to influence average fitness or transmission parameters. These feedbacks are the source of 'flow' in the infinite hierarchy of parameters in an effective theory as the scale of observation is changed from the short term to the long term [26].

The robust phenomena that cause changes of symmetry representation, or scale-dependent parameter changes, are collective waves of type-change by macroscopic collections of agents. They are caused by type changes of individual agents, but they do not depend on the behavior of any single agent or even of a definite number of agents[25].

We capture these collective fluctuation effects by working not with probability distributions ρ_n, but with the *Laplace transforms* of these distributions, which we introduce starting in section 4.3.1. The Laplace transform, like the Fourier transform, is a dual representation to the basis of integer lattice-points $\{n\}$, in which low-order terms correspond to the low-order cumulants of dynamical changes in ρ_n. In the same way as symmetry representations are not sensitive to parameter changes within convex regions of parameter space, collective fluctuations are not sensitive to many details of interaction at the individual/event scale, which may be hard to identify or model.

2.3.4 Symmetry breaking by short-term dynamical states and restoration by time averaging

The symmetries of a game model are the transformations that leave its fitness and transmission coefficients unchanged. These may be sub-groups of discrete permutations among agent types (for example, mirror reflection, or cyclic permutation) known as *point groups*, or they may come from displacing the time coordinate by any of a continuous range of values, forming what is known as a *Lie group* [30].

Symmetries imposed on individual-level transition coefficients, as part of the definition of a game, may become hidden if population states that determine

[24] The sub-set of these elementary events which actually change the distribution of agent types are separated by *waiting times* which are Poisson-distributed relative to the birth–death period. In some representations of population processes the waiting time distribution may be used directly, rather than the discrete-time model that tracks all births and deaths including those that do not change the population.

[25] Metaphorically, the straw may break the camel's back, but only if the camel has already been overloaded. Conversely, when it is overloaded, some straw will inevitably be the last. Which it will be may not be predictable, but for the robust outcome, this prediction also is not essential.

frequency-dependent fitness become asymmetrical. Such asymmetric states may be rest points, limit cycles, or more complex attractors, onto which the population state converges. If asymmetric population states cause agents to experience asymmetric forces of selection, even though the defining parameters of the game are symmetric, then the possible population states themselves must be *degenerate* as an expression of the hidden symmetry. For example, if the fitness function has a reflection symmetry under exchange of two types Left (L) and Right (R), but the population state converges to a rest point that is asymmetric on either the left or the right, then there must be two rest points that are identical under the reflection that either rest point alone masks. The same is true for continuous symmetries, which lead when they are broken to a one-dimensional manifold of solutions along a limit cycle or attractor. Which state is occupied then becomes an 'accident', of initial conditions, or of the observation time sampled over a long run in which the population occasionally (in the case of discrete rest points) or through a slow random walk (in the case of limit cycles) changes its state.

If asymmetries in population states that form through fluctuations decay rapidly—over timescales comparable to the elementary birth–death time—then the system dynamics is said to be *ergodic*[26]. If spontaneously formed asymmetries in population states persist over times much longer than the individual generation time, the population state is said to *spontaneously break* the defining symmetries of the underlying process[27].

The fact that hidden symmetry implies *strict* degeneracy of population states ensures autonomous dynamics at the population level[28]. Emergent population

[26] The technical definition of ergodicity is complicated to express, because it requires that samples from a dynamical trajectory accumulate to a distribution that is statistically equivalent to some reference distribution representing an ensemble. Here we will be concerned only with the ensembles that reflect the defining symmetries and beyond this we will not require a more precise criterion for ergodicity.

[27] Symmetry breaking is familiar from (second-order) phase transitions [28, 31, 32] in either equilibrium or non-equilibrium stochastic processes. Spontaneous symmetry breaking *sensu stricto* is defined formally by the existence of a large-system limit in which, asymptotically, asymmetric population states persist forever. Because we can always recover the asymptotic property (if it pertains) as a regular limit of finite-time persistence, we consider all forms of hidden symmetry in finite-time as well as infinite-time samples. The more general case includes many important forms of timescale separation and the scaling approach to the asymptote contains additional information about the symmetry-breaking transition beyond that present in the limiting distribution itself.

[28] This becomes a *non-trivial* property when the timescale separation between population state-changes and the generation time becomes very large. At a small timescale separation, bifurcations between asymmetric states and those between symmetric states may not seem qualitatively different. Both cases have basins of attraction and the non-symmetric case may simply have different occupancy times in different attractors. However, if the population dynamics separates exponentially (in some parameter such as selection strength) from the generation scale, in an asymmetric bifurcation the relative state occupancies will separate exponentially as well. In other words, all but the most frequently occupied state will come *never* to be occupied and the population state will effectively become non-dynamical. The observation that many properties of stochastic processes are fragile under exponential separation of scales if they are not protected by symmetries is known as the *hierarchy problem*. It is the origin of the statement in [33] that stochastically stable equilibria will 'typically' be unique, suggesting that exceptions involve some form of fine tuning and should occur rarely. The one class of exceptions in which this violation of typicality is not regarded as finely tuned is the case where it is ensured by symmetries.

dynamics will generally be stochastic and the stochasticity drives convergence to the ergodic distribution—which restores the expression of the underlying symmetries—under sufficiently long time-averaging. In such long-time averages, fluctuations in the population state become an autonomous variable with fitness consequences at the individual level, which then feed back to re-enforce the stability of population states. Spontaneous symmetry breaking is therefore an important source of multiple units of selection, and of dynamics that is coupled over multiple scales.

2.4 The major categories of symmetries and symmetry breaking

Symmetry breaking and restoration of an ergodic distribution by a time or ensemble average are very general phenomena [28, 29, 31, 32]. The particular symmetry groups affected, however, and the ways they are broken by sets of ordered population states, can vary widely. Each type of symmetry and type of symmetry breaking leads to a qualitatively distinctive kind of dynamics at the population scale. Whether a discrete or a continuous underlying symmetry is broken determines whether the characteristic timescale for population dynamics grows exponentially or only polynomially (or not at all) with population size, relative to the generation scale. This difference affects the convergence of asymptotic expansions and thus the computational methods that can be used. More complex forms of symmetry breaking introduce a new kind of macroscopic dynamics known as 'creep', and require qualitatively new calculation methods developed in the theory of glasses [34, 35].

2.4.1 Four symmetry examples

In chapters 5–8 we derive the stochastic dynamics for four standard game models representing different major categories of symmetry and symmetry breaking, and demonstrate the gross phenomenology produced by collective fluctuations in each class. Here we briefly summarize and contrast the distinct behaviors. When we have introduced methods for computing the distribution over population states and its generating functional in chapter 4, we will return in section 4.4 to show that each class of symmetry effects in chapters 5–7 draws from a different contribution to the large-deviations function. The example of breaking of a high-order permutation symmetry in chapter 8 is of a qualitatively different order of complexity than the other examples, for which we do not know how to present a systematic large-deviations analysis, so we are able to review only the qualitative properties of that case in an elementary correlation analysis.

2.4.2 Simple versus complex symmetry breaking: how does the number of ordered system states scale with the size of the state space?

The distinction between 'simple' symmetry breaking (which we will also term *classical*) and 'complex' symmetry breaking (which we will also term *glassy*) is made, not based on single models, but based on the behavior of population states in collections of models among which the number of agents, agent states, or some other variable is changed as a scaling variable. In simple symmetry breaking, the cardinality (if discrete) or topology (if continuous) of the set of ordered population states and the symmetry group that relates them do not change as the scale of the underlying system changes. The classical pitchfork bifurcation developed in chapter 5 and the Hopf bifurcation in chapter 6 are examples of this kind.

In complex symmetry breaking, the number of ordered population states grows (typically exponentially) with the scale of the underlying system. A model of linkage disequilibrium among heterotic loci on a chromosome introduced by Lewontin [11] and studied in chapter 8, in which the number of heterotic loci is the scale parameter, shows an example of complex symmetry breaking. Complex symmetry breaking in equilibrium systems was first worked out as a description of glasses and is now used to describe a number of other combinatorial optimization problems [34, 36] in which the constraints on a system cannot all be jointly satisfied, a condition known as *frustration*.

2.4.3 Continuous versus discrete broken symmetries and the consequences for fluctuations

When a broken symmetry is discrete, as occurs in the example of reflection symmetry for the pitchfork bifurcation, population-level dynamics occurs by hopping between basins of attraction, with hops becoming exponentially rare in a product of the selection strength with population size. Collective fluctuations must overcome kinetic 'barriers' and the problem for the fluctuation mechanics of chapter 4 to solve is to compute the rates and trajectories of population-dynamic escapes as functions of the parameters of the individual-level interaction.

A very different behavior results when the broken symmetry is continuous, as occurs for the supercritical Hopf bifurcation from a fixed point to a limit cycle. In this example the ordered states must form a continuum, along with the group of symmetries that transform one ordered state into another. No 'barrier' impedes the action of collective fluctuations and these therefore accumulate as Brownian motion along the limit-cycle trajectory [37, 38]. The diffusivity for this Brownian motion scales as a polynomial (in our examples, it is constant) in the population size, rather than exponentially as for the case of discrete symmetry breaking.

The Hopf bifurcation in our example shows a special feature of non-equilibrium stochastic processes, including stochastic evolutionary games: the space of states itself is discrete and its symmetry group is the discrete group of cyclic permutations. The Hopf bifurcation, however, does not break this discrete group, but rather breaks the continuous symmetry of *time translation*. The continuity of time translation implies the degeneracy of positions around the limit cycle through a topological argument, for both the mean population state and its fluctuations. This result would be difficult to derive directly from the fluctuation distribution, which would need to be computed at (or beyond) all orders in a small-parameter expansion. The proof that a form of barrier-free dynamics at the population level is protected by an underlying continuous symmetry is the stochastic version of *Goldstone's theorem* [29][29].

2.4.4 Spontaneously broken symmetry contrasted with externally imposed symmetries

Systems that produce degenerate states through symmetry breaking behave differently from systems in which degeneracy is imposed by symmetries built into the underlying process. We illustrate this by contrasting a coordination game with broken reflection symmetry, from chapter 5, against a deterministic RPD game in chapter 7, in which a continuum of population states is rendered degenerate by a symmetry between certain strategy pairs. The symmetry is built into the parameters of the prisoners' dilemma game and is essential to its function, which is to use repetition to support strategies in the long term that would be strictly dominated in the short term. Whereas the hidden symmetry in the coordination game results in discrete two-scale dynamics consisting mostly of small fluctuations in an average population state with rare large regime shifts, the imposed symmetry of the RPD leads to constant large fluctuations of the instantaneous population state about a unique average state. Because the latter large fluctuations are common rather than rare, they can cause the best estimator for the long-term average state to differ from any limit suggested by the naïve MFA.

As we remarked in chapter 1, externally imposed degeneracies are an important cause of *neutrality* in genotype–phenotype maps [39–44]. Whereas the phenomenology associated with broken symmetry is robust hierarchies of multiscale dynamics, the phenomenology associated with neutrality is the addition of *evolutionary entropy* terms that convert the microscopic fitness function into what might be called a population-level 'free fitness', just as the addition of entropy corrections changes internal energy to free energy in classical thermodynamics.

Bibliography

[1] Mayr E 1985 *The Growth of Biological Thought: Diversity, Evolution, and Inheritance* (Cambridge, MA: Harvard University Press)

[2] Gould S J 2002 *The Structure of Evolutionary Theory* (Cambridge, MA: Harvard University Press)

[3] Hartl D L and Clark A G 1997 *Principles of Population Genetics* (Sunderland, MA: Sinauer Association)

[4] Ewens W J 2004 *Mathematical Population Genetics* 2nd edn (Heidelberg: Springer)

[5] Provine W B 2001 *The Origins of Theoretical Population Genetics* (Chicago, IL: University of Chicago Press)

[6] Fisher R A 2000 *The Genetical Theory of Natural Selection* (London: Oxford University Press)

[29] In equilibrium statistical mechanics, Goldstone's theorem ensures the masslessness of particles brought into existence by symmetry breaking of the vacuum. In the stochastic version it ensures the existence of noisy clocks.

[7] Frank S A 1995 George Price's contributions to evolutionary genetics *J. Theor. Biol.* **175** 373–88
[8] Price G R 1972 Fisher's 'fundamental theorem' made clear *Ann. Human Genetics* **36** 129–40
[9] Dawkins R C 1976 *The Selfish Gene* (New York: Oxford University Press)
[10] Mayr E 1997 The objects of selection *Proc. Natl Acad. Sci. USA* **94** 2091–94
[11] Lewontin R C 1974 *The Genetic Basis of Evolutionary Change* (New York: Columbia University Press)
[12] Gerhart J and Kirschner M 1997 *Cells, Embryos, and Evolution* (New York: Wiley)
[13] Ethier S N and Kurtz T G 1986 *Markov Processes: Characterization and Convergence* (New York: Wiley)
[14] Kurtz T G 1981 *Approximation of Population Processes* Society for Industrial and Applied Mathematics, Philadelphia, PA, CMBS-NSF Regional Conference Series in Applied Mathematics vol 36
[15] Gillespie J H 2004 *Population Genetics: a Concise Introduction* (Baltimore, MD: Johns Hopkins University Press)
[16] Fudenberg D and Tirole J 1991 *Game Theory* (Cambridge, MA: MIT Press)
[17] Ratliff J 2010 A folk theorem sampler pp 1–34 http://www.virtualperfection.com/gametheory/Section5.3.html
[18] Frank S A 1998 *Foundations of Social Evolution* (Princeton, NJ: Princeton University Press)
[19] Cressman R 2003 *Evolutionary Dynamics and Extensive Form Games* (Cambridge, MA: MIT Press)
[20] Frank S A 1997 The Price equation, Fisher's fundamental theorem, kin selection, and causal analysis *Evolution* **51** 1712–29
[21] Hofbauer J and Sigmund K 1998 *Evolutionary Games and Population Dynamics* (New York: Cambridge University Press)
[22] Kirkpatrick M, Johnson T and Barton N 2002 General models of multilocus evolution *Genetics* **161** 1727–50
[23] Lehmann L and Rousset F 2009 Perturbation expansions of multilocus fixation probabilities for frequency-dependent selection with applications to the Hill–Robertson effect and to the joint evolution of helping and punishment *Theor. Population Biol.* **76** 35–51
[24] Smith J M and Price G R 1973 The logic of animal conflict *Nature* **246** 15–18
[25] Ken-iti S 2001 Basic results on Lévy processes *Lévy Processes: Theory and Applications* eds T Mikosch, O E Barndorff-Nielsen and S I Resnick (Boston: Birkhäuser) pp 3–37
[26] Polchinski J G 1984 Renormalization group and effective lagrangians *Nucl. Phys.* B **231** 269–95
[27] Weinberg S 1995 *The Quantum Theory of Fields* vol 1 (New York: Cambridge)
[28] Wilson K G and Kogut J 1974 The renormalization group and the ε expansion *Phys. Rep.* **12** 75–199
[29] Coleman S 1985 *Aspects of Symmetry* (New York: Cambridge)
[30] Georgi H 1999 *Lie Algebras in Particle Physics* 2nd edn (New York: Perseus)
[31] Ma S-K 1976 *Modern Theory of Critical Phenomena* (New York: Perseus)
[32] Goldenfeld N 1992 *Lectures on Phase Transitions and the Renormalization Group* (Boulder, CO: Westview)
[33] Young H P 1993 The evolution of conventions *Econometrica* **61** 57–84
[34] Mezard M, Parisi G and Virasoro M A 1987 *Spin Glass Theory and Beyond* (Singapore: World Scientific)

[35] Fischer K H and Hertz J A 1991 *Spin Glasses* (New York: Cambridge University Press)

[36] Krzakala F and Zdeborová L 2008 Phase transitions and computational difficulty in random constraint satisfaction problems *J. Phys.: Conf. Ser.* **95** 012012

[37] Boland R P, Galla T and McKane A J 2008 How limit cycles and quasi-cycles are related in systems with intrinsic noise *J. Phys. A: Math. Theor.* **41** 435003

[38] Boland R P, Galla T and McKane A J 2009 Limit cycles, complex Floquet multipliers, and intrinsic noise *Phys. Rev.* E **79** 051131

[39] Fontana W, Wagner G and Buss L W 1994 Beyond digital naturalism *Artificial Life* **1** 211–27

[40] Gruener W, Giegerich R, Strothmann D, Reidys C, Weber J, Hofacker I L, Stadler P F and Schuster P 1996 Analysis of RNA sequence structure maps by exhaustive enumeration: I. Neutral networks *Monatsh. Chem.* **127** 355–74

[41] Gruener W, Giegerich R, Strothmann D, Reidys C, Weber J, Hofacker I L, Stadler P F and Schuster P 1996 Analysis of RNA sequence structure maps by exhaustive enumeration: II. Structures of neutral networks and shape space covering *Monatsh. Chem.* **127** 375–89

[42] Reidys C, Stadler P F and Schuster P 1997 Generic properties of combinatory maps: neutral networks of RNA secondary structures *Bull. Math. Biol.* **59** 339–97

[43] Ancel L W and Fontana W 2000 Plasticity, evolvability and modularity in RNA *J. Exp. Zool. (Mol. Dev. Evol.)* **288** 242–83

[44] Fontana W 2002 Modeling 'evo-devo' with RNA *Bioessays* **24** 1164–77

IOP Publishing

Symmetry and Collective Fluctuations in Evolutionary Games

Eric Smith and Supriya Krishnamurthy

Chapter 3

Extensive-form games: from genomes to genes

The previous chapter introduced general formalisms for modeling evolutionary population processes. Natural selection incorporates information in population distributions and development subsumes all processes within the life course that constructs phenotype and fitness from genotype in a population context. In this chapter we pass from general frameworks to the problems of characterizing particular systems. We show how genes reflect covariance in heredity and how extensive-form games reflect the sub-structure of events within ontogeny. Covariance in heredity draws from modularities in the developmental program. A natural association therefore results between extensive-form games and projections of complete genotypes onto genes. Genes capture components of covariance in the dynamics of individuals, induced by the fact that they interact through a shared and structured game. We show how projections from genotypes to genes, based on the extensive forms of infinitely repeated games, may be used to regulate strategic complexity. Again, symmetry plays a central role, as strategies of similar complexity but different symmetry may still have very different evolutionary functions because they link the ontogenetic to the generational timescale in different ways.

3.1 The extensive form: a taxonomic system for developmental programs

The Darwinian algorithm—*ex ante* random variation filtered by *ex post* selection—is an extremely general mechanism for imprinting information from boundary conditions onto distributions. Its formalization in population genetics, while imposing a further restriction of vertical descent, is nonetheless applicable to a wide range of phenomena. Fisher's fundamental theorem, that the rate of change of fitness is proportional to the genetic variance in fitness (from (2.9)), is a strong and general theorem as long as a full genotype space is available from which to compute

doi:10.1088/978-0-7503-1137-3ch3

the variance. The expansion of the genotype × population → fitness map in k-player normal-form games is a tautology.

The existence of all these general results, presented in chapter 2, provides a formal basis for evolutionary theories. Their very generality, however, ensures that by themselves they are not theories of anything in particular. The work of evolutionary analysis consists of finding the appropriate abstractions for the various units and interactions that govern the dynamics of each evolving system. In population genetics, the main tool for representing structure is the decomposition of genotypes into *genes*, heritable components that multiple genotypes may have in common, with similar alleles causing related events in development or in patterns of interaction. In game theory, the main tool for representing structure is the *extensive-form game*, a refinement of the normal form that explicitly represents the sequence of events in the course of play. In this chapter we show that, within a general theory of evolutionary games, the gene decomposition and the extensive form are related.

Our formulation of the Price equation in chapter 2 required only genotypes that are passed down through generations, possibly with mutations. There we did not introduce genes either as a necessary simplification when the full genotype space is too large to represent or for natural populations to sample, or to accommodate more complex transmission models involving crossover, in which entire genomes may not be passed down directly, while smaller units (the genes) still may be. We therefore begin this chapter by introducing a general notation to represent genotypes as collections of genes and derive the resulting expression for fitness models as combined contributions from genes and gene interactions.

We then introduce the extensive form and show how it induces natural notions of the genic concepts of locus and allele, in terms of sets of nodes on a game tree and the moves that different strategies assign to them. As an illustration of the use of genic descriptions, we consider the case of repeated games, which will furnish several examples of important symmetry and collective-fluctuation effects in later chapters. Strategies on infinitely repeated games can become indefinitely complicated, giving rise to unboundedly large genotype spaces and raising fundamental questions about how much complexity is required for a game solution concept to solve certain problems. A decomposition into finite collections of genes, which approximate unbounded strategies with recursive strategies of finite depth and complexity, provides a taxonomy in which symmetry is related to fundamental strengths and weaknesses of strategies in different classes.

3.2 Development and the decomposition of genotypes into genes

The way population genetics formalizes evolutionary dynamics has been heavily driven by biological applications. Secondary applications of evolutionary ideas to non-biological areas are often treated as analogy. They may be fraught with difficulties when it is not clear whether abstractions that have become conventional in biology have well-defined counterparts in other areas [1, 2].

The presumption of a gene concept is a source of much of this difficulty. In mathematical population genetics it has become standard to take the existence of genes for granted—to such an extent that genes have even been claimed to be the unique elementary units of selection [3, 4][1]—even as efforts to frame a molecular basis for the gene concept in biology are becoming increasingly problematic [5].

Here we emphasize the genotype as the elementary abstraction, because it is the minimal unit required to pass through ontogeny from birth to death or reproduction[2]. This is in keeping with the way normal-form game theory treats fully defined strategies (which we associate with genotypes) as elementary entities, because they are the minimal units required to define the play of a game. For us the proper empirical foundation for a gene concept is necessarily a statistical one, as it was for Mendel [6]. Genes exist only when summary statistics exist, which compress the description of the dependence of fitness on genotypes and which are preserved under transmission enough of the time that they permit causally explanatory evolutionary models. Appendix 3.5 gives a brief review of the molecular basis for genes in biology and argues that they arise as such summary statistics because they reflect a strong tendency for development to factor into modules.

The statistical (compression/factorization) motivation for genes then translates naturally to a motivation for the use of extensive-form games as a general framework for development. Game trees, possibly with additional structure reflecting information conditions, provide a broadly useful representation of statistical dependencies in processes of development or interaction when these processes factor into smaller functional units, as we argue in the appendix will generally happen for complex processes.

3.2.1 Genes, data-compression codes, and multilevel selection

Most of the models in this and the following chapters will not involve variation by recombination. For these cases we will regard the types introduced in the formal framework of chapter 2 as complete genotypes, modified during descent only through mutation. Although mutation is an elementary operation, mutations of different genotypes may still decompose into shared modes of variation and these are what we wish to capture with summary statistics that we regard as genes. Genes are thus *partitions* of the space of genotypes, where all genomes in the same partition take the same value of some summary statistic.

The problem of identifying good gene-partitions for genotypes has much in common with the problem of finding good data-compression codes. When agents of different types respond similarly to the same population contexts because they make some common moves in an extensive-form game, grouping them together may provide a simplified coarse-grained type space that still captures much of the relevant variance in agent behavior. However, the same partitions that capture variance efficiently at the group level may leave out rare but important effects, which in some conditions require keeping track of the agent's membership in multiple groups,

[1] Both in chapter 2 and in chapters 7 and 8 we review a few of the constructions that show why, even within standard assumptions, this interpretation is unsupportable.

[2] This need not be the genotype of a diploid or higher organism; it may be a chromosome or a region within a chromosome.

leading to a more complex description. From experience with variable-length data-compression codes [7, 8], we expect that this property, which arises for partitions motivated by the game tree, will remain even as a property of optimal codes.

3.3 Genes in the Price equation and in game models

Up to this point the indices $\{i\}$ for genotypes have been treated as integers $i \in 1, \ldots, D$. To produce a 'genic' description we first replace this general-purpose (but meaningless) integer index with a multicomponent index in which different components represent different partitions, which are the 'genes'. Because the gene indexing is multicomponent, individual components will have smaller ranges than $|\{i\}| = D$. Hence linear regression models of fitness on the gene indices (conventionally termed *additive*) will contain fewer terms. Such additive models may provide a more compact representation and more statistical power if a few genes can be found that account for a large part of the covariance of fitness with the observables of interest. However, additive models are coarse-grained and hence lose information from the original regression in indices $\{i\}$. Therefore, non-linear contributions to fitness (termed *interactions*) are required to recover all information potentially in the fitness function. The notation for additive models and interactions is introduced in this section, and examples showing why each is required are developed in chapters 7 and 8.

A gene decomposition can provide a compressed representation of the statistics of a population process only if structure is present in the fitness coefficients $\{\alpha_i\}$ and $\{a_{ij}\}$ in (2.16) that can be captured by groupings of agent types. In the following section, we turn to extensive-form game models to show how a common developmental program, shared by the agents in a population as a reflection of their common descent, may force the types to group in a way that makes genes useful.

3.3.1 The space of gene decompositions: disjunction, conjunction and compression

We will formalize the notion of genes in our models in terms of *discrete partitions* on the space $\{i\}$ of genotypes. We require that the gene description be invertible, so that it changes the representation of genotypes but does not lose information or the Markovian property of completeness[3]. A partition is also known as an *equivalence relation*. The sets in the partition, called *equivalence classes*, correspond to unions of genotypes. In order for a representation by partitions to be invertible, we require multiple equivalence relations, so that each genotype can be uniquely identified by the intersection of the sets from different equivalence relations in which it appears. We will refer to any equivalence relation as a *locus* and to the equivalence classes in that partition as the *alleles*. A *gene* is a particular allele at a particular locus. We further restrict, without loss of generality, to sets of equivalence relations that are *minimal*, meaning that no partition consists entirely of sets that may be identified by

[3] These two criteria alone permit much more general models than discrete partitions, including continuous *partitions of unity*. We restrict ourselves to discrete partitions for simplicity.

intersections of sets from the other partitions. We do not, however, require that the decomposition into genes reconstructs the genotype space as a product space, or even that all individuals in the population share the same loci.

A good set of partitions will be one in which many equivalence classes contribute little to the variance that drives population dynamics and, of those that remain, most of the dynamically significant variance is captured in additive models. We now consider the expansion of game parameters in indicator functions for the presence of genes.

3.3.1.1 Disjunction and conjunction systematically approximate type dependencies with associations of genes

Introduce capital indices I, J, K,... to indicate genes, and indicator functions σ_{Ii},... to map genotypes i into equivalence classes I. Then each indicator function σ_{Ii} is a *disjunction* (logical OR) of indicator functions for the genotypes $\{i \mid i \in I\}$, and each genotype i is a *conjunction* (logical AND) of indicators σ_{Ii} for some collection of genes $\{I \mid i \in I\}$.

Each coefficient in the polynomial expansion of frequency-dependent fitness then has its own expansion in terms of gene-indicator functions. For the coefficients in (2.16), this expansion may be written

$$\begin{aligned} \alpha_i &= \sum_I \sigma_{Ii}\left(\alpha_I^{(0)} + \sum_K \sigma_{Ki}\left(\alpha_{KI}^{(1)} + \sum_M \sigma_{Mi}\alpha_{MKI}^{(2)} + \ldots\right)\right) \\ a_{ij} &= \sum_{IJ} \sigma_{Ii}\sigma_{Jj}\left(a_{IJ}^{(0,0)} + \sum_K \sigma_{Ki}\left(a_{KI,J}^{(1,0)} + \sum_M \sigma_{Mi} a_{MKI,J}^{(2,0)} + \sum_L \sigma_{Lj} a_{KI,LJ}^{(1,1)} + \ldots\right)\right). \end{aligned} \tag{3.1}$$

The leading coefficients $\{\alpha_I^{(0)}\}$ in the expansion of α_i give the additive genetic contribution to (inherent) fitness. In population genetics, the higher-order products of indicator functions are known as *associations* and non-zero coefficients for associations represent epistasis[4].

Note that the series for payoff a_{ij} contains an expansion in associations for both the focal individual (i) and the opponent (j). We use superscripts $a^{(p,q)}_{I_1,\ldots,I_p,J_1,\ldots,J_q}$ to indicate the orders p and q of 'epistatic' interaction in each player.

Just as the polynomial expansion in frequency has uniformly mixed k-player normal-form games as a default interpretation[5], the joint-gene expansion of payoff

[4] Note that, if we are using players to model haplotypes and two-player games to model development in a diploid organism, then inherent fitness is a transmitted property of chromosomes (such as meiotic drive) and epistasis refers specifically to *cis* interactions among genes within a chromosome. *Trans* interactions appear in the payoff matrix. *Dominance* appears as a combination of an inherent fitness from the dominant allele and a payoff from pairing with any opponent having the dominant allele. A model for dominance with saturation, in a pairing of exactly two chromosomes of types i and j, might be $f_i = a_i + a_i a_j/(a_i + a_j)$, where both $a_i, a_j \in \{a, A\}$. The interaction term then requires a slowly converging polynomial series to approximate.

[5] This interpretation is only a 'default' in the sense that it assumes random matching as a parameter-free model. However, random matching depends on the representation of types for its meaning. Therefore, like any Bayesian prior, it should be regarded as a particular assumption which is a part of model selection [9].

coefficients, for suitably chosen partitions, has a default interpretation in terms of interactions on an extensive-form game that refines the normal form, with payoffs that are additive over internal nodes on the game tree. We return to develop this interpretation in the following sub-section.

3.3.1.2 Compression and the competition between obtaining additive variance and minimizing multilevel selection

If partitions of the type space may be chosen so that the number of genes $|\{I\}|$ grows only logarithmically in the number of genotypes $|\{i\}|$, then the truncation of the series (3.1) to fixed order provides an exponentially compressed representation of the dependence of fitness on genotype. Whether this truncated approximation is good depends on the rate at which the coefficients $\alpha^{(p)}$ and $a^{(p,q)}$ fall off with increasing p or q.

We expect as the general case, even for genes defined purely for the sake of compression, that a typical gene will be pleiotropic[6] with differing contribution to the additive genetic variance for different traits. That is, a partition I may account for most of the variance of one component of f_i as additive variance, while it contributes epistatically to other components of f_i, perhaps to varying degrees depending on population state. Maximizing additive variance favors the status of I as a gene, while pleiotropy and epistasis make any purely linear model involving I insufficient to explain population dynamics and require the retention of higher-order associations involving I. This competition is analogous to the problem of designing data-compression codes, which leads (through Fano's theorem [8]) to variable-length codes. We will present an example in section 7.3.3 with this property, for which the precise components of variance explained by each higher-order interaction in relation to the main additive term are worked out in section 7.3.3.3.

3.4 Extensive-form games and partitions that reflect the structures in development and interaction

The interpretation of extensive-form games in terms of genes is a large topic, partly developed in the monograph by Ross Cressman [10]. Here we consider only three aspects: (i) the gene partition induced by extensive-form games, which relates interaction structure to mutation and recombination; (ii) the problem of expanding the strategies on repeated games in a systematic hierarchy of complexity, memory and symmetry, in order to relate repeated games in evolutionary and rational-choice game theory; and (iii) the directionality of solutions on game trees from root to leaves and the asymmetry this creates between transient and recurrent states when strategies are mapped to finite-state automata. We develop examples of each of these points which are used in the evolutionary game models of later chapters.

[6] In standard usage, a gene is *pleiotropic* if it influences multiple phenotypic traits. In our discussion of the role of games, we have not introduced the abstraction of a 'trait'—a term already fraught with difficulty and ambiguity—but work instead with the direct model of development that determines fitness. Pleiotropy in a game model corresponds to a gene's (or a move's) influencing fitness through multiple interaction contexts, such as play against more than one kind of opponent.

3.4.1 The gene partition induced by an extensive-form game

An extensive-form game has three elements: (i) an underlying *game tree*; (ii) a collection of *information sets* telling which nodes on the tree are to be regarded as indistinguishable by players[7]; and (iii) an assignment of information sets to players, telling who moves at each information set. A *strategy* for a player on an extensive form game is a specification of a move on every information set assigned to the player by that game.

When the tree in an extensive-form game has a repeated structure of the same simultaneous or sequential moves by the players, the repeated units are called *plies*. The joint strategies of a collection of agents, who together are sufficient to play a game, is called a *strategy profile*. In an extensive-form game, which further decomposes strategies into sequences of moves, the joint moves of the players at their respective information sets within a ply is called a *move profile*.

If strategies in evolutionary game theory are to provide the operational notion of genotypes, then any mutation and recombination processes must respect the information-set structure on the extensive-form game. They may alter moves at information sets, singly or jointly, but they cannot introduce distinctions at nodes within an information set[8], and must under all conditions produce fully defined strategies.

Any extensive-form game induces a gene decomposition of genotypes in which the information sets are loci and the moves on each information set are alleles[9]. Mutation is free to change any move to any other at a single information set, while recombination acting on a set of existing strategies may shuffle the moves on different information sets to produce one or more new well-formed strategies. (Note that recombination draws from more than one strategy, so if it is used, the individual moves rather than whole strategies must become the elementary replicators to ensure that each offspring has a unique parent and retains a well-defined concept of fitness.) In some cases the topology of the game tree may define a useful notion of adjacency, so that recombination by 'crossover' switches all information sets on the sub-tree descended from the crossover point.

The gene decomposition induced by an extensive-form game in this manner does reconstruct the genotype/strategy space as a product space. For games with asymmetric player roles, a distinct product space exists for the sub-sets of players who can take each role.

3.4.2 Repeated games and strategic complexity

Two quantities of interest for a wide range of problems are the complexity and the symmetries of the ontogenetic events and interactions that govern a population's dynamics. Whether a game is interpreted biologically as a model of development, or

[7] Information sets may be used to compensate for deficiencies in the tree representation for handling simultaneous play, or they may reflect limits of complexity or memory imposed on the available strategies. See figures below for examples and graphical denotations.

[8] The purpose of the extensive form is to separate the 'reality' defined by the game from the strategic freedom and solution concepts of players. In the reality defined by information sets, distinctions among the nodes within them do not exist as strategic options for players.

[9] This 'natural' decomposition is also the one adopted in [10].

strategically as a structured competition, the complexity of play is often limited by resource costs and it is desirable to know how the complexity of interactions limits the kinds of problems evolution can use them to solve.

The treatment of repeated games in rational-choice game theory [11–14] has made extensive use of the notion of infinite (or indefinite) repetition and thus assumed essentially unbounded resources to support contracts. Unlimited repetition is invoked to permit the use of arbitrarily complicated trigger strategies to enforce compliance with prior agreements, or to preclude unraveling of Pareto-superior repeated strategies by backward induction, when they require moves that are strictly dominated in a finite game. One can argue that indefinite repetition is meant to abstract a variety of practical limits on memory or reasoning capacity that make long games 'effectively' infinite in practice, but it remains a problematic concept. Solutions in infinitely repeated games may be singular limits, which are not approached with the same solution concepts in finite games of any length. Infinite repetition is not a property of real interactions, and its use in strategic game theory often glosses over the difficult mathematical and empirical problems of what limits the complexity of play in real systems[10]. We are interested in using extensive-form games to regulate the complexity of play, to thus determine when and at what cost the properties of infinite games can be recovered.

Among strategies of comparable complexity, differences of symmetry can have a large effect on the outcomes of population dynamics and the interpretation of placing limits on strategies as a 'solution' to evolutionary problems. In chapter 7 we consider the repeated prisoners' dilemma (RPD) as an example. Narratives have been spun out for decades about the relative performance of different recursive strategies, which mix self-protection, punishment, or coordination in various ways. Strategies have been tested against errors in play, against innovations in complexity in a tournament setting [16, 17] and in computer simulations [18, 19]. The narratives, in addressing apparently unrelated dimensions of the problem of coordination, have no common basis. Yet a symmetry analysis of the ways different strategies are partitioned by the extensive form explains a variety of their differences in performance by showing that they connect different timescales: some strategies respond to events during the course of play, while others reference the whole course of play to events on the generation scale. Either of these can be a strength or a weakness, depending on context, but the underlying cause of both strength and weakness is the same.

Finally, the structure of the game, which determines the population dynamics, thereby determines what trade-offs are possible between the compression of additive models and the need to augment them with interaction terms.

3.4.3 Coarse-graining extensive-form games, and memory and symmetry of strategies

The limitation on the depth and symmetry of recursive strategies is carried out not on the strategies themselves, but on the game trees for which they are defined.

[10] The field of *combinatorial game theory* [15] is devoted to quantifying these problems of complexity as they relate to limits in human or computer memory and analysis and shows how complex and difficult (and interesting!) such quantification can be.

The information sets on game trees provide a way to make specific complex strategies *impossible*, by eliminating the distinctions among states of play that those strategies require.

We call this procedure coarse-graining of games. For any game in extensive form, a *coarse-graining* is another game with the same tree, in which the information sets are unions of information sets in the original game. The original game is then a *refinement* of the coarse-grained game. Coarse-graining is a many-to-one map, so multiple refinements will be possible for any coarse-grained game.

The following is a procedure that uses coarse-graining of extensive-form games to progressively extend the complexity of recursive strategies and to classify them according to the memory and symmetries of the minimal game tree on which each strategy can be defined. We illustrate with a binary, simultaneous-move, total information game. A sub-set of these strategies will be used to study the RPD in chapter 7, where we will use payoffs that are sums of values from the internal nodes visited in the play of a strategy profile.

Figure 3.1 shows an extensive-form game for two players, each given two simultaneous, pure-strategy moves L and R in each ply. This tree provides the largest possible complexity of strategies and memory consistent with simultaneity and pure moves. We will refer to it as the *total information* game.

For the game in figure 3.1 with $N_{\rm ply}$ plies, the number of information sets assigned to either player is $(2^{2N_{\rm ply}} - 1)/(2^2 - 1)$, so the number of possible distinct strategy types is

$$|\{i\}| = 2^{\frac{2^{2N_{\rm ply}}-1}{2^2-1}}. \tag{3.2}$$

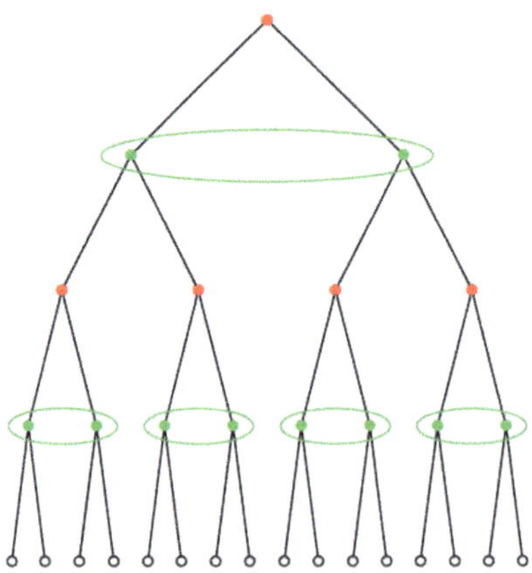

Figure 3.1. The two-player game with simultaneous binary moves and complete information. Information sets on alternating rows belong to the first (red) and the second (green) player. Information sets for the first player are single nodes, which require no special notation. Those for the second player include two nodes each and are denoted with loops (in the same color) surrounding the equivalent nodes. Two plies of the tree are shown; a tree with indefinitely repeated moves is obtained by extending this and all subsequent trees from their leaf nodes (open dots), recursively. Since all paths of play are distinguished by these information conditions, this game tree supports solution concepts such as sub-game perfection.

The most general additive normal form giving payoff a_{ij} to an individual of strategy type i when matched with an individual of type j is then

$$a_{ij} = \frac{1}{N_{\text{ply}}} \sum_{k=0}^{N_{\text{ply}}-1} \sum_{\alpha^k = L,R} \sum_{\beta^k = L,R} \sigma_{\alpha^k i} \sigma_{\beta^k j} a^{(k)}_{\alpha^k \beta^k}. \tag{3.3}$$

In (3.3) $a^{(k)}_{\alpha\beta}$ is the normal-form payoff in ply k. The α^k, β^k take values L or R at each information set and we use indicator functions $\sigma_{\alpha^k i}$ that take value 1 for those strategies i with move α^k at position k.[11]

Strategies are permitted to specify moves only at information sets, but payoffs need not be the same at all nodes within a single information set. By separating the control over an agent's strategic freedom (with information sets) and an agent's payoffs (with nodes of the underlying tree), the extensive form permits independent control over the complexity and memory of strategic opportunities and the payoff functions which form the basis for solution concepts [20].

The maximal coarse-graining of the total-information game in figure 3.1 is shown in figure 3.2. Each player has a single information set and therefore no memory and minimal complexity. The size of the type space is

$$|\{i\}| = 2. \tag{3.4}$$

This coarse-graining projects the extensive form at any N_{ply} onto the normal form for a single ply, and reduces the payoff a_{ij} of (3.3) to a one-ply effective normal form $a_{\alpha\beta}$, averaging the $\{a^{(k)}_{\alpha^k \beta^k}\}$ with equal weight.

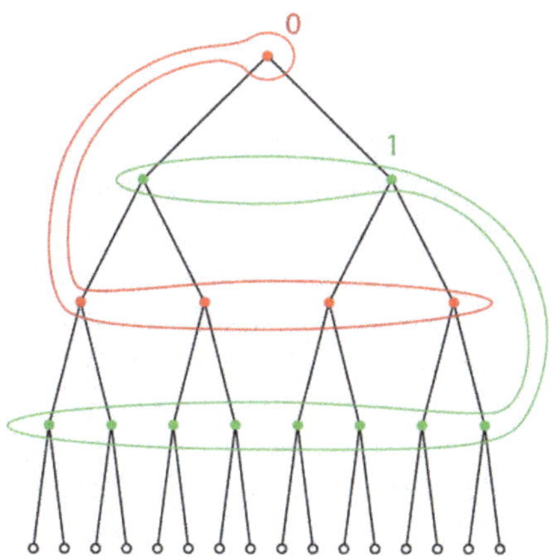

Figure 3.2. Maximal coarse-graining of the simultaneous-move game of figure 3.1, to one information set per player. The information sets 0 (red) and 1 (green) belong to the first and second player, and are understood to be extended as plies of the tree are added.

[11] The normalization by N_{ply} is arbitrary, but provides a useful way to handle scaling when we consider repeated games in chapters 7 and 8.

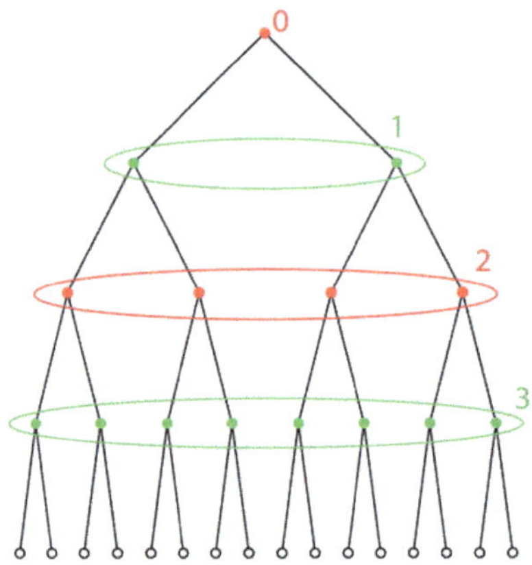

Figure 3.3. A coarse-graining of the total information tree 3.1 that distinguishes plies but not paths of play. Even-numbered information sets (red) belong to the first player and odd-numbered information sets (green) belong to the second player. In this tree, as plies are added, new information sets are introduced with each ply.

An intermediate coarse-graining of the total-information tree shown in figure 3.3 distinguishes plies and permits strategy complexity to grow with tree size, but precludes memory, because nothing in the history of play is distinguished by the information sets. The number of strategy types is

$$|\{i\}| = 2^{N_{\text{ply}}}. \tag{3.5}$$

The adjacency relation between plies on the tree is the same as that between loci on a pair of linear chromosomes which are the 'players'. This tree will be used as the basis for a model that maintains linkage disequilibrium in chapter 8. In the simplest such model, the payoff contributions from each ply are independent, corresponding to independent (additive) contributions to fitness from each of a set of loci on paired chromosomes. There is no strong motivation from the biological viewpoint to assume strict repetition of the plies in generating the extensive form: the payoffs in each ply might as well be assigned independently and non-identically. However, in the model we will develop, an identical anticoordination game will be strictly repeated. The added symmetry under arbitrary interchange of plies, induced by strict repetition, will allow us to focus on adjacency structure and the way selection can coordinate play across nearby plies through feedback from the population state.

3.4.4 Recursion and repeated games: systematic approximation of strategies

Recursive strategies with increasing memory are given variable-length descriptions by assigning each strategy to the simplest coarse-grained game on which it can be defined. More symmetric strategies can be defined on more coarsely grained games. Strategies with less symmetry require refinements of the coarse-grained game, which are more complex to specify[12]. An illustration is provided by refining the maximally coarse-grained tree of figure 3.2 to admit moves that respond to the immediately preceding round of play.

3.4.4.1 Strategies with a memory depth of one ply

Recall that on the tree of figure 3.2, the only definable strategies, L and R, are equivalent to moves on a single 2×2 normal form. The maximally coarse-grained tree permits three distinct refinements having a memory depth of one ply and equal numbers of L and R moves, shown in figures 3.4–3.6. Each tree has three information sets per player and thus

$$|\{i\}| = 2^3 \tag{3.6}$$

possible strategies. However, the first two strategies on each refined tree, which assign either L or R to all information sets, coincide with those on the tree of figure 3.2 and are not counted again. Therefore each refined tree brings six new strategies.

The refinement shown in figure 3.4 describes memory of the player's own previous move. The two new strategies defined on the recurrent moves either repeat the player's previous move (which we denote by REP), or do the opposite, leading to alternation (denoted by ALT).

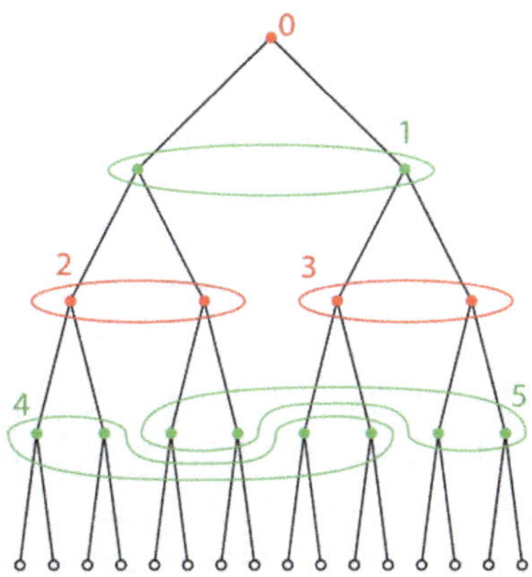

Figure 3.4. A tree with memory of the player's own previous move. Here information sets 0 and 1 are opening moves for the first and second player. Information sets 2, 3 are the *recurrent* moves for the first player and 4, 5 are the recurrent moves for the second player. Information sets representing recurrent moves are to be extended recursively as plies are added to the tree. Strategy REP moves L on information sets 2 or 4 and R on 3 or 5, while strategy ALT moves R on 2 or 4 and L on 3 or 5, thus repeating or alternating the player's previous move.

[12] In matching a strategy's complexity to its description length, we are following the *minimum description length* approach of Rissanen [21] for representing algorithms. Strategies on the total-information tree of figure 3.1 occupy a space with a product structure, since moves are specified independently in each ply. Strategies on coarser games occupy spaces in which alternatives for moves can only be selected jointly at multiple nodes. The decomposition of the space of all strategies into sub-groups on trees of varying coarseness is similar to the decomposition of a direct-product group into its direct-sum components [22].

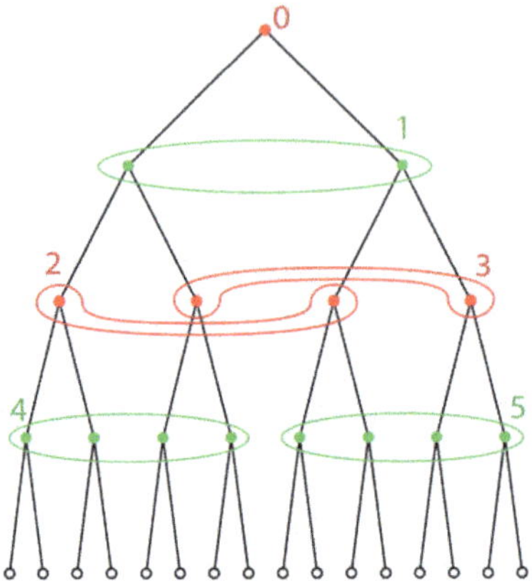

Figure 3.5. The coarse-graining of the total-information game of figure 3.1 that defines one-period memory for the opponent's move in the previous round. Strategies 0 and 1 are again opening moves, while 2–5 are recurrent moves. Strategy TFT moves L on sets 2 or 4 and R on sets 3 or 5, and attempts to achieve coordination between players by opening with L on sets 0 or 1.

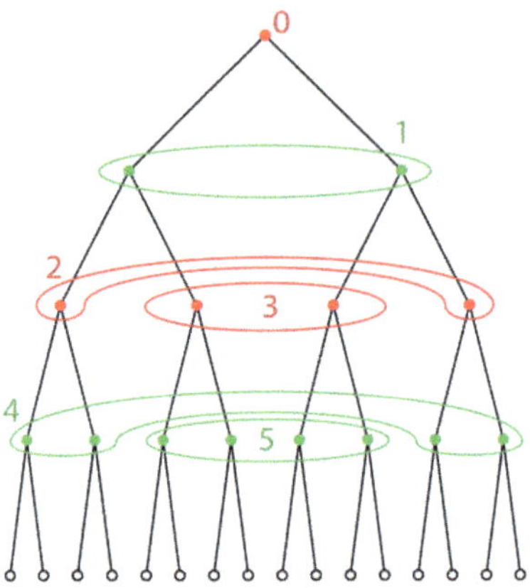

Figure 3.6. A tree that distinguishes parity of the players' previous-round moves. Sets 0 and 1 are opening moves and 2–5 are recurrent. The strategy WSLS assigns L to recurrent sets 2 or 4, and R to sets 3 or 5. It therefore breaks parity with the recurrent move in every ply and its opening move is not needed for coordination and does not matter in long repeated games.

The refinement shown in figure 3.5 describes memory of the opponent's previous move. The two new strategies defined in the recurrent stage of this tree are tit-for-tat (TFT), which repeats the opponent's last move, and its opposite, anti-TFT (ATFT).

3.4.4.2 Game trees that preserve or violate exchange symmetries between agent types and strategies

Both of the previous refinement trees have the property that their non-trivial recurrent moves (those that are not all L or all R) are *symmetric* with respect to joint exchange of responses L and R together with reflection of the information sets through the vertical. That is, by taking in asymmetric moves from the previous round and generating asymmetric responses, all four of REP, ALT, TFT, and ATFT make symmetrically defined *responses* to their inputs.

The refinement shown in figure 3.6 describes memory of the *parity* between the player's and opponent's previous move. Its information sets are therefore symmetric under reflection. The two non-trivial recurrent-move strategies on this tree *break* the response symmetry by mapping even or odd information sets to L or R moves. The widely explored strategy of win–stay/lose–shift [18, 19] (WSLS, also called Pavlov [23]) maps even parity to L and odd parity to R. Its opposite, which we term anti-WSLS (AWSLS), does the reverse.

3.4.5 Finite-state automata and behavior strategies

Game trees were introduced [24] to represent a particular form of sequential play of actual games and some forms of interaction among genes in development can occur for which a tree is not a well-suited representation. The tree representation is given some flexibility by the use of information sets, as is shown by the representation of simultaneous moves in figure 3.1. Memory can even be removed altogether and with it the direction of time, as in figure 3.3.

When memory is used, however, it inherits a direction from the branching of the underlying tree in the extensive form[13]. This directionality need not entail temporal sequence, but it does entail one-way contingency in some form. In order to classify the forms of strategic irreversibility that can result, we map finitely defined strategies on an indefinitely repeated extensive form game tree to their corresponding finite-state-automata (FSA) [25].

Any finitely defined strategy on a tree of infinite depth must correspond to a coarse-grained tree in which infinitely many nodes are assigned to some information set or sets. Because information sets are Markovian, we may always collapse the tree onto a finite-state graph. For any such graph, a minimal form exists [26]. The states in the FSA may be divided into *transient states,* which are not revisited asymptotically, and *recurrent states.* In the simple examples above, the transient and recurrent states correspond to the opening-move and recurrent-move information sets on the game tree.

Figure 3.7 shows the recurrent FSA for the tree of figure 3.5. The nodes are labeled with the move profiles that arrive at them. This FSA will be the basis for the RPD example in chapter 7.

Strategies defined by moves on the states of a FSA are known as *behavior strategies* [27]. Depending on the structure of the game tree, behavior strategies may or may not be intrinsically reversible. We may isolate behavior strategies as models of development by using long extensive-form games in which payoffs from opening moves are negligible, though transient states may still coordinate the recurrent moves.

[13] This one-way branching during interaction is reminiscent of the one-way branching of population processes described by population genetics.

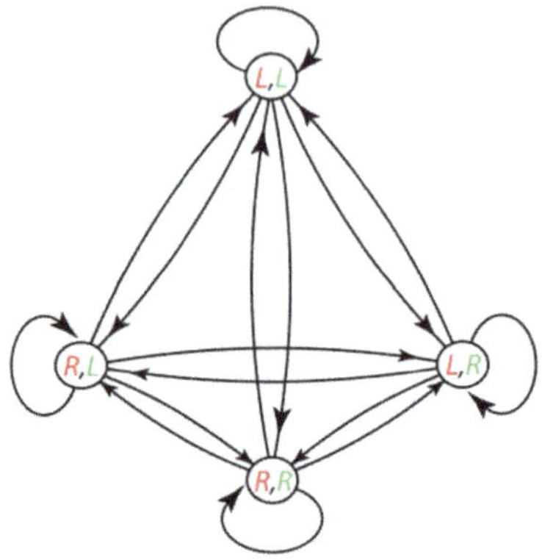

Figure 3.7. Possible moves in the recurrent stage of the repeated game of figure 3.5.

An example using behavior strategies in which correlations are removed with trembles is given in section 7.3.1, using the recurrent moves from figures 3.4–3.6.

3.5 Appendix. Genes in genomes and in physiology and development

Our discussion of genes in relation to extensive-form games is formally straightforward. We could have presented it without preamble and studied its consequences as an applied mathematics problem. However, we wish to avoid the interpretation that the gene description generated by an extensive-form game is somehow an analogy to a biological gene concept that is simply a primitive of life. That interpretation gives a false simplicity to a biological problem that is actually complex and difficult, it obscures the origin of genes which is not limited to cellular living systems and it robs the game of its proper justification, not as analogy to something else, but as a formalization of the kind of statistical construction that is properly the basis for biological genes.

Therefore, this appendix makes a short digression to point out that the biological gene concept was never a given, unique, material digital system. Rather, it is a reflection of modularity in development that becomes expressed in heredity. It derives from the same kinds of modularity in development that we wish to capture in the structure of extensive-form games.

3.5.1 The limited role of DNA in creating the gene concept for cell biology

The gene concept is complicated because it is related to structure both in development and in heredity. From the 1950s onward, genes have increasingly come to be associated with DNA or RNA sequences. It is not uncommon to find leading-edge papers in population genetics that *define* the gene as 'a particular copy of a non-recombining sequence at some locus in some individual' [28]. A hazard of this materialist interpretation of genes (apart from a growing list of recognized empirical problems with it [5]) is that it obscures the essentially *statistical* role of genes in population genetics and it masks the important role of development as the source of highly conserved units that can carry the interpretation of genes.

It is possible to understand the special place of contiguous sequences of DNA known as *open reading frames* (ORFs)[14], without losing sight of their source and their statistical nature. Empirically, viable recombination occurs much more readily between ORFs than within them. As a result, these regions are replicated either faithfully or with substitution mutations over more generations than longer regions containing multiple ORFs, between which crossover may occur. This tendency is the basis of the argument for genic selection [3, 4].

Although the conservation of ORFs is expressed as a pattern in the transmission of DNA sequences, it originates largely through a combination of constraints from the translation system and protein function.

3.5.2 The origin of biological genes in development

Many ORFs code for protein-folding domains. They depend on the preservation of the three-base reading frame in order for translation to process codon meanings and this frame is readily lost under crossover. Even in cases where crossover preserves a reading frame, the translated folding domain has generally evolved to satisfy a large number of constraints among multiple residues distributed along the polypeptide sequence, making the fold and its function fragile against insertion or deletion of residues. For these regions, the difficulty of obtaining viable recombinants is a very indirect reflection of the mechanics of translation, together with the complex constraints on protein folding and function.

The existence of a natural domain size that precludes most recombination is a problem of geometry, chemistry and physics [29]. Sequences much shorter than 50 residues do not usually fold stably into domains, while sequences much longer than 100 residues create very complex webs of constraint and may be difficult to evolve compared to assemblies of multiple smaller domains. A natural domain size implies a trade-off between the rate of recombination (which may facilitate adaptation) and the fraction of the germ line that is lost due to non-viable crossovers. The factorization of entire genomes, whether under crossover or with respect to transposable element insertions, into sequences of contiguous residues that resist insertion or recombination is a consequence of the interaction of all these effects.

Since the DNA substrate is not the source of the characteristic scales and constraints that lead to conserved domains of roughly similar size that can serve as hereditary particles, the absence of close analogs to DNA in behavioral or institutional systems need not preclude the existence of a gene concept in these systems. The proper question to ask is whether the mechanisms responsible for modularity lead to preferred levels of conservation in other processes as well.

3.5.3 Arguments from robustness and evolvability, that modularity in system architecture should be common

Modularity in the architecture of complex systems appears common, even though proximate causes such as those in cell physiology and development are system-specific.

[14] ORFs are regions between start codons and transcription pause sites. Generally they contain one or more contiguous translation regions for proteins, terminated by stop codons.

Arguments that modularity is a general requirement to permit hierarchical complexity have been put forward on grounds of robustness [15, 16] and evolvability [28]. Most such claims at present are descriptive. While they suggest causes in dynamical or evolutionary stability, or in limiting the information requirements of control systems, they are pre-formal in the sense that they propose explanations for what is observed, but do not attempt to make precise predictions. However, we find the general premise both a convincing and an important interpretation of very widespread empirical patterns.

Because our treatment of the role of extensive form games in evolutionary dynamics is primarily a technical derivation, it is important that the use of these models be based on appropriate and general abstractions of empirical phenomena. We believe that anchoring both gene decompositions and their counterparts in the extensive-form decomposition of strategies into moves, in the statistics of functional modularization is the correct empirical foundation for these abstractions.

Bibliography

[1] Nelson R R and Winter S G 1985 *An Evolutionary Theory of Economic Change* (Cambridge, MA: Balknap Press)

[2] Young P H 1998 *Individual Strategy and Social Structure: An Evolutionary Theory of Institutions* (Princeton, NJ: Princeton University Press)

[3] Williams G C 1966 *Adaptation and Natural Selection* (Princeton, NJ: Princeton University Press)

[4] Dawkins R C 1976 *The Selfish Gene* (New York: Oxford University Press)

[5] Stadler P F, Prohaska S J, Forst C V and Krakauer D C 2009 Defining genes: a computational framework *Theory Biosci.* **128** 165–70

[6] Mendel G 1901 Experiments on plant hybridization *J. R. Horticultural Soc.* **26** 1–32 (Engl. transl.)

[7] Shannon C E and Weaver W 1949 *The Mathematical Theory of Communication* (Urbana, IL: University of Illinois Press)

[8] Cover T M and Thomas J A 1991 *Elements of Information Theory* (New York: Wiley)

[9] Gelman A and Shalizi C R 2013 Philosophy and the practice of bayesian statistics *British J. Math. Stat. Psychol.* **66** 8–38

[10] Cressman R 2003 *Evolutionary Dynamics and Extensive Form Games* (Cambridge, MA: MIT Press)

[11] Aumann R J and Shapley L S 1994 Long-term competition—a game-theoretic analysis in *Essays in Game Theory: In Honor of Michael Maschler* ed N Megiddo (Heidelberg: Springer) pp 1–15

[12] Rubinstein A and Wolinsky A 1995 Remarks on infinitely repeated extensive-form games *Games Econ. Behavior* **9** 110–5

[13] Fudenberg D and Maskin E 1991 The folk theorem in repeated games with discounting or with incomplete information *Econometrica* **54** 533–54

[14] Fudenberg D and Tirole J 1991 *Game Theory* (Cambridge, MA: MIT Press)

[15] Nowakowski R J 1998 *Games of No Chance* (Cambridge: Cambridge University Press)

[16] Axelrod R 1997 *The Complexity of Cooperation* (Princeton, NJ: Princeton University Press)

[17] Axelrod R 2006 *The Evolution of Cooperation* (New York: Perseus)

[18] Nowak M A 1993 A strategy of win-stay, lose-shift that outperforms tit-for-tat in prisoner's dilemma *Nature* **364** 56–8
[19] Imhof L A, Fudenberg D and Nowak M A 2007 Tit-for-tat or win-stay, lose-shift *J. Theor. Biol.* **247** 574–80
[20] Shubik M 1984 *Game Theory in the Social Sciences: Concepts and Solutions* (Cambridge, MA: MIT Press)
[21] Rissanen J 1989 *Stochastic Complexity in Statistical Inquiry* (Teaneck, NJ: World Scientific)
[22] Georgi H 1999 *Lie Algebras in Particle Physics* 2nd edn (New York: Perseus)
[23] Kraines D and Kraines V 1995 Evolution of learning among Pavlov strategies in a competitive environment with noise *J. Conflict Resolution* **39** 439–66
[24] von Neumann J and Morgenstern O 1944 *Theory of Games and Economic Behavior* (Princeton, NJ: Princeton University Press)
[25] Hopcroft J E and Ullman J D 1979 *Introduction to Automata Theory, Languages, and Computation* (Reading, MA: Addison-Wesley)
[26] Crutchfield J P and Young K 1989 Inferring statistical complexity *Phys. Rev. Lett.* **63** 105–8
[27] Koller D and Megiddo N 1990 The complexity of two-person zero-sum games in extensive form, *Games Econ. Behavior* **4** 528–52
[28] Kirkpatrick M, Johnson T and Barton N 2002 General models of multilocus evolution *Genetics* **161** 1727–50
[29] Denton M 2002 The protein folds as platonic forms: new support for the pre-darwinian conception of evolution by natural law *J. Theor. Biol.* **219** 325–42
[30] Simon H A 1962 The architecture of complexity *Proc. Am. Phil. Soc.* **106** 467–82
[31] Simon H A 1973 The organization of complex systems *Hierarchy Theory: The Challenge of Complex Systems* ed H H Pattee (New York: George Braziller) pp 3–27
[32] Gerhart J and Kirschner M 2007 The theory of facilitated variation *Proc. Natl Acad. Sci. USA* **104** 8582–9

Eric Smith and Supriya Krishnamurthy

Chapter 4

Symmetry and collective fluctuations: large deviations and scaling in population processes

The first three chapters defined the scope of the population processes and developmental models that we will consider. They introduced the essential structural elements of both evolutionary dynamics and games that may be used to form both stochastic models and coarse-grained approximate descriptions such as the replicator equation. This chapter introduces methods from the theory of generating functions to pass back and forth between stochastic processes and the dynamics of the observables they produce. We make extensive use of a scaling relation known as the large-deviations property, which simplifies fluctuation probabilities and defines the equivalent of thermodynamic limits and the appropriate concept of entropy for stochastic evolutionary dynamics. Large-deviations scaling will be the property that makes possible robust inference about population processes from incomplete observations, as long as the defining symmetries of the process can be correctly deduced. We argue that this is not only fortuitous for theoretical inference; separation of dynamical scales for the sake of using the simplifications of the large-deviations limit is one way natural systems achieve both complexity and robustness. We close the chapter by listing a series of approximation methods within the large-deviations framework which capture dynamics associated with different classes of symmetry and its representations. Each method will be used in a worked example in the subsequent chapters.

4.1 Large-deviations scaling and collective fluctuations

Chapter 2 introduced the essential problem of unpacking the black box of development in evolutionary models and reviewed the Price equation as a universal organizing framework that captures the essential constraints common to population processes with well-defined vertical descent. Chapter 3 considered some of the methods offered by game theory—particularly the extensive form—to characterize the dynamics of development and interaction.

doi:10.1088/978-0-7503-1137-3ch4

These two inputs together encompass a fairly rich range of population processes and models, and capture several properties now regarded as important to the structure of evolutionary dynamics, including neutrality and repetition. A general framework for population processes, however, does not address the quantitative question of how the descriptions of any particular system at multiple scales are related to one another. That subject is developed in this chapter.

4.1.1 Multiscale processes: properties and the problems of theory

The relation between the individual and the aggregate enters evolutionary modeling from two directions. First, it subsumes many of the results we would like to predict for evolving systems. Second, it defines a large part of the problem of robust theory building, because our inferences about cause in evolutionary dynamics is often very indirect and drawn from observations that may be at different scales than those at which the causes act.

Frequently the purpose of an evolutionary dynamical model is to answer questions such as: how do interactions of individuals in single events relate to the population-level behaviors they generate? Conversely, to what extent do robust features of aggregate population dynamics permit us to infer the individual-level interactions responsible for them? By what mechanisms do new levels of selection or new forms of individuality emerge from aggregation and coupled dynamics? Computing aggregate behaviors in toy models is (to some extent) straightforward and a collection of very good reviews survey a large body of such results [1–3].

It is a more difficult problem to move beyond the use of models as toys for proof of concept and to show that they are justified by incomplete observations. A related problem, which requires committing to and justifying a model, is predicting statistics for a population which are not simply functions of those statistics that are observed. For instance, from summary statistics on the mean composition of a population we may wish to predict the moments and correlations of its fluctuations. Or, given a fit to short-term evolutionary dynamics, we may wish to know whether apparently open-ended degrees of freedom are really open-ended in the evolutionary long term (and not merely transients that will encounter a bound we cannot directly observe). More generally, we wish to know under what conditions the predictions from a model guessed by incomplete observations qualify as a *theory* of the actual phenomenon.

The emergence of multilevel dynamics or selection in particular requires a delicate balance of opposing tendencies: we ask when aggregation confers sufficient robustness to a population-level property that we have a hope of correctly capturing its cause in simplified and therefore highly incomplete models. Yet at the same time aggregation is expected to preserve the most delicate of conditions: the capacity for variation that permits either dynamics or filtering by selection at a level that must be much larger than the individual.

Problems of this kind have been the major area of advance in statistical mechanics in the 20th century, and it has been learned that the co-occurrence of robustness and multiscale dynamics is a subtle consequence of aggregation in which symmetry plays a major role [4, 5].[1] In section 2.4 we summarized a few major classes of symmetry and the ways aggregation may lead to changes in the *representation* of symmetries and the qualitative system-level properties to which they correspond. In this chapter we develop the key concepts from combinatorial properties of large numbers, and computational tools, to extract these qualitative behaviors systematically and to fill in their quantitative parameters.

We will use models throughout, because the treatment of every empirical case is a labor for specialists. However, implicitly we always have *two* models in mind: one that is a proxy for the 'hidden' generator of the natural process and the source of the observables from which we draw hypotheses, and another that is our attempt at a theory to account for the observables and approximate the hidden model. This chapter proposes a sequence from a generating process to observables to theory, in which robustness enters in each of the three stages.

4.1.1.1 Three stages of robust theory building

We suppose that a natural system evolves under a stochastic process satisfying the axioms of population genetics[2], but which is otherwise not directly observable to us and is probably complicated.

1. The first step is to understand, within the class of such processes, which are likely to be their robust observables and why. This is the problem of moment closure introduced in section 2.2. In stochastic processes, lower-order moments tend to be more stable at a given sample size than higher-order moments and so the dynamics of the mean is the most used observable. The attempt to fit models for the dynamics of the mean in terms of its own value leads to the mean-field approximation (MFA), an unregulated but frequently useful first-moment closure approximation. The replicator equation is usually written as an MFA. From model summaries of robust observables, such as fits to a replicator equation, we wish to extract particularly the defining symmetries of the real process. For many cases these can be correctly described despite weaknesses of the MFA.
2. Having captured the symmetries of a system in some rough characterization such as a set of mean-field equations, we then wish to construct a model stochastic process that reproduces the mean behavior but also allows us to predict properties that are not simply functions of the mean (either in the model or, we suppose, in reality). The problem of doing this in a least-biased

[1] Hence, while the descriptive account of the cause of ferromagnetism had appeared already in the work of Pierre Curie before the turn of the century and an initial formalization in mean-field theory was offered by Pierre-Ernest Wiess in 1907 [6], a proper understanding of critical phenomena and its generalization to become *the* theory of the hierarchy of structure in matter occupied much of the remainder of the century and led to the reconceptualization of some of our most fundamental concepts of object and interaction.

[2] For organisms or genes, obtaining a reliable answer to this question from a study of reproductive mechanisms is relatively straightforward. If the same property is to be claimed for a deme [7, 8] or for a species [9] as a proper unit in an evolutionary population process, the determination may become more difficult.

way has led to the extensive theory of maximum-entropy methods in thermodynamics [10, 11]. In the maximum-entropy approach to out-of-sample prediction, a maximum-entropy distribution for a set of constraints is the least-committal distribution consistent with those constraints [12, 13]. Moment properties extracted from the maximum-entropy distribution are predictions that aim to impose the necessary consequences of the observed data but nothing more [14]. We adopt a variant on maximum-entropy modeling, which is to construct a stochastic process that assumes as little as possible about higher moments of the generator beyond those required to produce the behavior in the MFA.

3. Having selected a least-committal model, the third step is to systematically extract its robust implications. These may include the presence of multiscale dynamics (chapter 5 and chapter 6), or multiple levels of selection that admit coarse-grained descriptions (chapter 8), or fluctuation effects that cause the interactions among individuals in a model to aggregate to population behaviors that, in the MFA, can appear quantitatively different from what pairs of individuals experience in their direct interactions (chapter 7). To the extent that the implications of the model are robust consequences of its symmetries, we may derive useful inferences about the actual generating process despite having had limited ability to estimate most of its fine-scale parameters.

The remainder of this chapter defines our approach to model inference and introduces the set of approximation methods that extract phenomena associated with different symmetries. The remaining chapters then illustrate each approximation method in a representative model.

4.1.2 The large-deviations property

4.1.2.1 Levels of selection, buffering and robustness

The premise of vertical transmission *at some sufficiently fine scale*, which defines classical population-genetic processes, does not ensure that when we have coarse-grained the type space we will be left with any uniform level of interaction or any tractable dynamical system. Yet empirically we find that much of selection operates on numerous copies of entities that share a comparable level of organization, whether they are organisms, cells, chromosomes, or genes. When multiple levels of selection emerge and interact, they do not usually form a chaotic, coupled continuum, but rather resolve into well-separated scales, many of which seem to arise repeatedly in parallel systems.

A number of related arguments have been put forth [15–17] that modular rather than continuous architecture is a helpful and perhaps necessary property for the emergence, maintenance and regulation, and evolution of hierarchical complex systems. In one way or another, all of these arguments rely on the concept of *buffering* of the dynamics within different sub-systems, for the sake of simplifying and reducing the diversity of the control or regulatory signals that pass between them[3]. It seems likely that the reason uniform levels of interaction and well-separated levels of selection are common in natural systems is that only this architecture permits the kinds of persistence and stable variation that can appear in a population process as a form of individuality or agency.

Buffering of the observed states of a module or sub-system confers a kind of robustness on the interactions that integrate it within a larger assembly. One of the most common mechanisms of buffering is aggregation which simplifies the behavior of coarse-grained observables compared to those at the microscale. For example, the sample mean of independent draws of a random variable converges to the mean of the underlying distribution, reducing sample dispersion, a result known as the law of large numbers [11]. By the *central-limit theorem* [20–22], the residual distribution for fluctuations of the sample mean about its average attracts to a member of the *Lévy α-stable family*[4], meaning that as more independent samples are included in the average, higher-order moments of the distribution for that average attenuate compared to the leading asymptotic behavior of the tail, which is governed by the parameter α. Related results exist for convergence of the distributions of extreme-value statistics to functions within the Fisher–Tippet family [23].

What these results have in common is that the properties of the combinatorics of large sums or extrema come to dominate all but a few low-order properties of microscopic event distributions when these events are aggregated. The shared combinatorial factors make the aggregate behaviors less diverse—both more robust and in a system context more predictable—than the diverse microscopic phenomena that produce them.

4.1.2.2 Scaling of fluctuation distributions

This simplification due to aggregation under the combinatorics of large numbers will be the basis for our discussion of multiscale dynamics and multilevel selection,

[3] The concept of buffering is formalized in the domain of optimal control theory, following Ross Ashby's notion of *requisite variety* [18]. Control systems, including evolutionary feedback, must contain internal models [19] with at least as much diversity as the system states they attempt to control. Yet controllers, as aggregate entities, often cannot have as many microscopic states as their targets. The difference, therefore, between microstate diversity and the diversity of signals that can come from a controller, must be buffered so that most internal degrees of the controlled system are self-stabilizing, relying on only lower-dimensional signals from its interaction with its environment.

[4] The Lévy α-stable distributions make up a four-parameter manifold, which are the unique solutions to the problem that if X_1 and X_2 are random variables drawn from one such distribution, and a and b are real numbers, then $aX_1 + bX_2$ may be written $cX + d$, where X is sampled from the same distribution as X_1 and X_2 and c and d are real numbers. The α-stable family does not have closed-form expressions at most values of α, but their characteristic functions have the form

$$\varphi(t) = e^{i\mu t - |ct|^{\alpha}(1 - i\beta\,\mathrm{sgn}(t)\tan(\pi\alpha/2))}.$$

In this monograph we will consider only the solutions with $\alpha = 2$, which are the Gaussian or *normal* distributions.

and also for the criteria of robust inference and analysis in population models. As justification for this premise we invoke the empirical observation of separated levels of selection and the foregoing arguments that only systems which can employ these as buffering mechanisms have evolved both complexity and also robustness.

For stochastic processes our concern is not only mean values, but the entire distribution of fluctuations in observables at multiple scales. The reason robust theories can be inferred and analyzed is that these probability distributions, when they describe the dynamics of aggregate entities, do not remain complex and arbitrary but collapse onto a restricted class of functional forms. Their collapse is characterized by a scaling relation known as the *large-deviations property* [24] for probability distributions of fluctuations. The large-deviations property holds when log probabilities separate into a *scale factor*[5] that characterizes all fluctuations in a system—in our examples this will be the population size N—and a factor called the *rate function,* which depends on the structure of fluctuations in a scale-independent way. We will see below, for instance in (4.84) and (4.86), that the probabilities for macroscopic composition fluctuations in the state of a population with N individuals converge on the form

$$P_{\text{fluct}} \sim \mathrm{e}^{-N\hat{S}}. \tag{4.1}$$

The notation $\sim$ indicates leading exponential order at large N.[6] More formally, there exists a function $\hat{S}$, asymptotically independent of N, for which

$$\lim_{N\to\infty}\frac{1}{N}\log P_{\text{fluct}} + \hat{S} = 0. \tag{4.2}$$

The rate function $\hat{S}$ is also referred to in large-deviations theory as the *entropy*. We will take the existence of a large-deviations property, and the rate function that characterizes it, to *define the appropriate entropy* for a general stochastic process.

$\hat{S}$ is a function only of the fractional changes in the population composition, which furnish a description of what we have termed the 'structure' of fluctuations independent of overall system scale[7]. The large-deviations property requires that it

[5] Large-deviations scale factors may come from any form of aggregation over degrees of freedom with some conditional independence. In all examples in this volume, the number of agents N in a population will define one such scale factor. In many applications, however, the length of an observation time may provide a different scale factor [25]. In such cases the total scale factor can be a product of multiple factors reflecting independent aspects of aggregation.

[6] Sub-leading factors absorbed in $\sim$ include terms that scale more slowly than N in the exponential and prefactors that are typically powers of N, which normalize the distribution appropriately for its characteristic width of $O(1/N)$.

[7] As noted in [24], the large-deviations principle does not imply that the scale factor will be linear in N, or that population size N will be the only scaling variable, but in the models solved here this scaling will always pertain.

first be possible to *describe* the structure of fluctuations in an invariant way across a range of scales, which means that aggregation can change the system quantitatively but not qualitatively. Therefore, large-deviations scaling is the defining characteristic of a thermodynamic level of description separate [26] from the detailed underlying stochastic dynamics of individuals.

Both the laws of large numbers (which characterize the mean) and the central-limit theorem (which in our examples will characterize small Gaussian fluctuations around the mean) may be understood in the context of large-deviations theory as noted in [24]. The mean or most probable values occur where the rate function $\hat{S}$ in a properly normalized probability distribution takes value zero corresponding to a law of large numbers for the system. The central limit theorem (in its simplest form) arises in the context of large-deviations theory if $\hat{S}$ is expandable to second-order about its minimum value (at $\hat{S} = 0$) leading to a Gaussian approximation for P_{fluct}. Large-deviations scaling is however a more general property than both the above, since the rate function $\hat{S}$ contains information not only about small fluctuations about the mean but also about rare large excursions.

4.1.2.3 Concentration in the support of fluctuations

Systems with large-deviations scaling and non-vanishing rate functions progressively concentrate the support of the fluctuation distribution with an increasing scale of aggregation [27, 28], as occurs for example when we increase the size of a well-mixed population whose members interact through the play of the same game. Concentration of the support of fluctuations can occur either in space or in time. As we will show for a coordination game with discrete broken symmetry (via a pitchfork bifurcation), the ordered population can exist in two states, whose separation does not diminish with large population size. The trajectories that take the system from one of the two states to the other cannot be made small and their associated rate function is bounded below by a positive number. Therefore the support for fluctuations concentrates *in time*, as events of escape from basins of attraction become exponentially rare[8].

In contrast, a game that breaks a continuous symmetry to form a limit cycle, such as rock–paper–scissors (RPS) (via a supercritical Hopf bifurcation), must possess a continuous set of ordered limit-cycle histories and an expansion of the rate function to Gaussian order with a spectrum that is continuous down to zero. As a result of this continuity, Gaussian-order fluctuations whose magnitude does

[8] It is common that large-deviation forms for fluctuations will apply to many different partitions among events in a stochastic process, so that rate functions with different offsets may characterize conditional versus unconditional distributions. In the example with a pitchfork bifurcation, the conditional distribution for trajectories that begin in one state and end in another has a rate function bounded above zero, resulting in concentration of large macroscopic fluctuations in time. In the corresponding *unconditional distribution*, the trajectories that change state are given a weight that becomes exponentially small, so that the lower bound on the overall rate function approaches zero exponentially in N, reflecting the existence of a stable macrostate for the population at almost all times.

not decrease as a power of $1/N$ remain dense in time in the directions where the spectrum vanishes. Since their support in all other directions decreases as $O(1/N)$ with increasing population size N, macroscopic fluctuations attract onto a lower-dimensional manifold than the full state space and so concentrate their support in space.

4.1.2.4 Concentration in space or in time, in relation to symmetries

Many results in the large-deviations theory of non-equilibrium stochastic processes, including evolutionary games, are mathematically analogous to familiar results in equilibrium thermodynamics or field theory[9]. The non-equilibrium cases are generally more complicated to represent, because their large-deviation rates are *functionals* of histories rather than *functions* of states, as the equilibrium entropy is. However, for many of the game models we will illustrate, the configuration spaces are sufficiently simple that we may sketch the qualitatively distinct behaviors of different classes of fluctuations here by way of introduction, and the remainder of this and the subsequent chapters will introduce a sufficient notation to express them fully.

The two important analogies to develop (because they reflect shared mathematical roles for symmetries) are that (i) fluctuations about discrete equilibria correspond (in condensed matter) to excitations in a system with an energy gap, while fluctuations about backgrounds that spontaneously break continuous symmetries include gapless modes, and (ii) for cases with discrete symmetry breaking and a finite number of degenerate equilibria, the escape trajectories that are local minima of the action resemble solitonic excitations, in having a gap but also being indexed by a time which is a coordinate of symmetry, along which they are rare compact events, much as solitons are separate compact forms in space[10].

Small diffusive fluctuations

To consider the simplest case as a reference case, suppose $\bar{n}$ is a classical rest point of the equations for the mean of some population process, and let $\hat{\Delta}n$ denote some direction of deviation in the population composition n.[11] Let $\delta n_{\omega|\Delta n}$ denote the Fourier component at frequency ω of some fluctuation history in the direction $\hat{\Delta}n$ away from $\bar{n}$. Then the fluctuation probability for such a component, evaluated over a time interval T, letting $\hat{\Delta}n$ be one of the principle directions of relaxation

---

[9] Readers not familiar with these domains need not worry. We include the points of contact because some readers may appreciate the familiarity, but all claims made in these introductions are demonstrated with full calculations in later sections and chapters.

[10] The mathematical analogy of escape trajectories to solitons has long been understood, and led to the adoption of the term 'instanton'—coined by Gerard t'Hooft [29]—to refer to escape trajectories, first for use in Euclidean field theories of the vacuum or equilibrium [5], and more recently in non-equilibrium rare-event statistics as well [30].

[11] More formally, we think of $\hat{\Delta}n$ as a *ray* in the population-configuration space. How we may choose to represent that ray with a normalized vector is unimportant to this discussion.

(eigenvectors of the diffusive Laplacian matrix), and restricted to small fluctuations, generally has the form appropriate to simple linear-diffusive relaxation[12]

$$P_{\text{fluct}}\left(\delta n_{\omega|\hat{\Delta}n}\right) \sim \mathrm{e}^{-\alpha_{\hat{\Delta}n} T\left(\omega^2+\omega_{\hat{\Delta}n}^2\right)\left(\delta n_{\omega|\hat{\Delta}n}\right)^2/N}. \tag{4.3}$$

Every stationary point is surrounded by a cloud of such fluctuations, in which the mean regression rate $\omega_{\hat{\Delta}n}$ may depend on the direction $\hat{\Delta}n$ of the fluctuation (dispersion and mean regression may be anisotropic in the state space), and the prefactor $\alpha_{\hat{\Delta}n}$ depends on details of the system. The common feature of these fluctuations is that $\langle (\delta n_{\omega|\hat{\Delta}n})^2 \rangle$ scales as N (appropriate to Poisson statistics), and as $1/T$ (appropriate to central-limit theorem aggregation for a single frequency component). The presence of a non-zero $\omega_{\hat{\Delta}n}^2$ is the stochastic process counterpart to an energy gap in condensed matter physics; it ensures that fluctuations even at zero frequency regress to the mean. This simple diffusion provides an example of what we mean by concentration of the support *in space*—namely the probability distribution becomes more and more concentrated around the mean value as N gets larger.

Fluctuations about broken symmetry backgrounds

In systems such as RPS with its Hopf bifurcation and limit cycle, the new feature is that the regression rate $\omega_{\hat{\Delta}n} \to 0$ for one of the directions of fluctuation ($\delta n_{\omega|\hat{\Delta}n}$ locally tangent to the limit cycle)[13]. Fluctuations in that direction accumulate as a random walk without mean regression. Other fluctuations at frequency ω remain concentrated in the *transverse* directions, but this form of spatial concentration has a co-dimension that is one lower than the co-dimension of attraction to a rest point. Modes with $S \propto \omega^2$ in the large-deviations formula (4.3) are the stochastic process counterparts to gapless excitations (in condensed matter) or massless particles (in vacuum field theory) and their origin in symmetries is of the same kind.

Rare large excursion trajectories

The Gaussian fluctuations in the cloud around a rest point generally have exponentially larger probability than the large and generally non-linear excursions that lead to escapes into other basins of attraction. In other words, in the unconstrained space of histories, the small fluctuations have lower values of S than the escapes.

---

[12] This can be derived as the behavior of one degree of freedom in a simple diffusive system, for example from the action (4.79) in section 4.3.4.1 below. The full algebra to produce an action functional for such systems, first demonstrated by Onsager and Machlup [31], is presented in equation (4.108) of [25]. Near rest points and in converging directions toward stable paths, all of our stochastic processes converge on behavior indistinguishable from simple-diffusive relaxation, in an appropriate eigenbasis of the Laplacian matrix that determines the independent diffusive modes for $\hat{\Delta}n$.

[13] This result may be derived by taking as a fluctuation the limit cycle offset by an oscillatory time displacement $\mathrm{e}^{-\mathrm{i}\omega t}\delta t$, rather than a constant time displacement δt, in the derivations leading to (4.115) of section 4.4.2.3 below. If desired, the Onsager–Machlup construction can then be carried out as in the previous example.

However, the escape trajectories are configurations of locally least action, and in the *conditional distribution* restricting to histories that dwell in different rest points at different times, they are the *global* least-action configurations[14]. These trajectories are not generally simple to represent in the frequency domain, but they have compact support in time and may be indexed by some center point t_0 of the transition. If we denote such histories by n_{t_0} (remembering that this denotes an entire function!), their probability of occurrence in a time series has the form[15]

$$P_{\text{fluct}}(n_{t_0}) \sim e^{-S_{\text{esc}}^{(0)}}. \tag{4.4}$$

The action $S_{\text{esc}}^{(0)}$ does not depend on the time t_0 because time translation is a symmetry of the underlying dynamics, which is merely hidden by any particular excursion trajectory. $S_{\text{esc}}^{(0)} \propto N$, so whereas the temporal profile of these trajectories is roughly independent of N, they become exponentially rare in large N, and in that sense their support is compressed in time. These large excursions behave like isolated particles or solitons. They are like excitations with an energy gap in that $S_{\text{esc}}^{(0)} > 0$, but more analogous to solitons in that they are indexed by a coordinate t_0 which is a direction of symmetry, rather than by a frequency ω which affects their probability as for the ordinary Gaussian fluctuations described in (4.3).

4.1.3 Generating functionals and Hamiltonian dynamical systems

4.1.3.1 Generating functions and sources in equilibrium large-deviations theory

Large-deviations scaling is familiar from equilibrium thermodynamics, in which multiple quantities (e.g. particle number, internal energy, volume) scale in proportion to each other *and in proportion to the entropy* when aggregate systems are made by combining many sub-systems with the same intensive state variables (respectively, chemical potential, temperature, or pressure). Any of these may be used as the scaling variable. If particle number is chosen as the scaling variable, the rate function for fluctuations in the thermodynamic (large-deviations) limit is then the specific entropy [11].

The most common way to extract properties of the equilibrium entropy is with a variety of moment-generating functions known as *partition functions.* The logarithms of partition functions, which are cumulant-generating functions, are the *free energies* of classical thermodynamics.

The generating function for a probability distribution introduces the concept of a *source* with which to probe the moments or cumulants of the distribution. In the simplest example, for ρ_n a distribution over some integer index n, and a weight

[14] In the unconditional distribution, the value $S^{(0)} = 0$ distinguishes the classical background from escape trajectories with $S^{(0)} > 0$ and we will retain these constant offsets because they also distinguish different approximation regimes for our solution methods. In the projection onto the conditional distribution for escapes, the normalization factor subtracts the positive constant from $S^{(0)}$, so that, in keeping with the status of the escape trajectory as a global minimizer, the shifted action likewise takes value zero on this path.

[15] This form is introduced in section 4.4.1 below and an example is developed in chapter 5.

parameter z that we may write $z \equiv e^q$, the moment-generating function Ψ and the cumulant-generating function $\Gamma = \log \Psi$ are related to ρ_n as

$$\begin{aligned} \Psi(z) &\equiv \sum_n z^n \rho_n \\ &\equiv e^{\Gamma(q)} = \sum_n e^{nq} \rho_n. \end{aligned} \tag{4.5}$$

We refer to q as a 'source' for n because

$$\frac{d\Gamma}{dq} = \langle n \rangle_q, \tag{4.6}$$

the expectation of n in the distribution distorted by q. Higher-order derivatives of Γ with respect to q generate higher-order cumulants in n. If the density ρ_n, regarded as a fluctuation probability for n away from its equilibrium value, has the large-deviations form (4.1) at leading exponential order, then the moment-generating function $\Psi(z)$ is likewise exponential, and its logarithm $\Gamma(q)$ is related to the logarithm of ρ by a Legendre transform [24].[16] This relation is known as the Gärtner–Ellis theorem.

In equilibrium thermodynamics, ρ_n (and its generalizations) are often pure combinatorial terms that count the number of ways the state n can be realized *within the system*, and the weight may take a form such as $z = e^{-E/k_B T}$, with E the energy associated with an increment of n, T temperature and k_B Boltzmann's constant. Such weights are yet-further large-deviations expressions, for the probability of fluctuations in an *environment* that permit the system energy increments by E, and $E/k_B T$ is their corresponding rate function.

4.1.3.2 Collective fluctuations, the Laplace transform and hydrodynamic modes

We argued in section 2.3 that just as symmetry is the property that unifies the representations of a process among different scales, collective fluctuations rather than rare events of individual interaction are the predominant dynamical entities that intermediate between representations at different scales. Generating functions provide particularly convenient and effective ways to treat collective fluctuations because they are the Laplace transforms of probability distributions over the discrete basis of counts of the agent types.

Like Fourier transforms, Laplace transforms provide a basis for the collective motion of many agents in which the lowest-order terms more directly approximate so-called 'hydrodynamic limits' [32, 33] than does the basis of single-agent changes. Second, the way the generating function extracts fluctuation moments is by first *deforming* the probability density of interest into a skewed density, in which the mean plus the fluctuation becomes the new typical value, and then differentiating the skewed distribution from the original. Rather than measure the probability of a single exact instance of a fluctuating population history, the generating function

[16] In the case where $\log(\rho)$ is not convex, the Legendre transform is generalized to a *Legendre–Fenchel* transform.

measures the weight of this skewed distribution, which is the proper probability to assign to a macroscopic fluctuation that may be compatible with many slightly distinct exact histories[17].

Collective fluctuations are the mediators of scale transformation in equilibrium systems [4, 34–36] whether or not they are modeled as explicitly dynamical entities. The simplification afforded by generating functions in passing to hydrodynamic limits becomes even greater, however, in passing to non-equilibrium systems, where fluctuations are time-dependent deviations of histories.

4.1.3.3 Generating functions and large-deviation methods for non-equilibrium systems

For time-dependent population processes, large-deviation properties may be derived much as they are for equilibrium systems, but they generally apply to correlation functions at multiple times as well as at single times, and to time-dependent as well as stationary observables. The observables for which fluctuation probabilities are computed are generally elevated from states to *histories*. Large-deviation limits on spaces of histories are constructed, not from generating functions with sources at single times, but from generating functionals that probe time-dependent probability distributions with sources at multiple times. The resulting rate functions, which are the appropriate non-equilibrium generalizations of specific entropies, therefore must be understood as functionals of histories rather than functions of states.

A remarkable property which we will derive and demonstrate in this and later chapters is that the rate functions for Markov processes over extended time have the structure of Hamilton–Lagrange *action functionals* for Hamiltonian dynamical systems constructed from the transition probabilities of their associated Markov processes [24]. We refer to them as *stochastic effective actions* [25], because they are the stochastic process counterparts to functionals known as *quantum effective actions* [37], which play a large role in renormalization of quantum field theories. A sub-set of the solutions to these Hamiltonian equations, in which the 'conjugate momentum' plays no specific role, contains the ordinary first-moment conditions for relaxation in Markovian systems. Much more surprising and at first counter-intuitive, a different sub-set of solutions, describing escapes, captures the first-moment dynamics of *conditional* distributions for rare events, with equations that are both deterministic and have a key role for the Hamiltonian conjugate momentum, even though they are derived from stochastic processes with no notion of kinematic momenta.

4.2 From robust structure estimation based on low-order moments, to minimal underlying stochastic models

We now propose a systematic way to perform the two necessary tasks of first choosing and then solving an appropriate stochastic model for a given population process, with an understanding that both steps should separate robust outcomes

[17] The way this skew is introduced, and the way it distinguishes probabilities for macroscopic fluctuations from those for microscopic fluctuations, are discussed in detail in section 3 of [25].

from properties sensitive to fine details including modeling assumptions. Our starting point will be the classical replicator equation (2.15), which in structure is a mean-field equation. We construct a stochastic process which gives us back the replicator equation when evaluated in its MFA and which, beyond the mean-field limit, satisfies criteria of least-additional assumptions. When the stochastic process has been specified, it is of course no longer necessary to limit its solution to the MFA and we can elaborate the fluctuation effects that result in interesting or important departures from the naïve predictions of the replicator equation.

From the discussion of general population processes in chapter 2 let $\mathrm{n} \equiv (\mathrm{n}_1, \ldots, \mathrm{n}_D)$ denote the counts of agents of types $1, \ldots, D$ in a population of size $N \equiv \sum_{i=1}^{D} \mathrm{n}_i$. We consider general branching-and-replacement processes that leave type number unchanged[18]. From the derivation of the Price equation [38, 39], we know that for processes satisfying the assumptions of population genetics, these may always be parametrized in terms of a vector of fitnesses (f_i) and a stochastic matrix $[\mu_{ij}]$ whose coefficients we generically term 'mutation' rates. Both the $\{f_i\}$ and $\{\mu_{ij}\}$ will generally be functions of the population state. We use roman indices n for discrete sample values of population states, and italic n for the corresponding continuous-valued variables which arise as mean values from distributions over samples n. Probability density functions for the stochastic process are denoted as ρ_{n} and we will use overbars ($\bar{n}$ etc) for averages under ρ_{n}, either exact or in various stationary-point approximations.

4.2.1 The association of minimal stochastic process models with first-moment recursions in the MFA

As long as all types in a population have fitnesses that differ in relative magnitude by more than $O(1/N)$, the population diverges exponentially from equal proportions and towards a well-defined asymptotic equilibrium (although which equilibrium may depend on initial conditions). In this situation the MFA for closure of first-moment conditions is often valid and therefore it is adopted as the starting point to specify structure in most discussions of evolutionary games[19]. Although the methods we develop here become most useful when convergence to a steady state is *not* exponential due to degeneracies in fitness—whether they are externally imposed or result from broken symmetry—the deterministic replicator provides a useful starting point to choose the model structure in many cases.

We assume that the population process is *Markovian* so that the distribution ρ_{n} is well defined on a complete space of types $\{\mathrm{n}\}$. A general Markovian process is specified by its *master equation*,

$$\frac{\mathrm{d}}{\mathrm{d}t}\rho_{\mathrm{n}} = \sum_{\mathrm{n}'} T_{\mathrm{nn}'}\rho_{\mathrm{n}'}. \tag{4.7}$$

[18] In examples, we will reduce these to Moran processes, as a kind of minimality assumption explained below.

[19] Fitness differences lead to exponential divergence from the uniform population state because they appear as multipliers of the numbers of different types in the replicator equation. The persistence of such fitness differences in neighborhoods of rest points, in the MFA, underlies the definition [40] of the evolutionary stable state (ESS) through linear stability analysis.

We assume that we may pass back and forth between discrete- and continuous-time formulations, and will usually write the continuous-time form for simplicity[20]. $T_{\mathrm{nn}'}$ is called the *transfer matrix* of the stochastic process. It is a stochastic matrix (meaning $\sum_{\mathrm{n}} T_{\mathrm{nn}'} \equiv 0, \forall \mathrm{n}'$) because it must conserve probability: $\mathrm{d}(\sum_{\mathrm{n}} \rho_{\mathrm{n}})/\mathrm{d}t = 0$.

The time-dependent first-moment of the distribution ρ is denoted as $\bar{n}$, with components

$$\bar{n}_i = \sum_{\mathrm{n}} \mathrm{n}_i \rho_{\mathrm{n}}. \tag{4.8}$$

Recall from chapter 2 that the dynamics for $\bar{n}_i$ delivered by the Price equation (2.9) generally require knowledge of higher-order moments of n, creating the problem of finding a *closure* for the equations of motion to some finite order. The MFA attempts to close the equations using functions only of the first moment $\bar{n}_i$. If a good approximation of this form can be found, then we can require that the transfer matrix induces mean population dynamics which may be written in the form

$$\begin{aligned} \frac{\mathrm{d}\bar{n}_i}{\mathrm{d}t} &= \sum_{\mathrm{n},\mathrm{n}'} \mathrm{n}_i T_{\mathrm{nn}'} \rho_{\mathrm{n}'} \\ &= \bar{n}_i \left(f_i - \phi \right) + \sum_{j=1}^{D} \mu_{ij} \bar{n}_j. \end{aligned} \tag{4.9}$$

The second line has the form of the *replicator equation with mutation*, in which the MFA consists of writing f_i and μ_{ij} as functions of $\bar{n}$ but not of higher-order correlations. If all relative fitness differences are $>O(1/N)$, the functions f_i and μ_{ij} used to define the model will be close to the values obtained by running regressions of instances of $\mathrm{dn}_i/\mathrm{d}t$ against the state n in populations with dynamics generated by that model. In other words, the regression-fits of the population-level rate equations to fitness and mutation functions of the forms (4.9) will yield coefficients close to those that describe fitness and mutation values experienced by the individual agents as a result of their interactions. If some fitness differences are smaller than $O(1/N)$, then corrections from higher-order moments of the distribution over fluctuations in population composition may lead to significant differences between the microscopic interactions that define the model and population-level regression coefficients used to fit the rate equations to a model with the same functional form. We provide an example of such corrections in (7.5) below.

The population number is exactly preserved,

$$\frac{\mathrm{d}}{\mathrm{d}t} \sum_{i=1}^{D} \bar{n}_i \equiv 0, \tag{4.10}$$

[20] Much literature in population genetics follows a notation that Price introduced, in which n′ represents population number in an offspring generation, when n refers to the parent's generation. For historical continuity we adopted this notation in chapter 2 where the Price equation was introduced, but in these sections where the master equation is developed, n′ simply represents an alternative index of summation.

if we choose

$$\phi = \frac{1}{N}\sum_{i=1}^{D}\bar{n}_i f_i . \tag{4.11}$$

It is important that the conditions (4.9–4.11) amount to a stronger assumption than the use of the replicator equation implies on its own. A fixed absolute population number implies fixed fluctuation strength. Although nothing in the methods below requires that we assume fixed fluctuation strength, doing so simplifies the examples we present.

In order to specify a master equation, probabilities per unit time must be provided for all events in which agents change type. When agents switch both from a type l to a type j, and also from j to l, only the excess of one rate over the other results in a change of the population state on average. Some of the type switching balances and, because it does not affect average population composition, it is not reported in the dynamics of the mean $\bar{n}$ that we approximate with (4.9). However, this canceling motion is an important source of variance and higher-order moments in the distribution of population states and therefore must be assigned somehow.

In the language of stochastic differential equations or the diffusion equations that they generate on distributions, one says that the deterministic replicator equation captures only the antisymmetric part of population change, known as *drift*. It is insensitive to the symmetric part of population fluctuations, which leads to *diffusion*[21]. Expressed in terms of probabilities, if p_{jl} (n) is the probability per unit time of a type conversion from type l to type j, as a result of reproduction and mutation in the context of a population whose composition is n, then the antisymmetric combination $(p_{jl} - p_{lj})/2$ determines the net conversion, while the symmetric combination $(p_{jl} + p_{lj})/2$ describes mutual exchanges that leave the net composition unchanged. The rate equations only constrain mean drift; higher-order antisymmetric moments are left undetermined by (4.9).

Readers who are new to some of these properties of master equations are encouraged to pause at this point and work through a small warm-up example developed in appendix 4.5. The example omits all of the consequences of selection and almost all of the consequences of mutation in a population with only two types. It demonstrates in microcosm the relation of generating processes, statistical estimators and minimal models summarized in section 4.1.1.1. The example in appendix 4.5 shows how parameters in the generating process affecting dispersion and higher-order moments can fail to be reported in the dynamics of the mean and thus require that the modeler make specific assumptions about their form.

---

[21] Collision of terminologies seems inescapable when population genetics encounters more general stochastic processes. 'Drift' in the terminology of neutral theory in population genetics corresponds to the 'diffusion' term in the stochastic process nomenclature. The quantity known as 'drift' in stochastic processes is associated with 'selection' in population genetics, though it can also result from regularity in the transmission process.

4.2.1.1 The notion of a minimal stochastic model

In much the same way as the method of maximum entropy [14] may be used to assign a maximum-ignorance distribution to any system of linear constraints (however simple), we may assign a 'maximum-ignorance' stochastic process to a chosen set of rate equations. Just as adding further constraints to the specifications of a system's boundary conditions refine the resulting maximum-entropy distribution, higher-order polynomial expansions of fitness, or other constraints on fluctuation moments, may be incorporated with the addition of terms in the stochastic process description. The particular parameters that define the stochastic process will appear as cumulants in a fluctuation probability in the field theory defined below.

We do not prove formal relations between our minimal stochastic processes and maximum-entropy conditions for the distributions they produce, although a formalism of path entropies exists within which such relations could be developed [25]. Rather, we define minimality through the two conditions that: (i) both drift and diffusion terms in the stochastic process introduce no new parameters beyond those in the mean-field replicator equation, and (ii) the generator of the Markov chain may be written as a Moran branching process [41] on a sufficiently short, discrete timescale. In the Moran process, only one birth or death occurs within a time step, so all events of replication and replacement are independent, and we may write the transfer matrix $T_{\mathrm{nn}'}$ with only diagonal and one-index off-diagonal elements[22].

To separate drift from diffusion, we divide a general transfer matrix into its antisymmetric and symmetric terms as

$$T_{\mathrm{nn}'} = T_{\mathrm{nn}'}^{\mathrm{Mean}} + T_{\mathrm{nn}'}^{\mathrm{Dispersion}}. \tag{4.12}$$

We have labeled the antisymmetric drift term 'Mean' because we will constrain it by the mean dynamics given in the replicator equation. The symmetric term is labeled 'Dispersion' as a reminder that it is the leading source of fluctuations.

If we now suppose linear frequency-dependent fitness[23]

$$f_i = \sum_{j=1}^{D} a_{ij}\left(\bar{n}_j - \delta_{ij}\right), \tag{4.13}$$

corresponding to the second-order polynomial regression model in (2.16), with the intrinsic fitness $\alpha_i \equiv 0$, and isotropic diffusion (2.6), then the lowest-order part of the drift operator $T_{\mathrm{nn}'}^{\mathrm{Mean}}$ is uniquely specified by the dynamics (4.9). We show next how

[22] Alternative formulations such as the Wright–Fisher model, describing simultaneous reproduction of all individuals in a generation, are well known to produce different relations between actual population size and the *variance-effective population size* than the Moran process does [41], although they may have similar diffusion limits when appropriately rescaled. Subtleties in correctly calibrating models that describe strictly simultaneous type changes of more than one agent at a time are illustrated in [42] and may have biologically meaningful counterparts for organisms with synchronized reproduction. For the examples in this volume, we have not pursued a systematic treatment of the identification of models involving synchronization.

[23] Like the assumption that defines structure from the MFA, the retention of only affine-order frequency dependence is an assumption that the lowest non-trivial order expansion is the most robust. These linear fitness and mutation terms define the Gaussian order of fluctuations, so these robustness assumptions have a basis in the central-limit theorem.

the transfer matrix may be systematically computed for a minimal stochastic process defined by these conditions.

4.2.2 Systematic construction of transfer matrices

We illustrate the construction of transfer matrices for linear-order normal form games with two players and the Moran birth–death process. The derivation follows the general form reviewed in [25]. This construction readily generalizes to k-player normal form games. The approach can also be extended to processes incompatible with the Moran model, involving simultaneous birth/death events, by the addition of higher-order 'contact terms', as explained in appendix 4.5.

First the payoff matrix is written as a sum of antisymmetric and symmetric parts,

$$a_{ij} \equiv a_{ij}^{A} + a_{ij}^{S}. \tag{4.14}$$

Then (4.9) is expanded in terms of these as

$$\begin{aligned}\frac{\mathrm{d}\bar{n}_i}{\mathrm{d}t} = & \sum_{j=1}^{D} \bar{n}_i a_{ij}^{A} \bar{n}_j \\ & + \sum_{j=1}^{D} \bar{n}_i a_{ij}^{S} \left(\bar{n}_j - \delta_{ij}\right) - \frac{\bar{n}_i}{N} \sum_{k,j=1}^{D} \bar{n}_k a_{kj}^{S} \left(\bar{n}_j - \delta_{kj}\right) \\ & + \sum_{j=1}^{D} \mu_{ij} \bar{n}_j.\end{aligned} \tag{4.15}$$

(The matrix μ is stochastic, so that its transpose vanishes in the sum over j.) The exchange antisymmetry between the (summed) index j and the (not summed) index i is further exposed by introducing a factor of $\sum_{k=1}^{D} n_k / N \equiv 1$ into the first term on the second line, to give

$$\begin{aligned}\frac{\mathrm{d}\bar{n}_i}{\mathrm{d}t} = & \sum_{j=1}^{D} \bar{n}_i a_{ij}^{A} \bar{n}_j \\ & + \frac{\bar{n}_i}{N} \sum_{j,k=1}^{D} \bar{n}_j \left[a_{ik}^{S} \left(\bar{n}_k - \delta_{ik}\right) - a_{jk}^{S} \left(\bar{n}_k - \delta_{jk}\right) \right] \\ & + \sum_{j=1}^{D} \mu_{ij} \bar{n}_j.\end{aligned} \tag{4.16}$$

The sum of (4.16) on i now vanishes manifestly by the antisymmetry of its argument.

Population size N regulates the strength of fluctuations, which we wish to hold at a fixed scale for ease of analysis. It is as easy to conserve N exactly as keeping it fixed in an ensemble average, so we will restrict attention to transfer matrices that conserve probability over configurations of separate N independently. All such matrices must transfer probability through equal numbers of death and reproduction events.

We express master equations compactly without needing to write index sums explicitly, by using a notational shortcut introduced in [25]. Density ρ is treated as a function of a continuous index, but only evaluated on integer (vector) values n. Off-diagonal matrix positions correspond to powers of the operators $e^{\pm\partial/\partial n_i}$, where each power shifts the corresponding n_i values by ± 1 wherever they appear.

For processes in which all population changes may be built up from single-individual type changes, $T_{nn'} \neq 0$ only on indices n, n′ that differ by ± 1 at two types j and i. The operator for the Moran process (type conversion of a single agent), which acts on any function of n to shift the index by $+1_i - 1_j$, is

$$e^{\partial/\partial n_i - \partial/\partial n_j}. \tag{4.17}$$

For continuous-time master equations, we require that all terms be built from single-agent shift operators of this form[24]. Because only net conversion from types i to j appears in the deterministic replicator, only the antisymmetric combination $\sinh(\partial/\partial n_i - \partial/\partial n_j)$ will be constrained by the first-moment equation (4.16).

Recalling the warm-up example developed in appendix 4.5, the master equation (4.122) conserves probability because it is written as a series of differences between terms at shifted indices. The shift acts to take index $n \to n + k_i - k_j$ in both the density and the multiplying rate terms. Such difference terms are simplified and the identical form of the terms in which probability enters and exits is more clearly expressed by replacing the shifted indices with powers of the operator (4.17). (Although we do not pursue k-player simultaneous shifts in what follows, these are generated by the kth power of the operator (4.17).)

The remaining task is to work backward from the mean-field rate equation to the set of terms in a master equation that will produce it. Since the replicator equation for linear frequency-dependent fitness is a polynomial of no more than third order, we may do this systematically by constructing the terms in probabilities $p_{ji}(n)$ that will produce each order of dependence of $d\bar{n}/dt$ on $\bar{n}$ in mean-field theory, and then matching coefficients with the corresponding term in (4.16).

Whether we assign rates to change events for agents sampled with-replacement or without-replacement can of course depend on the details of the system considered. For large, well-mixed populations it is usually a statistical fine point without consequences on average, and in models the choice may be made for mathematical convenience (or to reduce clutter in expressions). We will introduce terms $-\delta_{jl}$ at various points in what follows as corrections to type numbers n_j or n_l, so that all sampling is performed without replacement. In the game language, this amounts to excluding self-play from calculations of fitness; in the evolutionary dynamic it amounts to excluding self-replacement as a kind of 'change' event in the statistics. The reason this simplifies expressions will become apparent as we develop the representation in linear algebra in the following sections.

[24] Including multiple-agent type conversions that are correlated over arbitrarily short times is a kind of 'fine tuning'. It requires specific empirical motivation to be introduced at the individual level and in many cases we do not expect differences from the Moran process to be preserved under aggregation.

The correspondence is then constructed as follows. For any function $A_{\rm n}$ with the expectation $\sum_{\rm n} A_{\rm n}\rho_{\rm n}$ denoted $\langle A \rangle$, these shift operators, by shifting the index of summation, lead to the identity

$$\begin{aligned}
&\sum_{\rm n} {\rm n}_j \sinh\left(\frac{\partial}{\partial {\rm n}_i} - \frac{\partial}{\partial {\rm n}_j}\right) A_{\rm n}\rho_{\rm n} \\
&= \sum_{\rm n} A_{\rm n}\rho_{\rm n} \sinh\left(\frac{\partial}{\partial {\rm n}_j} - \frac{\partial}{\partial {\rm n}_i}\right) {\rm n}_j \\
&= \langle A \rangle. \qquad (4.18)
\end{aligned}$$

In particular, taking $A_{\rm n}$ to be the function $\sum_{j,l=1}^{D} {\rm n}_l a_{lj}^{A} {\rm n}_j$ defined from the decomposition (4.14), yields

$$\frac{1}{2}\sum_{\rm n} {\rm n}_i \sum_{j,l=1}^{D} \sinh\left(\frac{\partial}{\partial {\rm n}_l} - \frac{\partial}{\partial {\rm n}_j}\right) {\rm n}_j a_{jl}^{A} {\rm n}_l \rho_{\rm n} = \sum_{j=1}^{D} \left\langle {\rm n}_i a_{ij}^{A} {\rm n}_j \right\rangle. \qquad (4.19)$$

If correlated fluctuations are ignored, and $\langle {\rm n}_i {\rm n}_j \rangle$ is replaced by $\bar{n}_i \bar{n}_j$, (4.19) yields the first line of (4.16). A similar expansion with $A_{\rm n}$ set to $\sum_{j,k,l=1}^{D} {\rm n}_j a_{jk}^{S} ({\rm n}_k - \delta_{jk}) {\rm n}_l$ gives

$$\begin{aligned}
&\frac{1}{N}\sum_{\rm n} {\rm n}_i \sum_{j,k,l=1}^{D} \sinh\left(\frac{\partial}{\partial {\rm n}_l} - \frac{\partial}{\partial {\rm n}_j}\right) {\rm n}_j a_{jk}^{S} \left({\rm n}_k - \delta_{jk}\right) {\rm n}_l \rho_{\rm n} \\
&= \sum_{l,k=1}^{D} \left\langle {\rm n}_i \frac{{\rm n}_l}{N} a_{ik}^{S} \left({\rm n}_k - \delta_{ik}\right) \right\rangle - \sum_{j,k=1}^{D} \left\langle \frac{{\rm n}_i}{N} {\rm n}_j a_{jk}^{S} \left({\rm n}_k - \delta_{jk}\right) \right\rangle, \qquad (4.20)
\end{aligned}$$

which in the MFA is the second line in (4.16). Finally, the antisymmetric component of the combination $\mu_{ij} {\rm n}_j$ is extracted with the identity

$$\sum_{\rm n} {\rm n}_i \sum_{j,l=1}^{D} \sinh\left(\frac{\partial}{\partial {\rm n}_l} - \frac{\partial}{\partial {\rm n}_j}\right) \mu_{jl} {\rm n}_l \rho_{\rm n} = \sum_{j=1}^{D} \left\langle \mu_{ij} {\rm n}_j \right\rangle, \qquad (4.21)$$

giving the third line of (4.16).

The sum of the three terms (4.19–4.21) gives an un-symmetrized representation for the mean-field component of the transfer matrix,

$$\begin{aligned}
\sum_{\rm n'} T_{\rm nn'}^{\rm Mean} \rho_{\rm n'} \equiv &\sum_{j,l=1}^{D} \sinh\left(\frac{\partial}{\partial {\rm n}_l} - \frac{\partial}{\partial {\rm n}_j}\right) \\
&\times \left\{ {\rm n}_j \left[\frac{1}{2} a_{jl}^{A} + \frac{1}{N} \sum_{k=1}^{D} a_{jk}^{S} \left({\rm n}_k - \delta_{jk}\right) \right] + \mu_{jl} \right\} {\rm n}_l \rho_{\rm n}. \qquad (4.22)
\end{aligned}$$

Since only the antisymmetric part of the braces in (4.22) avoids cancellation against the sinh, and since it will be useful to have explicit antisymmetric forms to motivate the dispersion component of $T_{\mathrm{nn}'}$ in the text, we extract the antisymmetric part of the braces explicitly as

$$\sum_{\mathrm{n}'} T_{\mathrm{nn}'}^{\mathrm{Mean}} \rho_{\mathrm{n}'} = \frac{1}{2} \sum_{j,l=1}^{D} \sinh\left(\frac{\partial}{\partial \mathrm{n}_l} - \frac{\partial}{\partial \mathrm{n}_j}\right) \times \left\{ \mathrm{n}_j(\mathrm{n}_l - \delta_{jl}) a_{jl}^{A} + \frac{1}{N} \sum_{k=1}^{D} \mathrm{n}_j(\mathrm{n}_k - \delta_{jk}) a_{jk}^{S} \mathrm{n}_l + \mu_{jl} \mathrm{n}_l - \frac{1}{N} \sum_{k=1}^{D} \mathrm{n}_l(\mathrm{n}_k - \delta_{lk}) a_{lk}^{S} \mathrm{n}_j - \mu_{lj} \mathrm{n}_j \right\} \rho_{\mathrm{n}}. \tag{4.23}$$

Equation (4.23) implicitly includes self-play (and selection of a player against itself) in the combinations of n coefficients multiplying a_{jk}^{S} and a_{lk}^{S}. A pair of completely canceling terms, which remove double counting of any agent from both of these a^S coefficients, may be introduced to put the mean-field transfer matrix in a form that includes only interactions of distinct agents,

$$\sum_{\mathrm{n}'} T_{\mathrm{nn}'}^{\mathrm{Mean}} \rho_{\mathrm{n}'} = \frac{1}{2} \sum_{j,l=1}^{D} \sinh\left(\frac{\partial}{\partial \mathrm{n}_l} - \frac{\partial}{\partial \mathrm{n}_j}\right) \times \left\{ \mathrm{n}_j(\mathrm{n}_l - \delta_{jl}) a_{jl}^{A} + \frac{1}{N} \sum_{k=1}^{D} \mathrm{n}_j(\mathrm{n}_k - \delta_{jk}) a_{jk}^{S} (\mathrm{n}_l - \delta_{jl} - \delta_{kl}) + \mu_{jl} \mathrm{n}_l - \frac{1}{N} \sum_{k=1}^{D} \mathrm{n}_l(\mathrm{n}_k - \delta_{lk}) a_{lk}^{S} (\mathrm{n}_j - \delta_{lj} - \delta_{kj}) - \mu_{lj} \mathrm{n}_j \right\} \rho_{\mathrm{n}}. \tag{4.24}$$

The antisymmetric combination of shift operators $\sinh(\partial/\partial \mathrm{n}_l - \partial/\partial \mathrm{n}_j)$ in (4.24) describes net conversion from type l to type j and the remaining quantity in braces gives the rate of conversion per unit time. Although (4.24) appears cumbersome with its extra δ terms, it will yield the simplest form for the algebraic and field-theoretic generating functionals, derived in section 4.3.1.2 below.

The operator $T_{\mathrm{nn}'}^{\mathrm{Mean}}$ cannot define a full master equation, because the quantity in braces is not positive definite and cannot be interpreted as a probability. We may associate a unique symmetric operator $T_{\mathrm{nn}'}^{\mathrm{Dispersion}}$ with the deterministic replicator dynamic by requiring that the resulting $T_{\mathrm{nn}'}$ contain only positive-definite rate constants and *no additional parameters* beyond those appearing in (4.24). While the resulting transfer matrix is not the only one compatible with the first-moment conditions, it operationalizes our definition of a maximum-ignorance model choice.

We construct $T_{\mathrm{nn}'}^{\mathrm{Dispersion}}$ as follows: the lowest-order symmetric shift operator which conserves probability, counterpart to the antisymmetric $\sinh(\partial/\partial \mathrm{n}_l - \partial/\partial \mathrm{n}_j)$ in (4.24), is $[\cosh(\partial/\partial \mathrm{n}_l - \partial/\partial \mathrm{n}_j) - 1]$. Similarly, the unique symmetric hopping rate that neither introduces nor suppresses parameters is the symmetric counterpart (under exchange of j and l) to the antisymmetric braced terms in (4.24). If we adopt the product of these two symmetric operators as the dispersion component of the transfer matrix,

$$\sum_{n'} T^{\text{Dispersion}}_{nn'} \rho_{n'} = \frac{1}{2} \sum_{j,l=1}^{D} \left[\cosh\left(\frac{\partial}{\partial n_l} - \frac{\partial}{\partial n_j} \right) - 1 \right]$$
$$\times \left\{ n_j \left(n_l - \delta_{jl} \right) a^S_{jl} + \frac{1}{N} \sum_{k=1}^{D} n_j \left(n_k - \delta_{jk} \right) a^S_{jk} \left(n_l - \delta_{jl} - \delta_{kl} \right) \right.$$
$$\left. + \mu_{jl} n_l + \frac{1}{N} \sum_{k=1}^{D} n_l \left(n_k - \delta_{lk} \right) a^S_{lk} \left(n_j - \delta_{lj} - \delta_{kj} \right) + \mu_{lj} n_j \right\} \rho_n, \tag{4.25}$$

then the right-hand side in the master equation (4.7) becomes

$$\sum_{n'} T_{nn'} \rho_{n'} \equiv \sum_{j,l=1}^{D} \left(e^{\partial/\partial n_l - \partial/\partial n_j} - 1 \right)$$
$$\times \left\{ \mu_{jl} n_l + n_j \frac{a_{jl}}{2} \left(n_l - \delta_{jl} \right) + \frac{1}{N} \sum_{k=1}^{D} n_j a^S_{jk} \left(n_k - \delta_{jk} \right) \left(n_l - \delta_{jl} - \delta_{kl} \right) \right\} \rho_n. \tag{4.26}$$

As long as all a_{jl} and all off-diagonal components of μ_{jl}, are non-negative, the quantity in braces in (4.26) is non-negative and has the interpretation of the probability per unit time for single-agent conversions from type l to type j, through any combination of replacement with replication or mutation.

For master equations in which the shift operators are written this way (as functions of $\exp(\partial/\partial n_l - \partial/\partial n_j) - 1$), it is particularly easy to extract the diffusion approximation known as its *Fokker–Planck* equation. The second-order Taylor expansion of the exponentials in $\partial/\partial n_l - \partial/\partial n_j$ directly yields this diffusion approximation. Proper proofs of convergence in probability to the diffusion equation are provided in the probability literature [43–45]. The Taylor's series explanation given here may be regarded as a notational short-cut to the formal derivation, if desired.

The hopping rates in braces in (4.26) are the defining quantities for the stochastic process. We will not solve dynamics from them directly in this representation, though that can be done [46, 47] using eigenvalue and eigenvector decompositions of the transfer matrix. Instead, we will convert the transfer matrix representation into a sequence of equivalent representations, first on the generating function for the probability distribution ρ in which the transfer matrix is replaced by a differential *Liouville* operator, and then in a field-theoretic representation for the generating functional, in which the transfer matrix is replaced by a Lagrange–Hamilton *action* functional. The hopping terms will re-appear as parameters both in the Liouville operator in equations (4.33, 4.52) and in the action (4.85) of the Doi–Peliti field-theoretic construction.

Master equations such as (4.7), with transfer matrices of the form (4.26), may be written more compactly as

$$\frac{d}{dt}\rho_{n} = \sum_{j,l=1}^{D} \left(e^{\partial/\partial n_l - \partial/\partial n_j} - 1 \right) p_{jl}(n) \rho_{n}, \tag{4.27}$$

in which $p_{jl}(n)$ denote the population-composition-dependent probabilities to change type from l to j, referred to previously. We will sketch the systematic derivation of the Liouville operator and field action for the sake of completeness, but in the final field theory that results, the form (4.85) for the action functional may be written down directly from (4.27), with p_{jl} given by the braced term in (4.26).

4.3 The Doi–Peliti method to construct the thermodynamics of non-equilibrium stochastic processes

The treatment of discrete-state stochastic processes with operator methods originally due to Masao Doi [48, 49], followed by the conversion to a field-integral representation introduced by Luca Peliti [50, 51], provide one instantiation of a very general framework for computing properties of perturbed dynamical systems[25]. The Doi–Peliti functional integral construction is equivalent to several approaches in field theory that have been developed for particular application domains. These include equilibrium [52] and non-equilibrium [53, 54] finite-temperature quantum mechanics, stochastic differential equations [55] and systems with quantum superposition but classically definable entropy [56]. The field theories fall into the general class derived by Martin, Siggia and Rose [57], and their common properties are reviewed in [46, 47, 58]. The use of these methods to compute large-deviation limits for fluctuation and escape probabilities often goes under the name of *Freidlin–Wentzell theory*, after the exposition of [55].

Evolutionary dynamics in single, well-mixed populations are described by a vector-valued composition variables at a single 'point', which corresponds to the population. Similar composition vectors evolving in multiple, interacting populations—for example, island models in bio-geography, or locally interacting populations diffusing on a spatial landscape—would occupy a state space with a product structure. The set or space of states within each sub-population is a copy of the same configuration space, while the population indices make up a second set or space. The product set (or product manifold, if these sets are continuous) is called a *fiber bundle*, in which the set of configurations is the fiber and the set of indices for populations is the base space. Migration or diffusion of individuals among populations (but without changing types) defines the projection of the total space onto the base space for each population state. A parallel literature exists for lattice gases, hydrodynamics and similar cases, in which the base space is continuous-valued, and diffusion on the base space as well as in the space of configurations occurs [32, 33, 59–61].

[25] A somewhat more thorough review is provided in [25].

We will note, where they arise in the following chapters, the conversions that can be performed from functional-integral representations of the time-dependent generating function to the van Kampen expansion [62–64] and the Langevin equation, two approximations that are often taken as starting points in the treatment of perturbed dynamical systems or non-equilibrium stochastic processes. As part of the development of the basic method, in this chapter we include perturbative methods to expand beyond Gaussian fluctuations and the method of instantons [5, 30, 65].[26]

Theories of Freidlin–Wentzell or Martin–Siggia–Rose form naturally yield probabilities for macroscopically specified *histories*, which generalize the fluctuation theorems of equilibrium thermodynamics [11] to incorporate information about kinetics and which may be interconverted with Jaynes's *maximum-caliber* approach to non-equilibrium statistical mechanics [25, 68–71]. In the large-population limits where field-theoretic methods become most useful, the leading large-deviations scaling of the probabilities generated by the stochastic evolutionary dynamic is expressed directly through a rescaling of field variables[27].

4.3.1 From master equations to Liouville equations; from function spaces to abstract Hilbert spaces

We provide a slightly unconventional introduction to the Doi–Peliti construction, by starting with the usual representation [72] of generating functions as analytic functions of a complex variable. We do this to make clear the stage at which the discrete indices of the master equation are replaced by continuous variables (obscured in the more conventional operator representation) and also to show why generating functions count identical particles much as they are counted in quantum mechanics, *without* any implied interpretation in terms of quantum state spaces. We then convert the analytic representation of generating functions—which is often used only as a formal power series anyway [72]—to a representation by abstract operators and vectors in a linear space that is much more efficient for the treatment of time dependence.

4.3.1.1 The ordinary-power-series generating function of complex arguments

The generating function of a stochastic process replaces discrete number indices in the master equation (4.7) with continuous-valued complex arguments. The

[26] Terminology for ray methods in stochastic processes is not as uniform as the counterpart for quantum systems. Ray methods in general may correctly be referred to as *eikonal approximations* [30, 66, 67] because of the manner in which the leading exponential contribution to the probability is calculated. Occasionally the word 'instanton' is found used as a synonym for eikonal [30], but we will reserve the word 'instanton' for the sub-set of eikonals that begin and end in rest points of the stochastic process and so can exist as local least-action solutions in a system's autonomous dynamics, not anchored by source terms in generating functions. See section 4.4.1 for more on this distinction. This usage is most consonant with the original coinage of the term in Euclidean field theory [5] to refer to a time-bounded, soliton-like solution to the equations of motion.

[27] This simple approach to extracting large-deviations scaling is similar to the extraction of the Fokker–Planck equation by mere Taylor's series expansion in the previous section and in fact the two approaches are related under the conversion from the discrete master equation to the equivalent field theory performed here.

ordinary-power-series generating function [72] for distribution ρ is defined as the polynomial

$$\Psi(z) \equiv \sum_{\mathrm{n}} \left(\prod_{m=1}^{D} z_m^{\mathrm{n}_m} \right) \rho_{\mathrm{n}}. \tag{4.28}$$

Here D complex arguments are components of a vector $z \equiv (z_1, \ldots, z_D)$. The monomials in the z_m are basis functions for Ψ and differentiation with respect to the z_m yields the various moments of ρ. The master equation (4.7) induces dynamics for Ψ, which are also linear and may be written in terms of a differential *Liouville equation* of the form

$$\frac{\mathrm{d}}{\mathrm{d}t}\Psi = -\mathcal{L}\left(z, \frac{\partial}{\partial z}\right)\Psi. \tag{4.29}$$

The map from ρ to Ψ is invertible and the transfer matrix (4.26) determines the form for its associated Liouville operator $\mathcal{L}$, as the following construction shows.

4.3.1.2 Conversion between transfer matrices and Liouville operators

In the Liouville equation (4.29), the left-hand side $\mathrm{d}\Psi/\mathrm{d}t$ is made a function of $\mathrm{d}\rho/\mathrm{d}t$ by definition (4.28), with $\mathrm{d}\rho/\mathrm{d}t$ evaluated according to the master equation (4.7) and the form (4.26) for the transfer matrix. Factors of z and $\partial/\partial z$ are chosen on the right-hand side so that $\mathcal{L}$ will reproduce the terms in $T_{\mathrm{nn}'}$, multiplying each basis function $\prod_{m=1}^{D} z_m^{\mathrm{n}_m}$. The elementary terms that contribute to the transfer matrix (4.26) are factors of n_i, and the single-agent shift operators $\exp(\partial/\partial \mathrm{n}_l - \partial/\partial \mathrm{n}_j) - 1$.

To extract these, again let A_{n} be any function of the components of n defining the context in which an index appears. The combination

$$\sum_{\mathrm{n}} \left(\prod_{m=1}^{D} z_m^{\mathrm{n}_m} \right) \mathrm{n}_i A_{\mathrm{n}} \rho_{\mathrm{n}} = z_i \frac{\partial}{\partial z_i} \sum_{\mathrm{n}} \left(\prod_{m=1}^{D} z_m^{\mathrm{n}_m} \right) A_{\mathrm{n}} \rho_{\mathrm{n}} \tag{4.30}$$

simply extracts the number index. When multiple derivatives act in turn, they recover number indices with corrections so that no factor of z_m is multiply counted,

$$\sum_{\mathrm{n}} \left(\prod_{m=1}^{D} z_m^{\mathrm{n}_m} \right) \mathrm{n}_i \left(\mathrm{n}_j - \delta_{ij} \right) A_{\mathrm{n}} \rho_{\mathrm{n}} = z_i z_j \frac{\partial}{\partial z_j} \frac{\partial}{\partial z_i} \sum_{\mathrm{n}} \left(\prod_{m=1}^{D} z_m^{\mathrm{n}_m} \right) A_{\mathrm{n}} \rho_{\mathrm{n}}. \tag{4.31}$$

The shift operator in any context is even more simply obtained as

$$\sum_{\mathrm{n}} \left(\prod_{m=1}^{D} z_m^{\mathrm{n}_m} \right) \left(\mathrm{e}^{\partial/\partial \mathrm{n}_l - \partial/\partial \mathrm{n}_j} - 1 \right) A_{\mathrm{n}} \rho_{\mathrm{n}} = \left(\frac{z_j}{z_l} - 1 \right) \sum_{\mathrm{n}} \left(\prod_{m=1}^{D} z_m^{\mathrm{n}_m} \right) A_{\mathrm{n}} \rho_{\mathrm{n}}. \tag{4.32}$$

Repeated application of suitably ordered z and $\partial/\partial z$ as in (4.30, 4.31), followed by (4.32), shows that the transfer matrix (4.26) is produced by the Liouville operator

$$\mathcal{L}\left(z, \frac{\partial}{\partial z}\right) \equiv \sum_{j,l=1}^{D} (z_l - z_j)\left(\mu_{jl} + z_j \frac{a_{jl}}{2} \frac{\partial}{\partial z_j} + \frac{1}{N} \sum_{k=1}^{D} z_j z_k a_{jk}^{S} \frac{\partial}{\partial z_j} \frac{\partial}{\partial z_k}\right) \frac{\partial}{\partial z_l}. \tag{4.33}$$

The δ terms in the transfer matrix (4.26), which remove self-play or self-replacement, lead to a form for $\mathcal{L}$ with all z factors to the left and all $\partial/\partial z$ factors to the right. This ordering, known as *normal ordering*, allows derivatives in the Liouville operator to act on Ψ alone, without the complexity of evaluating their effect on internal factors of z within $\mathcal{L}$ itself. Any Liouville operator can be converted to normal order; (4.33) is again a *minimal* form because the coefficients in the normal-ordered expression are the same as those in the deterministic replicator equation.

The Liouville operator is conventionally defined with the minus sign as in (4.33). On the space of generating functions, it is a positive-semidefinite operator and thus behaves like a *Hamiltonian* in the field theories corresponding to equilibrium thermodynamics [5]. In the Doi–Peliti construction (and equivalent constructions such as Freidlin–Wentzell theory [55]), $\mathcal{L}$ *is* in fact a Hamiltonian and induces a volume-preserving time-evolution map on the space of probability distributions, familiar from eikonal approaches to solutions of the diffusion equation on curved manifolds [66, 67, 73]. The role of $\mathcal{L}$ as a Hamiltonian will connect the general domain of irreversible stochastic processes to the concept of a potential that governs escape from domains of attraction and to the more general representation of symmetries.

4.3.2 Abstracting from analytic functions to a representation in linear algebra

Doi observed [48, 49] that monomials $z_m^{n_m}$ forming a basis for Ψ obey the same laws of superposition as quantum mechanical states and that the derivatives in the Liouville operator act on these states just as raising and lowering operators act in quantum mechanics or quantum field theory[28]. In particular, generating functions of a probability distribution ρ_{n} count the histories of numbers of each type in a population, but do not track histories of individuals. Therefore the many constructions in linear algebra invented to facilitate quantum mechanical calculations may be applied to stochastic processes. As in [25], we take care here to distinguish the *operator algebra*, which merely represents counting conventions, from the choice of *Hilbert space* on which the operators act. The latter distinguishes stochastic from quantum applications.

The algebraic relation that defines z and $\partial/\partial z$ as operators on a space of complex functions is the commutator

$$\left[\frac{\partial}{\partial z_i}, z_j\right] = \delta_j^i. \tag{4.34}$$

[28] This observation is commonplace; the algebraic representation of harmonic oscillator states was originally constructed from differential solutions to the Schrödinger equation.

It may be abstracted as a relation among formal operators a and $a^{\dagger}$,

$$\left[a^{i}, a_{j}^{\dagger}\right] = \delta_{j}^{i}, \tag{4.35}$$

which we refer to conventionally as raising and lowering operators, respectively, under the map

$$z_j \leftrightarrow a_j^{\dagger}, \tag{4.36}$$

$$\frac{\partial}{\partial z_i} \leftrightarrow a^{i}. \tag{4.37}$$

4.3.2.1 The correspondence of elementary objects

The space of complex multinomials is 'constructed' by application of the raising operators, simply by multiplying factors of z_m on the number 1. Therefore, we identify the right *ground state* of the function space with the notation

$$1 \leftrightarrow |0). \tag{4.38}$$

Only a finite number of moments can be extracted from any basis polynomial. In the representation of the complex functions as a linear vector space this is ensured because all lowering operators a^i annihilate the ground state $|0)$,

$$0 = \frac{\partial}{\partial z_i} 1 \leftrightarrow a^{i}|0). \tag{4.39}$$

The requirement on an inner product is that the conjugate ground state $(0|$ be normalized against $|0)$ (written $(0\,|0) = 1$), and that the annihilation property $a^i|0) = 0$ be preserved under conjugation, so that $(0|a_j^{\dagger} = 0$. The unique operator on complex functions that satisfies these requirements is

$$\int \mathrm{d}^{D}z\, \delta(z) \leftrightarrow (0|, \tag{4.40}$$

for which we may check that

$$0 = \int \mathrm{d}^{D}z\, \delta(z) z_j \leftrightarrow (0|a_j^{\dagger}. \tag{4.41}$$

The basis monomials count the number of agents of each type without distinguishing them and are associated with right *number states*,

$$\left(\prod_{m=1}^{D} z_m^{\mathrm{n}_m}\right) \cdot 1 \leftrightarrow \prod_{m=1}^{D} \left(a_m^{\dagger}\right)^{\mathrm{n}_m}|0) = |\mathrm{n}), \tag{4.42}$$

on which the raising and lowering operators act by

$$a_j^{\dagger}|\mathrm{n}) = \left|\mathrm{n} + 1_j\right), \tag{4.43}$$

$$a^i|\mathrm{n}) = \mathrm{n}_i \left|\mathrm{n} - 1_i\right). \tag{4.44}$$

The number states are eigenvalues of the *number operator*, which corresponds to the operator $z_i \partial/\partial z_i$ that extracts the index n_i from complex functions,

$$a_i^{\dagger} a^i|\mathrm{n}) = \mathrm{n}_i|\mathrm{n}). \tag{4.45}$$

The state corresponding to a generating function is therefore the sum of number states weighted with their probabilities,

$$\Psi(z) \leftrightarrow \sum_{\mathrm{n}} \rho_{\mathrm{n}}|\mathrm{n}) \equiv |\Psi). \tag{4.46}$$

A unique inner product evaluates to unity on all number states, which is known as the *Glauber norm*, or also the inner product with the *Glauber state*. Its representation in analytic functions and abstract vectors are

$$\begin{aligned} 1 &= \int \mathrm{d}^D z\, \delta(z) \mathrm{e}^{\sum_m \partial/\partial z_m} \left(\prod_{m=1}^{D} z_m^{\mathrm{n}_m} \right) \cdot 1 \\ &\leftrightarrow (0|\mathrm{e}^{\sum_m a_m} \prod_{m=1}^{D} \left(a_m^{\dagger}\right)^{\mathrm{n}_m} |0) \\ &= (0|\mathrm{e}^{\sum_m a_m}|\mathrm{n}). \end{aligned} \tag{4.47}$$

From the expansion of the exponential e^{a_m} in (4.47), we can extract the definition of the normalized conjugate number state,

$$(\mathrm{n}| \equiv (0|\prod_{m=1}^{D} \frac{a_m^{\mathrm{n}_m}}{\mathrm{n}_m!}, \tag{4.48}$$

which has the interpretation of a moment-projection operator. From the commutation (4.35) and annihilation (4.41) relations it follows that number states are orthogonal and normalized,

$$(\mathrm{n}'|\mathrm{n}) = \delta^D_{\mathrm{n}'\mathrm{n}}. \tag{4.49}$$

As a consequence of the equal inner product that the Glauber norm (4.47) takes with each number state, the inner product on $|\Psi)$ in this space, preserved no matter what its time evolution, is simply the trace of the original probability distribution ρ_{n}.

$$\begin{aligned} 1 &= \int \mathrm{d}^D z\, \delta(z) \Psi(z+1) \\ &\leftrightarrow (0|\mathrm{e}^{\sum_m a_m}|\Psi) = \sum_{\mathrm{n}} \rho_{\mathrm{n}}. \end{aligned} \tag{4.50}$$

In the analytic representation, this corresponds to evaluation of the generating function at argument 1; that is, $\Psi(1) = 1$ for all distributions $\rho_{\mathbf{n}}$.

The Liouville equation (4.29) in this notation becomes

$$\frac{\mathrm{d}}{\mathrm{d}t}|\Psi) = -\mathcal{L}\left(a^{\dagger}, a\right)|\Psi). \tag{4.51}$$

The Liouville operator remains that of (4.33), with the appropriate change in notation,

$$\mathcal{L}\left(a^{\dagger}, a\right) = \sum_{j,l=1}^{D} \left(a_l^{\dagger} - a_j^{\dagger}\right)\left(\mu_{jl} + a_j^{\dagger}\frac{a_{jl}}{2}a_j + \frac{1}{N}\sum_{k=1}^{D} a_j^{\dagger} a_k^{\dagger} a_{jk}^{S} a_j a_k\right)a_l. \tag{4.52}$$

It can sometimes fail to be apparent in the operator representation that $a_j^{\dagger}$ is already, in some sense, a 'continuous' argument, since it may be represented by the variable z_j. Thus we have already replaced the discrete basis of the master equation with a continuous basis. The further conversion to an integral over field variables, which we will perform next, is only a small additional change.

4.3.3 Reduction to quadrature and the coherent-state generating functional of time-dependent correlations

Formally, the Liouville equation can be integrated in time as an operator equation, or 'reduced to quadrature'. Doing so provides a means to compute correlations between moments of the probability distribution $\rho_{\mathbf{n}}$ at different times. The bracket notation for inner products in the operator representation replaces the more cumbersome complex integral that would be required by analytic functions. The quadrature of (4.51) between times 0 and T is written formally as

$$\left|\Psi_T\right) = \mathrm{e}^{-\mathcal{L}\left(a^{\dagger},a\right)T}\left|\Psi_0\right). \tag{4.53}$$

Because $\mathcal{L}$ is not an explicit function of time, the exponential can be evaluated as a power series in which all terms commute. For more complex cases with explicit time dependence, it becomes necessary to refine (4.53) to a time-ordered exponential integral.

4.3.3.1 Coherent states: construction, inner product and action of the Liouville operator

It is possible to work directly with the eigenvalues of operators such as $\mathrm{e}^{-\mathcal{L}T}$ [46, 47]. We will find it more convenient, however, to adopt the method of Peliti [50, 51], which is an expansion in an overcomplete set of *coherent states*[29]. These are defined as eigenstates of the lowering operator (acting to the right),

$$a^i|\phi) = \phi^i|\phi), \tag{4.54}$$

[29] Recall from section 4.3.1.1 that a countable set of monomials is a complete basis for the ordinary-power-series generating functions. The coherent states, in contrast, are indexed by continuous-valued field variables and therefore are not orthonormal under the natural inner product (which we define below). Their inner product and measure, however, provide the necessary cancellations of cross-terms so that, like the monomials in the ordinary power series, they furnish a representation of unity. Hence we refer to them as 'overcomplete'.

or the conjugate raising operator (acting to the left),

$$\left(\phi^{\dagger}\right| a_j^{\dagger} = \left(\phi^{\dagger}\right| \phi_j^{\dagger}. \tag{4.55}$$

Here $\phi \equiv (\phi^1, \ldots, \phi^D)$ is a vector of complex numbers (which we will treat as a column vector), and $\phi^\dagger$ its Hermitian conjugate. Like the original variables z_i, the ϕ_i are continuous-valued arguments of generating functions.

The coherent states are constructed from the number states as

$$\begin{aligned} |\phi) &= \mathrm{e}^{-(\phi^{\dagger}\phi)/2} \mathrm{e}^{a^{\dagger}\cdot\phi} |0) \\ &= \mathrm{e}^{-(\phi^{\dagger}\phi)/2} \sum_{M=0}^{\infty} \frac{\left(a^{\dagger} \cdot \phi\right)^M}{M!} |0) \\ &= \prod_{m=1}^{D} \left(\mathrm{e}^{-(\phi_m^{\dagger}\phi_m)/2} \sum_{\mathrm{n}_m=0}^{\infty} \frac{\phi_m^{\mathrm{n}_m}}{\mathrm{n}_m!} a_m^{\dagger \mathrm{n}_m} \right) |0) \\ &= \sum_{\mathrm{n}} \prod_{m=1}^{D} \left(\mathrm{e}^{-(\phi_m^{\dagger}\phi_m)/2} \frac{\phi_m^{\mathrm{n}_m}}{\mathrm{n}_m!} \right) |\mathrm{n}). \end{aligned} \tag{4.56}$$

Because all $a_m^\dagger$ commute, the exponential may be expanded either as a simple sum on M or a multinomial over $\{n_m\}$. It is clear, then, that the right coherent state $|\phi)$ is simply the generating function for a *Poisson distribution* over discrete states n, with mean value ϕ.

The conjugate coherent states, indexed by Hermitian conjugate vector $\phi^\dagger$, and satisfying the conjugate eigenvalue relation (4.55), are then

$$\begin{aligned} \left(\phi^{\dagger}\right| &= (0| \mathrm{e}^{\phi^{\dagger}\cdot a} \mathrm{e}^{-(\phi^{\dagger}\phi)/2} \\ &= (0| \sum_{M=0}^{\infty} \frac{\left(\phi^{\dagger} \cdot a\right)^M}{M!} \mathrm{e}^{-(\phi^{\dagger}\phi)/2} \\ &= \sum_{\mathrm{n}} (\mathrm{n}| \prod_{m=1}^{D} \left(\left(\phi_m^{\dagger}\right)^{\mathrm{n}_m} \mathrm{e}^{-(\phi_m^{\dagger}\phi_m)/2} \right). \end{aligned} \tag{4.57}$$

The left coherent state $(\phi^\dagger|$ is the distinctive feature of Doi–Peliti and of all other 'two-field' functional-integral constructions of irreversible processes. It is a weighted shift operator, the Taylor's series expansion of which consists of the whole series of number states that act to project out moments of distributions in an inner product. In particular, the value $\phi^\dagger = 1$ recovers the Glauber state

$$(0| \mathrm{e}^{\sum_m a_m},$$

up to an overall normalization.

The asymmetry between left/right number states and left/right coherent states distinguishes the space of generating functions from the (symmetric) space of quantum states, which may share the same algebra of operators. The role of the left coherent states in propagating information *backward* in time, to identify most-probable paths associated with a final argument in a generating functional, is reviewed in [25].

To compute the quadrature (4.53) of the Liouville equation we need the inner product between two coherent states at adjacent times separated by an interval $\mathrm{d}t$. This inner product may be evaluated from the definitions (4.56, 4.57) as

$$\left(\phi^{\dagger}_{t+\mathrm{d}t}\middle|\,\phi_t\right) = \mathrm{e}^{\phi^{\dagger}_{t+\mathrm{d}t}\cdot\phi_t - \left(\phi^{\dagger}_t\cdot\phi_t + \phi^{\dagger}_{t+\mathrm{d}t}\cdot\phi_{t+\mathrm{d}t}\right)/2}. \tag{4.58}$$

The coherent states are readily checked to furnish a representation of the identity equivalent to that provided by the number operators,

$$\int \frac{\mathrm{d}^D\phi^{\dagger}\mathrm{d}^D\phi}{\pi^D}|\phi)\left(\phi^{\dagger}\right| = \sum_{\mathrm{n}}|\mathrm{n})(\mathrm{n}| = I. \tag{4.59}$$

The cancellation of phase integrals in (4.59) collapses two summations into one and the evaluation of the resulting Γ functions as integrals in $\phi^{\dagger}\cdot\phi$ cancels factorials, leaving only the number states.

To make use of the coherent-state representations of unity, the time-evolution operator in (4.53) is factored into a product of short-time evolution operators,

$$\mathrm{e}^{-\mathcal{L}\left(a^{\dagger},a\right)T} = \prod_{k=0}^{T/\mathrm{d}t} \mathrm{e}^{-\mathcal{L}\left(a^{\dagger},a\right)\mathrm{d}t}, \tag{4.60}$$

and a copy of the representation (4.59) of unity is inserted between each factor. This transformation replaces the operator equation (4.53) with a series of integrals over complex numbers ϕ and $\phi^{\dagger}$ at a sequence of times.

To evaluate the action of the Liouville operator, we begin with the expectation of the number operator,

$$\left(\phi^{\dagger}\right|\sum_{i=1}^{D} a_i^{\dagger}a^i|\phi) = \sum_{i=1}^{D}\phi_i^{\dagger}\phi^i \equiv \phi^{\dagger}\cdot\phi. \tag{4.61}$$

Because $\mathcal{L}$—like the number operator—is normal-ordered, the value from its insertion between coherent states is elementary to compute, as

$$\left(\phi^{\dagger}_{t+\mathrm{d}t}\right|\,\mathcal{L}\left(a^{\dagger},a\right)\,\left|\phi_t\right) = \mathcal{L}\left(\phi^{\dagger}_{t+\mathrm{d}t},\phi_t\right)\left(\phi^{\dagger}_{t+\mathrm{d}t}\middle|\,\phi_t\right). \tag{4.62}$$

The initial conditions for any instance of evolution under the master equation are represented in the state $|\Psi_0)$. It is convenient (and has become conventional) to choose $|\Psi_0)$ itself to be the generating function for a Poisson distribution with some mean value $\bar{n}$,

$$\left|\Psi_0\right) = \exp\left(\sum_j\left(a_j^{\dagger}-1\right)\bar{n}_j\right)|0). \tag{4.63}$$

For both systems with exponential convergence [27] and the weaker diffusive behavior that we will consider, information in the initial state rapidly decays to that of the asymptotic late-time distribution, so nothing computed below will depend sensitively on the choice of $|\Psi_0)$.

With these procedures for evaluation, the preceding steps may be assembled to convert the Doi operator expression [48, 49] for the generating function into its coherent-state equivalent, introduced by Peliti [50, 51]. The combination of the factorization (4.60) of the evolution operator with the insertion of coherent-state representations of unity (4.59), the choice (4.63) for the initial state and inner product with the Glauber norm (4.47), gives the integral representation for the time-evolved generating function

$$\left(0|e^{\sum_m a_m}|\Psi_T\right) = (0|e^{\sum_m a_m} e^{-\mathcal{L}(a^\dagger,a)T} e^{(a^\dagger-1)\cdot\bar{n}}|0) = \int_0^T \mathcal{D}\phi^\dagger\mathcal{D}\phi\, e^{1\cdot\phi_T} e^{-\int_0^T dt\left\{\phi^\dagger\cdot d\phi/dt+\mathcal{L}(\phi^\dagger,\phi)\right\}-\phi_0^\dagger\cdot\phi_0} e^{(\phi_0^\dagger-1)\cdot\bar{n}}. \tag{4.64}$$

The integration measure in (4.64) is defined as a continuum limit of the product measure over discrete-time values of ϕ and $\phi^\dagger$,

$$\int_0^T \mathcal{D}\phi^\dagger\mathcal{D}\phi \equiv \lim_{\delta t\to 0}\prod_{K=1}^{T/\delta t}\int\frac{d^D\phi_K^\dagger\, d^D\phi_K}{\pi^D}, \tag{4.65}$$

Collecting terms from the Liouville operator, the sequence of inner products of states at adjacent timesteps and surface terms, we may write (4.64) more cogently as

$$(0|e^{\sum_m a_m}|\Psi_T) = \int_0^T \mathcal{D}\phi^\dagger\mathcal{D}\phi\, e^{-S} e^{(\phi_0^\dagger-1)\cdot(\bar{n}-\phi_0)}. \tag{4.66}$$

(Recalling from (4.50) that the inner product with the Glauber state simply computes the trace of the probability density, the expression (4.66) evaluates to $\Psi_T(1)\equiv 1$ at all T.) The *action* S determines the behavior of the array of time-dependent correlation functions of ρ, or equivalently their moments extracted at each time through Ψ.

The left-most factor of $1\cdot\phi_T$ in (4.64) results from the inner product of the Glauber state with the left-most coherent state. It may be shifted to the right by absorbing a total derivative into the main integral in (4.64),

$$e^{1\cdot\phi_T} = e^{\int_0^T d(1\cdot\phi_t)/dt} e^{1\cdot\phi_0} \tag{4.67}$$

to put the action in (4.66) in the form

$$S \equiv \int_0^T dt\left\{\sum_{i=1}^{D}\left(\phi_i^\dagger - 1\right)\frac{d\phi_i}{dt} + \mathcal{L}\left(\phi^\dagger,\phi\right)\right\}. \tag{4.68}$$

The weights $\phi_i^\dagger$ in the left-coherent (moment-projecting) states have become the conjugate momenta to the fields ϕ_i which are the mean values indexing the basis of

Poisson distributions. The Liouville operator has become the *Hamiltonian* in S and takes the form

$$\mathcal{L}\left(\phi^{\dagger}, \phi\right) \equiv \sum_{j,l=1}^{D} \left(\phi_l^{\dagger} - \phi_j^{\dagger}\right)\left(\mu_{jl} + \phi_j^{\dagger}\phi_j\frac{a_{jl}}{2} + \frac{1}{N}\sum_{k=1}^{D}\phi_j^{\dagger}\phi_j a_{jk}^{S}\phi_k^{\dagger}\phi_k\right)\phi_l. \quad (4.69)$$

The normal-ordered form of the operator version (4.52) of $\mathcal{L}$ ensures that its functional counterpart (in which ordering no longer matters, because scalars commute) simply replaces coherent-state variables for raising and lowering operators.

4.3.3.2 Why use the abstract inner product in place of parallel complex integrals

The Doi–Peliti construction passes through several representations of the same quantities at increasing levels of abstraction. The use of generating functions and their motivation as Laplace transforms is by now fairly familiar, but the projection of analytic functions onto only their linear algebra is less familiar. Therefore before continuing further we pause to note what is gained by letting go of analyticity and keeping only the linear algebra of function spaces.

The most important reason to pass from analytic functions to an abstract representation of the Doi Hilbert space is that it simplifies the inner product. Since we will need to compute this inner product an infinite number of times in the functional integral, such simplifications are valuable. Also, the abstract Hilbert space captures properties of the *formal power series* representation of generating functions [72], which remain fundamental whether or not the power series converge to analytic functions.

The definition (4.40) of the left ground state $(0|$ as an integral, beyond being cumbersome to write, requires a separate variable of integration for each inner product to be taken. To see how the multiplicity of integration variables is simplified by the formal inner product, we may evaluate the coherent-state representation of unity (4.59) as it would appear directly in terms of differential operators. Expanding the exponentials in the coherent states, and using the phase integral to eliminate all but equal-power terms, the identity acting on inner products becomes

$$\begin{aligned}\int \frac{d^D\phi^{\dagger}d^D\phi}{\pi^D}|\phi)\left(\phi^{\dagger}\right| &= \int \frac{d^D\phi^{\dagger}d^D\phi}{\pi^D} e^{-\left(\phi^{\dagger}\cdot\phi\right)} e^{z_{t+dt}\cdot\phi} \int d^D z_t\, \delta(z_t) e^{\phi^{\dagger}\cdot\partial/\partial z_t} \\ &= \int d^D z_t\, \delta(z_t) e^{z_{t+dt}\cdot\partial/\partial z_t} \\ &= z_t \mapsto z_{t+dt}, \end{aligned} \quad (4.70)$$

that is, a map that removes the former integration variable and substitutes the latter.

4.3.4 A canonical transformation to number fields

There are now two standard approaches to using the field theory of the Doi–Peliti construction: either to expand in the coherent-state field coordinates, or to expand in fields corresponding directly to the number indices. In this section we note that the two approaches are related by a canonical transformation which has the form of a change of variables to *action-angle* coordinates [74].

The early literature from reaction–diffusion theory [46, 47, 75] emphasizes the coherent-state field variables ϕ and $\phi^\dagger$. The two virtues of these variables are that they provide a direct translation from the operator generating function, and that the Liouville operator (4.69) for finite-order interactions of individuals in the population[30] is itself a finite-order polynomial in fields. The disadvantage is that the directly observable quantities—the expected numbers of agent types—are bilinear forms in the fields $\phi^\dagger\phi$, and not elementary field variables. Thus, the most intuitive component in the Peliti construction—the mean of its Poisson distributions—cannot be directly interpreted as a particle number, but requires multiplication with the weights in the sampling operator to be assigned a meaning.

An alternative approach, finding increasing use [76–80], is to work directly with the characteristic function of the distribution in number operators. The resulting action contains transcendental functions and is formally infinite-order in the conjugate momenta to the number field. This is not usually a serious problem in practice, because the same low-order fluctuation calculations may be performed in either representation and the transformed variables provide many advantages for identifying and understanding the stationary paths of the action. They also lead directly to fluctuation expressions for the number field, which is observable, rather than for the coherent-state field ϕ which is an eigenvalue of the lowering operator.

The action-angle transformation from fields to number variables is given by

$$\begin{aligned} \phi_i^\dagger &\equiv e^{\eta_i}, \\ \phi_i &\equiv e^{-\eta_i} n_i, \end{aligned} \tag{4.71}$$

where n_i is a *number field* corresponding directly to the expectation of the number index n_i of section 4.2.1. An ordinary action-angle transformation [74] would be written $\phi_i \equiv e^{i\alpha_i} n_i$, so we see that η_i is the 'imaginary' coordinate of an angle variable analytically continued in the complex plane. The mean value and integration contour for η_i generally have both real and imaginary components, so we detail these in the contexts where they arise below. Under analytic continuation of η, the transformation (4.71) is equivalent to rotation from complex coordinates $(\phi^\dagger, \phi)$ to their equivalent polar coordinates $(n, i\eta)$. The transformation is canonical because it preserves the kinetic term in the action,

$$\frac{d\phi_i^\dagger}{dt}\phi_i = \frac{d\eta_i}{dt} n_i. \tag{4.72}$$

We show next that no problems arise in the functional measure (4.65) under transformation to action-angle variables by verifying that the resulting functional integral recovers the best Gaussian approximation to an exactly solvable

[30] In the terminology of population genetics, a k-individual interaction that determines the reproductive probability of the kth member would be said to describe 'frequency-dependent' fitness of order $k - 1$. Under the map that derives fitness from normal-form game payoffs in populations of randomly matched players, this interaction would correspond to a k-player game.

distribution: the case of neutral diffusion of a population between two types[31]. When we have verified this, we may perform Gaussian integrals by inverting the kernel in the action, equally well in coherent-state or action-angle variables.

4.3.4.1 Checking the measure in action-angle variables for the simplest case of drift between two types

The simplest check of the integration measure in action-angle variables is a comparison of the Gaussian integral to an exact solution for the problem of pure mutational drift between two types. We use the simplest version of the model of two types coupled by mutation, developed in appendix 4.5, with the two mutation rates both set equal to some probability p per unit time. As shown in the appendix, the total relaxation rate for the system becomes $2p$.

For this model the population structure is given by a vector $\mathrm{n} \equiv (\mathrm{n}_1, \mathrm{n}_2)$. The master equation for pure mutation becomes simply

$$\frac{d\rho_{\mathrm{n}}}{dt} = p\left[(\mathrm{n}_1 + 1)\rho_{\mathrm{n}+(1,-1)} - \mathrm{n}_1\rho_{\mathrm{n}}\right] + p\left[(\mathrm{n}_2 + 1)\rho_{\mathrm{n}+(-1,1)} - \mathrm{n}_2\rho_{\mathrm{n}}\right]. \tag{4.73}$$

Steady-state solutions to (4.73) separate to independent distributions along the diagonals $\mathrm{n}_1 + \mathrm{n}_2 = N$ for each value of N. Any one such diagonal solution is exactly solvable, to the binomial

$$\rho_{\mathrm{n}} = \frac{1}{2^N}\binom{N}{\mathrm{n}_1}. \tag{4.74}$$

The leading N dependence of the equal-time fluctuation correlation function in this distribution is

$$\left\langle\left(\frac{\mathrm{n}_1 - \mathrm{n}_2}{2}\right)^2\right\rangle = \frac{N}{4}, \tag{4.75}$$

displaying the N-linear scaling characteristic of either Poisson statistics or the central-limit theorem.

Following through the Doi–Peliti construction, the master equation (4.73) corresponds to the coherent-state Liouville operator

$$\mathcal{L} = p\left(\phi_1^\dagger - \phi_2^\dagger\right)\left(\phi_1 - \phi_2\right), \tag{4.76}$$

and to the action in the full generating functional

$$S = \int dt\left\{\phi_1^\dagger\frac{d\phi_1}{dt} + \phi_2^\dagger\frac{d\phi_2}{dt} + p\left(\phi_1^\dagger - \phi_2^\dagger\right)\left(\phi_1 - \phi_2\right)\right\}. \tag{4.77}$$

---

[31] This is equivalent to checking that the action-angle transformation generates no anomaly terms from the measure.

This action, representing pure diffusion, is *exactly quadratic*—that is, it has no second-order term in $\phi^\dagger$. By Noether's theorem it implies conservation of $\phi_1 + \phi_2 = N$, which we may make explicit with a basis rotation to write

$$S = \frac{1}{2}\int \mathrm{d}t \left\{ \left(\phi_1^\dagger + \phi_2^\dagger\right)\frac{\mathrm{d}}{\mathrm{d}t}\left(\phi_1 + \phi_2\right) + \left(\phi_1^\dagger - \phi_2^\dagger\right)\left(\frac{\mathrm{d}}{\mathrm{d}t} + 2p\right)\left(\phi_1 - \phi_2\right) \right\}, \quad (4.78)$$

Now we encounter the subtlety remarked by Kamenev [58] for free-field theories in coherent-state variables: in the continuum limit no Hessian term for $(\phi_1^\dagger - \phi_2^\dagger)^2$ appears to specify the correlation function for $(\phi_1 - \phi_2)$. This lacuna is partly a notational artifact of the continuum limit, in the sense that the fluctuation kernel can be correctly recovered by careful evaluation of the more precise discrete functional integral [56, 58]. It is also physical, however, in the sense that fluctuations are extensive when they are not regulated by terms larger than $1/N$ in the action.

The transformation to action-angle variables actually solves both of these problems, but only if it is performed in the original variables (n_1, n_2), so that it respects the discrete permutation symmetry of the problem. The form for S obtained by transforming (4.77) and then rotating to a diagonal basis is

$$\begin{aligned} S = \frac{1}{2}\int \mathrm{d}t \Bigg\{ &(\eta_1 + \eta_2)\frac{\mathrm{d}N}{\mathrm{d}t} + 2pN\left[1 - \cosh(\eta_1 - \eta_2)\right] \\ &+ (\eta_1 - \eta_2)\frac{\mathrm{d}(n_1 - n_2)}{\mathrm{d}t} + 2p(n_1 - n_2)\sinh(\eta_1 - \eta_2) \Bigg\}. \end{aligned} \quad (4.79)$$

In (4.79) we have substituted $N \equiv n_1 + n_2$ for the conserved total number. The nonlinear cosh creates an explicit second-order term in $(\eta_1 - \eta_2)$, where there was none in $(\phi_1^\dagger - \phi_2^\dagger)$. Note that such a term would not have been produced by introducing action-angle variables for the rotated linear combinations of coherent-state fields in (4.78).

The second-order expansion of the action (4.79) now contains all three terms in the continuum limit, which are needed to define the response and correlation functions. Inverting the Gaussian kernel to obtain the Green's function (demonstrated in detail below in section 4.4.2.1) recovers the result of the exact solution,

$$\left\langle \left(\frac{n_1 - n_2}{2}\right)^2 \right\rangle = \int_0^\infty \mathrm{d}t e^{-4pt}(Np) = \frac{N}{4}. \quad (4.80)$$

4.3.4.2 Rescaling the action to express large-deviations scaling of the entropy rate

In either coherent-state or action-angle variables, we immediately extract the leading large-deviations approximation by de-scaling the number field by the total population size N. In the polynomial expansion for fitness, $1/N$ then serves as a small parameter in which to expand correlation functions. In field variables, de-scaling is applied to ϕ while $\phi^\dagger$ is left unchanged. In action-angle variables, de-scaling is applied to the number fields, leaving the angle variables η unchanged. We introduce de-scaled number fields with the notation

$$\begin{aligned}\frac{n_i}{N} &\equiv \nu_i,\\ N\,\mathrm{d}t &\equiv \mathrm{d}\tau,\\ \frac{\mu_{jl}}{N} &\equiv \hat{\mu}_{jl}.\end{aligned} \tag{4.81}$$

In the deterministic replicator dynamic, the relative fitness $f_i - \phi$ may be expressed as a linear function of frequencies with a 'population-corrected' payoff matrix. In de-scaled variables, this adjusted payoff becomes independent of N and is written

$$\lambda_{jl} \equiv \frac{a_{jl}}{2} + \sum_{k=1}^{D} a_{jk}^{S} \nu_k. \tag{4.82}$$

The total mean rate of conversion from type l to type j, introduced above in (4.27), corresponding to the braced terms in (4.26), is then written in de-scaled variables as

$$\hat{p}_{jl} \equiv \left(\hat{\mu}_{jl} + \nu_j \lambda_{jl}\right)\nu_l. \tag{4.83}$$

$\hat{p}_{jl} = p_{jl}/N^2$ from (4.27). This quantity is non-negative for $j \neq l$ and contains both drift and diffusion contributions to type conversion. It does not generally vanish at either equilibrium or non-equilibrium steady states. In equilibrium systems, at steady states it satisfies a condition known as *detailed balance*, where $\hat{p}_{jl} = \hat{p}_{lj}$ between any two types l and j. At non-equilibrium steady states, only the weaker condition that total influx equals total efflux is required to hold at every node j, given by $\sum_l \hat{p}_{jl} = \sum_l \hat{p}_{lj}$.

The combined transformation from coherent-state to action-angle variables, and rescaling of the number fields and time, recasts the action (4.68) as

$$\begin{aligned}S &= \int \mathrm{d}t \left(\sum_{i=1}^{D} \phi_i^{\dagger} \frac{\mathrm{d}\phi_i}{\mathrm{d}t} + \mathcal{L} \right)\\ &= N \int \mathrm{d}\tau \left(\sum_{i=1}^{D} \eta_i \frac{\mathrm{d}\nu_i}{\mathrm{d}\tau} + \hat{\mathcal{L}} \right)\\ &\equiv N\hat{S},\end{aligned} \tag{4.84}$$

in which the rescaled Liouville operator $\hat{\mathcal{L}}$ takes the form

$$\begin{aligned}\hat{\mathcal{L}} \equiv \frac{\mathcal{L}}{N^2} &= \sum_{j,l=1}^{D} \left(1 - \mathrm{e}^{\eta_j - \eta_l}\right)\left(\hat{\mu}_{jl} + \nu_j \lambda_{jl}\right)\nu_l\\ &= \sum_{j,l=1}^{D} \left(1 - \mathrm{e}^{\eta_j - \eta_l}\right)\hat{p}_{jl}.\end{aligned} \tag{4.85}$$

We now note the relation between the field Liouville operator (4.85) and the original transfer matrix (4.26). Although a rather laborious sequence of transformations has been required to take one into the other, their forms are identical and the map from operators to field variables clarifies the meanings of the latter. The field η_j corresponds to the derivative $-\partial/\partial \mathrm{n}_j$. The term p_{jl} is the probability per unit time for all type conversions as a function of n from all processes, with the provision that simple products of *field n* correspond to products of discrete-index n in which agents are never redundantly counted. With these correspondences, one can write down the field functional integral for a given transfer matrix by inspection.

4.4 Three systematic approximation methods: a polynomial expansion of the effective action

All of the approximation methods we will develop for the functional integral (4.64) treat the variable n as a continuous field. The resulting equations for flow of the corresponding probability density ρ_n thus fall into the class of *diffusion equations*[32]. The large-deviations limit of such equations shares a property with high-frequency (also called 'optical') limits of wave equations: the leading values of a solution to the field equations at a point may be approximated by functionals defined along *rays* through that point which are appropriately connected to initial conditions or stationary sets. For diffusion solutions these rays are the *eikonals* of the diffusion equation [67, 73, 81]. The remarkable and non-intuitive property of diffusion eikonals is that they are solutions to Hamiltonian equations of motion, even though the underlying diffusion process is macroscopically irreversible.

The two-field methods, whether those of Freidlin–Wentzell, Doi–Peliti, or Martin–Siggia–Rose, share the feature that in the large-deviations limit they provide systematic constructions of ray approximations and the leading diffusive corrections to them. In the Peliti expansion in an overcomplete basis of elementary generating functions for discrete-state Markov processes, the resulting functional integral (4.66) may readily be approximated by the *stationary paths* of the action (4.68) corresponding to the eikonals, corrected by Gaussian fluctuations of the integration variables. This approximation recovers the mean-field replicator equations in a rather trivial way, but with almost equal ease it gives the leading large-deviations formula for escape trajectories and escape probabilities from basins of attraction of the stochastic process [82]. The ability to reduce quite complicated statistical escape probabilities in high-dimensional state spaces to elementary problems of Hamiltonian dynamics is the first in a sequence of remarkable simplifications provided by the Doi–Peliti construction for calculating complex correlation functions.

[32] These include, but are not limited to, the Fokker–Planck approximation to the original master equation (4.7), which corresponds to an approximation by Gaussian-order fluctuations about the mean background. The treatment of n as a continuous variable need not arise from an approximation at all. Two-field functional integrals with the same form as (4.64) arise in the Martin–Siggia–Rose construction [57], or in the treatment of stochastic differential equations by Freidlin and Wentzell [55], to evolve distributions over index variables that are continuous from the start.

The basis for approximating functional integrals is the expansion of the action S in stationary points and polynomial orders of residual fluctuations about them. In the same way that has now become familiar from quantum field theory [35], successive powers of fields or of time and space derivatives are often associated with powers of a small expansion parameter that measures the strength of corrections to the ray limit. In our examples this small parameter will be the inverse of the large-deviations scale factor, which will be provided by the population size N. Before decomposing the action, we note general properties that it must have by virtue of symmetries. When individual terms in the expansion respect these symmetries, the same properties must then hold order by order in the small expansion parameter.

The property of normalization of probability distributions leads to actions with minimum value $S = 0$, attained at all solutions of the classical diffusion equation[33]. (That is, the most likely solution must always have log-probability $\sim N^0$ in the large-deviations limit.) If the microscopic dynamics is invariant under time translation and if no explicitly time-dependent sources are added in the generating functional (4.66), then the Liouville operator (which is the generator of time translations) must satisfy $\hat{\mathcal{L}} = 0$, either as a classical expectation value, or as an operator identity in correlation functions. Finally, because we have restricted to models that conserve total number N, the population number appears as the conserved quantity in Noether's theorem for a global symmetry of the variable $\sum_j \eta_j$. We now proceed with approximation methods.

If we use overbar to denote stationary-path fields for S and expand the functional variables of integration $\phi \equiv \bar{\phi} + \phi'$, $\phi^\dagger \equiv \bar{\phi}^\dagger + \tilde{\phi}$ (or their counterparts in $\eta = \bar{\eta} + \eta'$, $n = \bar{n} + n'$), then S may be expanded in a functional Taylor's series in ϕ' and $\tilde{\phi}$ (or η' and n'). Using superscript (k) to denote the collection of all terms of order k in this expansion, the exponential e^{-S} may be systematically decomposed as

$$\mathrm{e}^{-S} = \mathrm{e}^{-S^{(0)}} \mathrm{e}^{-\left(S^{(1)} + S^{(2)}\right)} \left(1 - S^{(3)} + \frac{{S^{(3)}}^2}{2} + \cdots \right). \tag{4.86}$$

The first term, $\mathrm{e}^{-S^{(0)}}$, does not depend at all on the fluctuations of ϕ and $\phi^\dagger$ (or η and n) about their background values in the path integral (4.66). Its arguments provide a *deterministic* approximation to the expected values of these fields. The fields $\bar{\phi}$ and $\bar{\phi}^\dagger$ (or $\bar{\eta}$ and $\bar{n}$) are generally chosen to place $S^{(0)}$, including boundary terms at $t = 0$ and

[33] Our use of the term 'classical' here indicates the deterministic solutions to diffusion equations with a Laplacian and a drift term, in which fluctuations or the aggregate corrections they produce are not yet included. Classical diffusion equations need not be linear. Both the diffusivity and the drift rate may be functions of the population state, in which case we must adopt some procedure such as MFA to handle non-linearities. An established terminology distinguishing such 'classical' approximations from leading large-deviations approximations to limited fluctuation effects such as escapes (termed 'semi-classical'), or solutions for full distributions that retain all fluctuations, has become conventional in quantum mechanics, quantum field theory and condensed matter physics that uses the same mathematics presented here. We retain this terminology for consistency with the main body of literature.

$t = T$, at a local minimum[34]. Some of these are global minima, which may be discrete sets of states if the system has stationary fixed points, or could lie within a continuum if the stable set is a limit cycle. In addition to global minima where $S = 0$, local minima may also exist at paths such as escape trajectories, where $S > 0$.

About any of these minima of S, $S^{(1)} \equiv 0$, and the term $S^{(2)}$ in (4.86) defines a kernel for Gaussian fluctuations of $(\phi', \tilde{\phi})$, or (η', n'). Whenever the further terms $S^{(3)} + \cdots$ are multiplied by suitable powers of a small parameter such as $1/N$ to ensure convergence of the series, the Gaussian approximation dominates the exponential and the remaining corrections may be evaluated as moments of this distribution, as indicated by the large parenthesis in (4.86). When the series in $S^{(3)}$ is not small, a modified stationary-path approximation that goes beyond mean-field theory may be used to self-consistently incorporate Gaussian or higher-order fluctuations into the deterministic background, effectively incorporating entropy corrections into the definitions of fitness or mutation.

We will consider approximation methods in successive order of the terms in (4.86). Each approximation captures the leading behavior of games with characteristic symmetries and for each class we demonstrate a full calculation for a simple model. The stationary-point action itself has interesting structure, related to flips between two stable fixed points, for a coordination game with discrete symmetry breaking. The stationary-path action lacks interesting structure, but the second-order kernel for fluctuations becomes interesting in the case of hidden time translation in a RPS game with a limit cycle. Finally, the fluctuations feed back through third-order derivatives to alter the stationary paths themselves, for a canonical repeated prisoners' dilemma (RPD).

The next two sections show how stationary points are identified and how Gaussian fluctuations are evaluated and used.

4.4.1 Stationary-point expansions for mean-field flow equations and large-deviations formulae

Stationary paths of the action $S^{(0)}$ are of two types, which we will term *classical* and *non-classical*. Classical stationary paths satisfy $S^{(0)} \equiv 0$. They form a continuous space and comprise all solutions of the deterministic evolutionary game equation (4.9). The property $S^{(0)} \equiv 0$ implies that the probability of any such path is ~ 1 to leading exponential order. The path followed in a particular case is selected by the surface terms in the path integral (4.66) at $t = 0$ or $t = T$ stipulating the boundary conditions for that case. Classical stationary paths converge on the stationary sets of the replicator dynamic (rest points, limit cycles, or chaotic attractors) in the interior $0 < t < T$ given sufficiently long times.

The non-classical stationary paths are solutions with $S^{(0)} > 0$ of *locally* least action. They represent the most-probable paths for escape from the stationary sets of the classical replicator dynamic and correspond to the *eikonals* in the ray

[34] We will return below to a more general method of selecting backgrounds that may include fluctuation effects.

approximation to solutions of the Fokker–Planck equation [66]. The classical and non-classical stationary paths are also collectively referred to as *semiclassical trajectories*, emphasizing a different aspect of their relation to fluctuations. Whether characterizing mean values or escapes, they are interpreted as *macroscopic* rather than microscopic paths. Each macroscopic trajectory represents an entire distribution of micro-trajectories as described in section 4.1.3.2; the particular distribution is the one in which that macro-trajectory becomes typical[35]. The factor $e^{-S^{(0)}}$ gives the relative weight of the distribution within the full path ensemble.

A functional continuum of non-classical stationary paths exist, which may originate within the stationary set of the classical dynamics (rest points, limit cycle, etc), but diverge to points that are not within it. Almost all such trajectories never re-enter points in the stationary set and there is no way they can evolve smoothly forward in time to become classical trajectories in the $S = 0$ set. Therefore they cannot arise as dynamical solutions whose support lies entirely in the interior $0 < t < T$ (unless they are created by time-dependent source terms in the generating functional). In generating functions written as finite-time path integrals, these solutions are determined by the particular boundary condition at the final time in the path integral (4.66).[36]

An isolated, discrete set of non-classical stationary paths connects the rest points of the classical replicator dynamic. In the Hamiltonian description, these form the *heteroclinic network* [84, 85] of the dynamical system. The orbits in this network connect stable rest points (the ESS) to saddle points and form the *instantons* of the stochastic process. Because they terminate in configurations specified by the dynamics rather than by final-time boundary conditions—that is, they *do* evolve smoothly forward into classical solutions in the $S = 0$ set—any sufficiently long trajectory may have an unlimited number of such transitions. They are suppressed at leading exponential order with probabilities $\sim e^{-S^{(0)}}$, which provides the large-deviations probability for escape. Escapes have finite duration comparable to the time for diffusion toward stable rest points and therefore may occur as a (Poisson-distributed) 'dilute gas' in any long trajectory. The sum of probabilities from this distribution of possible escapes determines the characteristic lifetime of the stable rest points under perturbation from non-vanishing fluctuations [5]. We will consider the trajectories and probabilities of a single escape in this section and return in section 4.4.2.1 to the proper formulation of the Poisson distribution over escapes. The heteroclinic network of the dynamical system—if one exists—is important because it describes the autonomous dynamics of the aggregate population in the regime of symmetry breaking.

[35] See the discussion in [83] of typicality of sequences in distributions governed by the combinatorics of large numbers.

[36] These paths are the eikonals introduced by Freidlin and Wentzell [55] as the tool to compute boundary values for diffusion equations over entire surfaces bounding basins of attraction. In the absence of neutral directions, any such path will only grow exponentially within an effectively finite window of support before the final boundary and thus does not describe persistent, autonomous dynamics of the population.

The instantons of the stochastic process resemble those of equilibrium, but they differ in one important respect, which affects the interpretation of potentials in these theories. In equilibrium escape problems, trajectories which are mirror images under time-reversal have the same probabilistic suppression and the probability of a trajectory is therefore linked to its form. For irreversible stochastic processes this association need not hold. Escape trajectories will often approximate (in one dimension, they will *equal*) the time-reversed mirror images of classical diffusive paths. However, extra force terms in the Freidlin–Wentzell Hamiltonian serve to precisely cancel the probabilistic suppression of classical diffusion solutions, while approximately doubling the path contribution (squaring the probability) for escapes. A worked example of these properties of non-equilibrium escape trajectories and their probabilities is presented in section 5.2.3 and section 5.2.4.

4.4.1.1 Scaling properties of the stationary-point approximation: the large-deviations limit and thermodynamics

We noted in (4.84) that $S = N\hat{S}$, where $\hat{S}$ is approximately independent of N as $N \to \infty$. Thus the large-deviations scaling $e^{-N\hat{S}^{(0)}}$ of escape probabilities is exponential in population size, corresponding to the *extensive* scaling familiar from equilibrium thermodynamics. Escape represents a shift by a fixed *fraction* of the population to a saddle or unstable fixed point. Therefore, writing $\mathrm{e}^{-N\hat{S}^{(0)}} = (\mathrm{e}^{-\hat{S}^{(0)}})^N$, we see that we may interpret $\mathrm{e}^{-\hat{S}^{(0)}}$ as the conditional probability for each individual to follow the path to a saddle or boundary point, given the population state at each point, independent of the motion of other individuals. Although this scaling thus represents independent motion by individuals, because it is an essential singularity in the Taylor's series expansion in $1/N$ it cannot be approximated by any finite-order expansion of the action as in (4.86) [65]. In this respect, escapes are less probable than all orders of fluctuations about a rest point and are inherently missed by perturbations about the deterministic description.

4.4.1.2 'Kinematic' description of the stationary-point flow equations and the shape of the potential

The stationary point condition, if we do not attempt to incorporate fluctuation corrections, is $S^{(1)} = 0$, equivalent to vanishing of the first variational derivatives of S. In coherent-state fields, vanishing first derivative gives the Hamiltonian 'equations of motion'

$$\frac{\mathrm{d}\bar{\phi}_i}{\mathrm{d}t} = -\frac{\overline{\partial \mathcal{L}}}{\partial \phi_i^\dagger} \tag{4.87}$$

$$\frac{\mathrm{d}\bar{\phi}_i^\dagger}{\mathrm{d}t} = \frac{\overline{\partial \mathcal{L}}}{\partial \phi_i}, \tag{4.88}$$

while in the action-angle variables the same condition reads

$$\frac{d\bar{\nu}_i}{d\tau} = -\overline{\frac{\partial \hat{\mathcal{L}}}{\partial \eta_i}} \tag{4.89}$$

$$\frac{d\bar{\eta}_i}{d\tau} = \overline{\frac{\partial \hat{\mathcal{L}}}{\partial \nu_i}}. \tag{4.90}$$

Overbars on partial derivatives indicate that they are evaluated at arguments $\bar{\phi}^\dagger$, $\bar{\phi}$, or their equivalents $\bar{\nu}$, $\bar{\eta}$. The specific forms for derivatives of $\mathcal{L}$ (or $\hat{\mathcal{L}}$), used to compute properties of the examples, are provided in appendices in the chapters where they are used.

It is a general feature [58] of 'two-field' action functionals—whether derived using the Doi–Peliti construction or arrived at through alternative constructions such as Freidlin–Wentzell [55] or Martin–Siggia–Rose [57]—that $\phi^\dagger \equiv 1 \Leftrightarrow \eta \equiv 0$ are the stationary values for all classical stationary paths. From the forms (4.69) or (4.85) for the Liouville operator, it is also clear that these leave $S^{(0)} \equiv 0$.[37]

The structure of the Liouville operator reflected by the classical stationary trajectories is not directly apparent in coherent-state coordinates, but if we introduce the log-ratio of flow rates between types j and l,

$$r_{jl} \equiv \frac{1}{2} \log\left(\frac{\hat{p}_{lj}}{\hat{p}_{jl}}\right), \tag{4.91}$$

it is possible to recast the action-angle operator $\hat{\mathcal{L}}$ of (4.85) in the form

$$\begin{aligned}\hat{\mathcal{L}} &= 2\sum_{j=1}^{D}\sum_{l=1}^{j}\sqrt{\hat{p}_{jl}\hat{p}_{lj}}\left[\cosh(r_{jl}) - \cosh\left(\eta_j - \eta_l - r_{jl}\right)\right] \\ &\approx -\frac{1}{2}(\eta - \eta_0)^T m^{-1}(\eta - \eta_0) + V(\nu).\end{aligned} \tag{4.92}$$

The second line of (4.92) is an approximate expansion to second order in η, but keeping all orders in ν. It is applicable to any Liouvillian-cum-Hamiltonian with the form (4.85), compatible with having been generated on sufficiently short timescales by a Moran process.

[37] [58] shows that these properties are consequences of the structure of action functionals for stochastic processes and do not depend on other additional restrictions that may apply to different cases, such as Gaussian-order fluctuations ([55], chapter 4), discrete state spaces [48, 49], or whether the classical fields do or do not correspond to expectations of commuting observables [56]. The property $S^{(0)} \equiv 0$ applies even to the original near-equilibrium stochastic action of Onsager and Machlup [31], which arises from the more general two-field form after integration over the conjugate-momentum field $\phi^\dagger$ or η [25].

In (4.92) we have introduced, at each configuration-point ν, a function $\eta_0(\nu)$, defined as the maximizer of $\hat{\mathcal{L}}$ over η, and a *kinematic potential* $V(\nu)$ which is the maximizing value,

$$V(\nu) \equiv \hat{\mathcal{L}}\left(\nu, \eta_0(\nu)\right) \equiv \max_{\eta}\left(\hat{\mathcal{L}}(\nu, \eta)\right)$$
$$\approx \frac{1}{2}\eta_0^T m^{-1}\eta_0. \tag{4.93}$$

From the $\bar{\nu}$ equation of motion (4.89), it follows that $\mathrm{d}\bar{\nu}/\mathrm{d}\tau = 0$ at $(\nu, \eta_0(\nu))$, so we see that $\eta - \eta_0$ behaves as the kinematic momentum for motion in potential $V(\nu)$. The second line in (4.93) follows as a condition for consistency of the second-order approximation in (4.92), with the condition $\hat{\mathcal{L}} \equiv 0$ at $\eta \equiv 0$.

The final term in (4.92), which behaves as a matrix whose eigenvalues are the inverses of the masses in this system, will be the source of Langevin noise in action-angle coordinates,

$$m_{ij}^{-1}(\nu) \equiv -\left.\frac{\partial^2\hat{\mathcal{L}}}{\partial\eta_i\partial\eta_j}\right|_{\nu,\eta_0(\nu)}. \tag{4.94}$$

From the second line in (4.84) for the action and the approximation (4.92) for its Hamiltonian $\hat{\mathcal{L}}$, it is clear that this decomposition is the standard Hamilton–Lagrange decomposition familiar from mechanics, except that it gives rise to a free-standing term $\eta_0\partial_\tau\nu$, which arises because the *canonical* momentum η is offset from the kinematic momentum $\eta - \eta_0$. In systems where escapes occur between fixed points, along heteroclinic networks, this term often exercises little influence on trajectories (as we will demonstrate for a coordination game in section 5.2.3 below), but it is responsible for the asymmetry in the value of $S^{(0)}$ between the classical and non-classical solutions[38].

We return in section 5.2 below to develop the kinematic description of stationary solutions, their relation to the contours of integration for η in the path integral and the evaluation of the instanton probability $S^{(0)}$ on escape trajectories.

4.4.2 The Gaussian approximation for fluctuations, correlation functions and corrections to mean-field dynamics

We now show how the Gaussian fluctuation kernel $\mathrm{e}^{-S^{(2)}}$ is inverted to obtain the response and correlation functions, which are the *Green's functions* associated with the stochastic process. We first consider the distinctive form of this kernel associated with all two-field stochastic theories[39], which expresses causality and also determines

[38] This result does not necessarily extend to systems which escape along separatrices, for which see [86, 87].

[39] These include all of the Schwinger–Keldysh, Doi–Peliti, Freidlin–Wentzell and Martin–Siggia–Rose constructions.

the correct structure for Langevin approximations. We then show how the construction of the Green's function matrix defines the appropriate fluctuation-dissipation theorem for non-equilibrium stochastic processes.

4.4.2.1 The tri-diagonal form and causality

We observed in (4.94) that the Hessian of $\hat{\mathcal{L}}$ defines an effective-mass tensor for *real-valued* excursions of $\eta - \eta_0$ in the non-classical stationary paths. About any stationary path, the remaining fluctuations of η must generally be rotated to an imaginary contour, which is the stable contour of integration for the path integral (4.66). In this section we consider the effects associated with fluctuations along the rotated contour, whose imaginary part is analogous to the kinematic momentum in the real potential of a finite-temperature Hamiltonian system [5]. The imaginary parts may be integrated out to form an action involving only the fields ν and their time derivatives, known as the *Onsager–Machlup* action [31]. In section 4.4.2.2 below we develop an alternative formulation, which uses a *Hubbard–Stratonovich transformation* [36] to produce the Langevin equation of stochastic differential equation theory. Whatever approach is used, the quantities controlling the rate of injection of noise and its rate of damping are the second partial derivatives that make up $S^{(2)}$.

We will emphasize the distinctive 'tri-diagonal' form of $S^{(2)}$, whose meaning was first clearly understood by Keldysh [54] in the context of quantum mechanics, but which enforces certain constraints on all orders of fluctuation corrections in either classical or quantum Hilbert spaces [58]. We consider only expansions about the classical backgrounds, $(\phi^\dagger, \phi) = (1, \bar{\phi}) + (\tilde{\phi}, \phi')$, which are sufficient to demonstrate the method and are computationally simple. (Therefore in what follows both $S^{(0)} = 0$ and $S^{(1)} = 0$). The tri-diagonal form comes from the ability to organize all non-vanishing terms in $S^{(2)}$ into the bilinear form

$$S^{(2)} = \frac{1}{2}\int \mathrm{d}t \sum_{i,j=1}^{D} \begin{bmatrix} \phi_i' & \tilde{\phi}_i \end{bmatrix} \begin{bmatrix} \overline{\dfrac{\partial^2 \mathcal{L}}{\partial\phi_i \partial\phi_j}} & -\delta_{ij}\mathrm{d}_t + \overline{\dfrac{\partial^2 \mathcal{L}}{\partial\phi_i \partial\phi_j^\dagger}} \\ \delta_{ij}\mathrm{d}_t + \overline{\dfrac{\partial^2 \mathcal{L}}{\partial\phi_i^\dagger \partial\phi_j}} & \overline{\dfrac{\partial^2 \mathcal{L}}{\partial\phi_i^\dagger \partial\phi_j^\dagger}} \end{bmatrix} \begin{bmatrix} \phi_j' \\ \tilde{\phi}_j \end{bmatrix}. \tag{4.95}$$

The upper-left matrix element $\overline{\partial^2\mathcal{L}/\partial\phi_i\partial\phi_j} \equiv 0$ in (4.95) for all classical solutions, because $\bar{\phi}^\dagger \equiv 1$ causes vanishing of *all* terms at order $(\phi')^2$.[40] This property is preserved even if fluctuation corrections are incorporated back into the average dynamics to any order, because it is an expression of causality of the dynamics [58]. In any Gaussian

[40] The upper-left term will not generally vanish in expansions about non-classical stationary paths and indeed it is essential to preserve a zero-eigenvalue mode in the Green's function corresponding to the time derivative of the non-classical background. The existence of such a mode is a necessary expression of the fact that non-classical stationary paths generally hide time-translation symmetry and therefore must exist in continuous one-parameter families of solutions.

integral with such a tri-diagonal kernel, the expected covariance of fields $\tilde{\phi}$, ϕ' (the matrix inverse of the kernel), also has tri-diagonal form,

$$\left\langle \begin{bmatrix} \phi' \\ \tilde{\phi} \end{bmatrix}_t \begin{bmatrix} \phi' & \tilde{\phi} \end{bmatrix}_{t'} \right\rangle = \begin{bmatrix} D^K_{t;t'} & D^R_{t;t'} \\ D^A_{t;t'} & 0 \end{bmatrix}. \tag{4.96}$$

The D matrices are the Green's functions of the functional integral, which are the propagating waves of response to point-like sources. D^A and D^R describe, respectively, the time-advanced (A) and time-retarded (R) responses of ϕ' to $\tilde{\phi}$. D^K is the correlation function for fluctuations ϕ'. From standard methods of matrix inversion [56, 58], it follows that the retarded and advanced Green's functions in the continuous-time limit satisfy the equations

$$\left(\mathrm{d}_t \delta_{ij} + \overline{\frac{\partial^2 \mathcal{L}}{\partial \phi_i^\dagger \partial \phi_j}} \right) \mathcal{D}^R_{j,t;m,t'} \equiv \delta_{im} \delta(t - t') \tag{4.97}$$

$$\left(-\mathrm{d}_t \delta_{ij} + \overline{\frac{\partial^2 \mathcal{L}}{\partial \phi_i \partial \phi_j^\dagger}} \right) \mathcal{D}^A_{j,t;m,t'} \equiv \delta_{im} \delta(t - t'). \tag{4.98}$$

Their solutions are, respectively,

$$\mathcal{D}^R_{t;t'} = \theta(t - t') \mathcal{T} \exp\left\{ - \int_{t'}^{t} \mathrm{d}z \left. \overline{\frac{\partial^2 \mathcal{L}}{\partial \phi^\dagger \partial \phi}} \right|_z \right\} \tag{4.99}$$

$$\mathcal{D}^A_{t;t'} = \theta(t' - t) \mathcal{T}^{-1} \exp\left\{ - \int_{t}^{t'} \mathrm{d}z \left. \overline{\frac{\partial^2 \mathcal{L}}{\partial \phi \partial \phi^\dagger}} \right|_z \right\} = \left(D^{R^\dagger} \right)_{t;t'}. \tag{4.100}$$

Here $\mathcal{T}$ and $\mathcal{T}^{-1}$ denote forward and reverse *time-ordering* operators on their respective exponential integrals, defined so that for a general time-dependent matrix A_z,[41]

$$\mathcal{T} \exp\left\{ - \int_{t'}^{t} \mathrm{d}z \, A_z \right\} \equiv \lim_{\mathrm{d}z \to 0} \left(1 - \mathrm{d}z A_t\right)\left(1 - \mathrm{d}z A_{t-\mathrm{d}z}\right) \cdots \left(1 - \mathrm{d}z A_{t'}\right) \tag{4.101}$$

$$\mathcal{T}^{-1} \exp\left\{ - \int_{t}^{t'} \mathrm{d}z \, A_z \right\} \equiv \lim_{\mathrm{d}z \to 0} \left(1 - \mathrm{d}z A_t\right)\left(1 - \mathrm{d}z A_{t+\mathrm{d}z}\right) \cdots \left(1 - \mathrm{d}z A_{t'}\right). \tag{4.102}$$

[41] In (4.99, 4.100) and in what follows, it is necessary to explicitly time-order as it was not in the original quadrature (4.53) of the Liouville equation, because $\bar{\mathcal{L}}$ depends on the background $\bar{\phi}$, which may be a function of time.

The correlation function D^K, named after L D Keldysh, has a formally simple relation to the retarded and advanced Green's functions due to the tri-diagonal kernel. Its general solution can be written

$$\mathcal{D}^K_{t;t'} = \mathcal{D}^R_{t;t'} M_{t'} + M_t \mathcal{D}^A_{t;t'}, \tag{4.103}$$

in terms of a time-local matrix M satisfying

$$\frac{\mathrm{d}M}{\mathrm{d}t} + \overline{\frac{\partial^2 \mathcal{L}}{\partial \phi^\dagger \partial \phi}} M + M \overline{\frac{\partial^2 \mathcal{L}}{\partial \phi \partial \phi^\dagger}} = -\overline{\frac{\partial^2 \mathcal{L}}{\partial \phi^{\dagger 2}}}. \tag{4.104}$$

We have shown elsewhere [56] that (4.104) is an expression of the *fluctuation-dissipation theorem*, in which dissipation comes from the terms $\overline{\partial^2 \mathcal{L}/\partial \phi \partial \phi^\dagger}$ and its conjugate, and the fluctuation source from the term $\overline{\partial^2 \mathcal{L}/\partial \phi^{\dagger 2}}$.

Because the mixed partial derivatives $\overline{\partial^2 \mathcal{L}/\partial \phi^\dagger \partial \phi}$ in (4.99, 4.100), which give the proper population-dependent Laplacian matrix for the diffusion operator, may have zero eigenvalues, we must generally integrate (4.104) over finite rather than infinite times, so that the solution for M at two different times t'' and $t > t''$ obeys the relation

$$M_t = \mathcal{D}^R_{t;t''} M_{t''} \mathcal{D}^A_{t'';t} - \int_{t''}^{t} \mathrm{d}z \, \mathcal{D}^R_{t;z} \left. \overline{\frac{\partial^2 \mathcal{L}}{\partial \phi^{\dagger 2}}} \right|_z \mathcal{D}^A_{z;t}. \tag{4.105}$$

Then, completing the solution to (4.103), a similar relation for D^K at (t, t') both $> t''$, is

$$D^K_{t;t'} = \mathcal{D}^R_{t;t''} M_{t''} \mathcal{D}^A_{t'';t'} - \int_{t''}^{\min(t,t')} \mathrm{d}z \, \mathcal{D}^R_{t;z} \left. \overline{\frac{\partial^2 \mathcal{L}}{\partial \phi^{\dagger 2}}} \right|_z \mathcal{D}^A_{z;t'}. \tag{4.106}$$

In (4.105, 4.106), the positive eigenvalues of the kernel $\overline{\partial^2 \mathcal{L}/\partial \phi^\dagger \partial \phi}$ damp out correlations from earlier times, while the noise kernel $-\overline{\partial^2 \mathcal{L}/\partial \phi^{\dagger 2}}$ continuously contributes new fluctuations, as it does in the Langevin formulation derived later in section 4.4.2.2. For damped fluctuations, the finite-time integral (4.105) converges exponentially to the infinite-time integral. In the case where the diffusion kernels have zero eigenvalues, the undamped fluctuations accumulate linearly in time as Brownian motion.

The fluctuation at equal times, $\langle \phi'_t \phi'_t \rangle$, which will be useful for studying corrections to the mean replicator equations, is a degenerate case of the Keldysh correlation function,

$$D^K_{t;t} \to M_t. \tag{4.107}$$

4.4.2.2 Auxiliary fields and the Langevin equation

We will perform most of our calculations in the following chapters using the Keldysh decomposition of the Green's function. However, before considering more of its properties, it is useful to remark in passing how the Gaussian-order functional integral is converted to the older and more widely used Langevin equation. This mapping is especially helpful to convert between the discrete-state master equation and the stochastic differential equation that corresponds to it within the approximation of Gaussian fluctuations.

The tri-diagonal form (4.95) for $S^{(2)}$ in the main text results from splitting the mixed-partial derivatives in $\phi^\dagger$ and ϕ into two symmetric terms, whose inverses are the advanced and retarded response functions. If we do not split these terms, the same action, up to a total derivative, may be written

$$S^{(2)} = \int \mathrm{d}t \sum_{i,j=1}^{D} \left\{ \tilde{\phi}_i \left(\delta_{ij} \frac{\mathrm{d}}{\mathrm{d}t} + \overline{\frac{\partial^2 \mathcal{L}}{\partial \phi_i^\dagger \partial \phi_j}} \right) \phi_j' + \frac{1}{2} \tilde{\phi}_i \overline{\frac{\partial^2 \mathcal{L}}{\partial \phi_i^\dagger \partial \phi_j^\dagger}} \tilde{\phi}_j \right\}. \tag{4.108}$$

Its counterpart in action-angle coordinates reads

$$S^{(2)} = N \int \mathrm{d}\tau \sum_{i,j} \left\{ \eta_i' \left(\delta_{ij} \frac{\mathrm{d}}{\mathrm{d}\tau} + \overline{\frac{\partial^2 \hat{\mathcal{L}}}{\partial \eta_i \partial \nu_j}} \right) \nu_j' + \frac{1}{2} \eta_i' \overline{\frac{\partial^2 \hat{\mathcal{L}}}{\partial \eta_i \partial \eta_j}} \eta_j' \right\}. \tag{4.109}$$

The following construction may be carried out in either variables, but we will compute examples in later chapters using the coherent-state representation.

The Hessian for either $\tilde{\phi}$ or η', in equations (4.108, 4.109), is negative-semidefinite, requiring the rotation of the momentum variable to an imaginary integration contour. The consequences of this rotation are developed further in section 5.2.3. The direct use of the momentum variable as an 'imaginary Langevin field' is sometimes seen, but an older approach is to cancel the unstable Hessian term with an offset produced by an *auxiliary field*, which will become the conventional real-valued Langevin field. The procedure, known as the *Hubbard–Stratonovich transformation* [37] is to introduce a representation of unity as the Gaussian integral

$$1 = \det \left(\overline{\frac{\partial^2 \mathcal{L}}{\partial \phi^{\dagger 2}}} \right)^{-1/2} \int \mathcal{D}\zeta \mathrm{e}^{-S^{\mathrm{Aux}}}, \tag{4.110}$$

in which the 'auxiliary field action' is the non-dynamical quadratic form (meaning that it involves no τ-derivatives)

$$S^{\mathrm{Aux}} = -\frac{1}{2} \int \mathrm{d}t \sum_{i,j=1}^{D} \left(\zeta_i + \sum_{m=1}^{D} \tilde{\phi}_m \left(\overline{\frac{\partial^2 \mathcal{L}}{\partial \phi^{\dagger 2}}} \right)_{mi} \right) \left(\overline{\frac{\partial^2 \mathcal{L}}{\partial \phi^{\dagger 2}}} \right)^{-1}_{ij} \left(\zeta_j + \sum_{n=1}^{D} \left(\overline{\frac{\partial^2 \mathcal{L}}{\partial \phi^{\dagger 2}}} \right)_{jn} \tilde{\phi}_n \right). \tag{4.111}$$

(Equation (4.110) is normalized with the square root of a functional determinant because a single field, ζ, is introduced to correspond to $\tilde{\phi}$, rather than a complex-valued pair $(\zeta^\dagger, \zeta)$.)

When the expression (4.110) is inserted into the generating functional (4.66), the resulting integral remains Gaussian, with a kernel in which the original Hessian terms cancel, to be replaced with a convergent Hessian in the fields ζ,

$$S^{(2)} + S^{\mathrm{Aux}} = \int \mathrm{d}t \left\{ \sum_{i=1}^{D} \tilde{\phi}_i \left[\sum_{j=1}^{D} \left(\delta_{ij} \frac{\mathrm{d}}{\mathrm{d}t} + \overline{\frac{\partial^2 \mathcal{L}}{\partial \phi_i^{\dagger} \partial \phi_j}} \right) \phi_j' - \zeta_i \right] - \frac{1}{2} \sum_{i,j=1}^{D} \zeta_i \left(\overline{\frac{\partial^2 \mathcal{L}}{\partial \phi^{\dagger 2}}} \right)^{-1}_{ij} \zeta_j \right\}. \tag{4.112}$$

If $\tilde{\phi}$ is now integrated along an imaginary contour, it produces a functional δ-function,

$$\int \mathcal{D}\tilde{\phi} \exp \left\{ \int \mathrm{d}t \sum_{i=1}^{D} \tilde{\phi}_i \left[\sum_{j=1}^{D} \left(\delta_{ij} \frac{\mathrm{d}}{\mathrm{d}t} + \overline{\frac{\partial^2 \mathcal{L}}{\partial \phi_i^{\dagger} \partial \phi_j}} \right) \phi_j' - \zeta_i \right] \right\} = \prod_{i=1}^{D} \delta \left[\sum_{j=1}^{D} \left(\delta_{ij} \frac{\mathrm{d}}{\mathrm{d}t} + \overline{\frac{\partial^2 \mathcal{L}}{\partial \phi_i^{\dagger} \partial \phi_j}} \right) \phi_j' - \zeta_i \right], \tag{4.113}$$

which restricts field variations to a hypersurface within the remaining integral over ϕ' and ζ.

The argument of the δ-functional,

$$\sum_{j=1}^{D} \left(\delta_{ij} \frac{\mathrm{d}}{\mathrm{d}t} + \overline{\frac{\partial^2 \mathcal{L}}{\partial \phi_i^{\dagger} \partial \phi_j}} \right) \phi_j' - \zeta_i$$

is called the *Langevin equation* from stochastic differential equation theory. It expresses ϕ', through the inverse of the retarded Green's function, in terms of the Langevin field ζ, which has the time-local correlation function

$$\left\langle \zeta_{i_t} \zeta_{j_{t'}} \right\rangle = - \left(\overline{\frac{\partial^2 \mathcal{L}}{\partial \phi^{\dagger 2}}} \right)_{ij} \delta(t - t'). \tag{4.114}$$

The form of the Hessian $\overline{\partial^2 \mathcal{L} / \partial \phi^{\dagger 2}}$ is given in (4.136) of appendix 4.6, as a function of the original probabilities in the master equation. Thus, from the Gaussian approximation to the functional integral, we may derive the Langevin equation for an arbitrary stochastic process, with the correct noise structure and the correct damping. The Langevin equation augments the deterministic equations (4.87, 4.88) for the mean with a zero-mean stochastic fluctuation whose correlations are given by (4.114).

4.4.2.3 Time-dependent backgrounds and zero eigenvalues

A point that requires care in the evaluation of Gaussian fluctuations is that the second-order action $S^{(2)}$ may have eigenvectors of zero-eigenvalue (referred to as *zero modes*). For game models in which the constitutive parameters do not have explicit time dependence, a zero mode must arise for each time-dependent background, because such backgrounds hide the symmetry of the constitutive relations. Thus they must exist in continuous one-parameter families generated by time translation.

Formally, if $(\bar{\phi}_t, \bar{\phi}_t^{\dagger})$ is any solution to the stationary-path equations (4.88), then so must be $(\bar{\phi}_{t+\delta t}, \bar{\phi}_{t+\delta t}^{\dagger})$ at any δt. Expanding $S^{(1)} = 0$ in powers of δt generates a cascade of identities that must relate any stationary-path trajectory and the partial derivatives of $\bar{\mathcal{L}}$. The first-order term in δt is simply the first time derivative of the equations of motion (4.88), given by

$$\begin{aligned}
\frac{d^2\bar{\phi}_i}{dt^2} + \overline{\frac{\partial^2 \mathcal{L}}{\partial \phi_i^{\dagger} \partial \phi_j}} \frac{d\bar{\phi}_j}{dt} + \overline{\frac{\partial^2 \mathcal{L}}{\partial \phi_i^{\dagger} \partial \phi_j^{\dagger}}} \frac{d\bar{\phi}_j^{\dagger}}{dt} &= 0 \\
-\frac{d^2\bar{\phi}_j^{\dagger}}{dt^2} + \overline{\frac{\partial^2 \mathcal{L}}{\partial \phi_j \partial \phi_i^{\dagger}}} \frac{d\bar{\phi}_i^{\dagger}}{dt} + \overline{\frac{\partial^2 \mathcal{L}}{\partial \phi_j \partial \phi_i}} \frac{d\bar{\phi}_i}{dt} &= 0,
\end{aligned} \tag{4.115}$$

(in which repeated indices are summed). Taking the inner product of the first line with $d\bar{\phi}_j^{\dagger}/dt$ and the inner product of the second line with $d\bar{\phi}_i/dt$ and adding, we find that if $[\phi' \quad \tilde{\phi}]$ is set equal to $d_t[\bar{\phi} \quad \bar{\phi}^{\dagger}]$ in (4.95), the result is that $S^{(2)} = 0$.

Most of these zero modes will not concern us, as they result either from transient relaxation of initial conditions, or as transient departures if fluctuations in final conditions are imposed. In either case, the boundary conditions pin the time dependence of the solution. Two classes of cases, however, will be of great interest. The first are such time-dependent backgrounds as limit cycles or attractors that are (exactly or approximately) re-entrant on finite time intervals. Since both $d_t\bar{\phi}^{\dagger} = 0$ and $\overline{\partial^2 \mathcal{L}/\partial \phi_i \partial \phi_j} \equiv 0$ for these solutions, we must have $(d_t + \overline{\partial^2 \mathcal{L}/\partial \phi^{\dagger} \partial \phi}) d_t \bar{\phi} \equiv 0$ also[42].

The second class of solutions are the transient solutions of instantons in games with discrete symmetries and escapes, as in the coordination game illustrated. For these solutions, the non-vanishing of $\overline{\partial^2 \mathcal{L}/\partial \phi_i \partial \phi_j}$ is required for $S^{(2)} = 0$.

4.4.2.4 Fluctuations about non-classical stationary paths and the continuous sum over instantons

A zero mode of $S^{(2)} = 0$, if retained in the functional measure $\int \mathcal{D}\tilde{\phi}\, \mathcal{D}\phi'$ of the path integral (4.66), would appear to render the integral divergent because of the infinite range of integration. This apparent divergence would be a problem no matter what

---

[42] If a solution such as a limit cycle may be written in coordinates where the background advances linearly in time, these are zero-eigenvalue eigenvectors of $\partial^2 \mathcal{L}/\partial \phi^{\dagger} \partial \phi$. The small-radius limit cycle of the RPS game developed in chapter 6 will be an example of this kind.

contour of integration was chosen. The resolution of the problem is a combination of a contour choice, with a recognition that the zero-mode should *not* be integrated over an infinite range because it is the leading term in a non-linear fluctuation that is already included elsewhere in the integral. $\tilde{\phi}$ should be retained within a real contour of integration, but the naïve discrete sum over multiple instantons should absorb this mode to become an integral over the positions of the instanton centers as well as a sum over their counts.

The identification of the correct instanton sum goes as follows: to ensure a convergent expansion in Gaussian moments [5, 65], the separation of the action from (4.86) should be made about each stationary background $\bar{\phi}^{\dagger}$, $\bar{\phi}$, so that the generating functional (4.66) becomes a sum

$$(0|e^{\sum_m a_m}|\Psi_T) = \sum_{\bar{\phi}^{\dagger},\bar{\phi}} e^{-S^{(0)}} \int_0^T \mathcal{D}\tilde{\phi}\, \mathcal{D}\phi' e^{-\left(S-S^{(0)}\right)} e^{\left(\phi_0^{\dagger}-1\right)\left(\bar{n}-\phi_0\right)}. \tag{4.116}$$

Each $S^{(0)}$ is to be evaluated at the appropriate background, which may contain one or many instantons, and the integral is to be left with only convergent Gaussian moments about that background.

The formal sum in (4.116) includes a discrete count over the number $k \in 0, \ldots, \infty$ of escape events and for each k the full set of allowed solutions must contain the integral over all possible time-ordered positions $\{t_1, \ldots, t_k\}$ for the event centers. At the order to which $S^{(2)}$ accounts for Gaussian fluctuations, the incremental measure in the integral over instanton centers adds to each background $\bar{\phi}^{\dagger}$, $\bar{\phi}$ an infinitesimal amount of its time derivative. Hence the integral over backgrounds already includes the zero-eigenvalue mode of fluctuations, which therefore must not be redundantly included in the Gaussian integral. The time-derivative of the escape path is a finite and hence normalized mode, so that the individual integrals over t_i differ from the component $d\tilde{\phi}\, d\phi'$ for the zero-eigenvalue mode only by a normalization, which we denote t_{esc}.[43] Denoting the functional measure without the zero-eigenvalue mode as $\mathcal{D}'\tilde{\phi}\, \mathcal{D}'\phi'$, we have for a single escape

$$\int_0^T \mathcal{D}\tilde{\phi}\, \mathcal{D}\phi' = \int_0^T \frac{dt}{t_{\text{esc}}} \int_0^T \mathcal{D}'\tilde{\phi}\, \mathcal{D}'\phi'. \tag{4.117}$$

Because the functional measure was defined by (4.65) as a product, the normalization for removal of a fluctuation with k escapes is simply t_{esc}^k. Similarly, the value of $S^{(0)}$ for a k-escape trajectory is k times that for a single escape, which we denote $S_{\text{esc}}^{(0)}$.[44] The geometric restriction that escapes happen in order may be simplified by introducing the time-ordering operator $\mathcal{T}$, as was used to define Green's functions, this time applied to an unrestricted integral over the times of the escape events.

[43] For methods to compute this normalization constant, see [5], chapter 7, appendix 2.

[44] Here we assume for simplicity that the escape trajectories are all drawn from a symmetric set of possibilities, so that they all have the same value of $S^{(0)}$. The worked example will have this property. The more general case with different classes of escapes and values of $S^{(0)}$ can of course be treated with a more complicated expression.

One then divides by the factor $k!$ by which a product of unrestricted time integrals overcounts the configuration space for sequential escapes. The result is that a fully regular expansion of the generating functional including all escapes may be written

$$\left(0\middle|e^{\sum_m a_m}\middle|\Psi_T\right) = \sum_{k=0}^{\infty} \frac{e^{-kS^{(0)}_{\mathrm{esc}}}}{k!} \mathcal{T} \int_0^T \frac{dt_1}{t_{\mathrm{esc}}}, \ldots,$$
$$\int_0^T \frac{dt_k}{t_{\mathrm{esc}}} \int_0^T \mathcal{D}'\tilde{\phi}\, \mathcal{D}'\phi' e^{-\left(S-S^{(0)}\right)} e^{\left(\phi_0^{\dagger}-1\right)\left(\bar{n}-\phi_0\right)}. \qquad (4.118)$$

The mean-regressing fluctuations in the remaining Gaussian integral of (4.118) will be finite, and away from any critical points will lead to at most polynomial corrections to the exponential prefactors. Therefore the factors $e^{-S^{(0)}_{\mathrm{esc}}}$ and integrals over t_i collapse to an exponential sum of the form

$$e^{-r_{\mathrm{esc}} T},$$

in which the escape rate is given by

$$r_{\mathrm{esc}} = \frac{e^{-S^{(0)}_{\mathrm{esc}}}}{t_{\mathrm{esc}}}. \qquad (4.119)$$

The correction of the Gaussian measure is the step needed to relate the dimensionless probability term $e^{-S^{(0)}_{\mathrm{esc}}}$ to a rate by defining the characteristic timescale t_{esc} over which fluctuations sample the basin of attraction. Because t_{esc} depends at most polynomially on N, $S^{(0)}_{\mathrm{esc}} \equiv N\hat{S}^{(0)}_{\mathrm{esc}}$ defines the leading exponential dependence of the escape rate on N.

4.4.2.5 Breaking and restoration of ergodicity

The functional-integral approach that we adopt toward non-equilibrium stochastic processes lends itself to a somewhat different treatment of broken symmetry than one finds in many treatments in equilibrium thermodynamics. 'Broken symmetry' is more accurately characterized as *broken ergodicity organized according to symmetries.* This kind of broken ergodicity is the failure of a system's dynamics to be capable of sampling its state space in a manner equivalent to sampling in a symmetric ensemble. It is an inherently dynamical concept and as such inherently depends on the timescale of observation. Rather than replace an *a priori*, symmetric ensemble with a 'broken' ensemble as a proxy for failure of sampling, our approach is to compute observables involving multiple field values at multiple times.

Systems whose dynamics break ergodicity develop very long-term dependencies among field values; in those that break symmetry these dependencies factor into representations of the symmetry group. In systems with finite large-deviations scale factors, however, intertemporal correlations are never indefinitely long-lived. Fluctuations eventually render observations at sufficiently distant times independent of each other and dependent only on the symmetries of the underlying dynamics. Thus we never need to invoke more than one ensemble to describe a system, as long as we require the ensemble of multi-time correlation functions rather than of single-time observables. A single ensemble contains both representations that break symmetries on short timescales and representations that respect them at long timescales.

The background field expansion in the functional integral is by default an expression of an infinite-order correlation among the field values at all points along a history. Time dependent backgrounds express serial autocorrelation in the moments of the distribution ρ for the stochastic process. Expansion about any single background 'breaks ergodicity' in the measure by implicitly inserting a projection operator that makes expectations conditional on that background. The restoration of symmetry in the long term is by means of fluctuations that are also represented explicitly in the functional integral. Both escapes from domains of attraction, in systems whose population states break a discrete symmetry, and Brownian motion around limit cycles, cause conditional expectations on the short term to relax towards the ergodic distribution over population states at late time, restoring the expression of the hidden symmetry.

If we shift the scale of observation from individuals to populations by forming a coarse-grained description, either escapes or Brownian motions become the elementary moves of the populations-as-entities. The presence of a hidden symmetry is the precondition that lifts a restricted sub-set of individual-level dynamics to become the aggregated dynamics of emergent entities, at aggregation scales that may be arbitrarily larger and slower than the microscale of individual birth/death events.

4.4.3 Fluctuation-induced corrections to average dynamics and the evolutionary entropy

Our last fluctuation effect concerns corrections to the mean stationary paths, in models where some fluctuations are not suppressed by powers of $1/N$.

$S^{(2)}$ is *only* the correct kernel for Gaussian fluctuations, on the background identified by $S^{(1)} = 0$, in the case that $S^{(3)}$ and higher-order terms in the Taylor's series for the action do not *on average* produce values linear in $\bar{\phi}$ or ϕ'.[45] If this approximation is not good enough, rather than use (4.86), e^{-S} in the generating functional (4.66) can be expanded as

$$e^{-S} = e^{-\left(S^{(0)}+S^{(2)}\right)}\left(1 - S^{(1)} - S^{(3)} + \cdots\right). \tag{4.120}$$

The expansion is initially performed about a *general* background, which is then chosen so that the linear terms in the large parenthesis vanish on average. For corrections to classical stationary points where $\bar{\phi}^{\dagger} \equiv 1$, the identification of self-consistent backgrounds is particularly easy. For all backgrounds $\bar{\phi}$, $S^{(0)} \equiv 0$, and

[45] It is important to remember that the distribution producing these fluctuations, as well as the reference values for ϕ and $\phi^{\dagger}$, depend on the background about which the action is expanded.

the only non-vanishing term in $S^{(3)}$ leads to the correction to the equation of motion for $\bar{\phi}$,

$$\frac{d\bar{\phi}_i}{dt} + \overline{\frac{\partial \mathcal{L}}{\partial \phi_i^{\dagger}}} + \frac{1}{2} \sum_{m,n=1}^{D} \overline{\frac{\partial^3 \mathcal{L}}{\partial \phi_i^{\dagger} \partial \phi_m \partial \phi_n}} \left\langle \phi'_n \phi'_m \right\rangle = 0. \tag{4.121}$$

The fluctuation strength $\langle \phi'_n \phi'_m \rangle$ is given by the equal-time correlation function in equations (4.96, 4.105, 4.107), computed about the background $\bar{\phi}$.

The third-order partial-derivative term has the interpretation of the gradient of an entropy correction to the fitness or mutation parameters that define the game. For games with rest points, this entropy may be regarded as a property of the state of the population at the rest point, so it becomes a similar quantity to the constitutive parameters f_i and μ_{ij}. For more general treatment of relaxation trajectories, or in the case of time-dependent backgrounds, the correction must be computed as an *entropy functional of histories*. In this form it is an instance of the quantity termed 'caliber' by Jaynes [68–71].

Entropy corrections become particularly important in the context of evolutionary neutrality. We develop an example using the RPD game in chapter 7. Algebraic forms for the derivatives of $\mathcal{L}$ required to evaluate (4.121) are given in appendix 4.6.1.

4.5 Appendix. From distributions to moments and back

In this appendix we sketch the relation between the dynamics of distributions and the dynamics of moments, and the passage between the two which is the main project of section 4.2. Distributions (which may require complex master equations to specify) generate entire hierarchies of moment equations, which are not generally closed (cannot generally be solved recursively working upward from lower to higher orders). When the MFA is used to close the first moment equations, a corresponding minimal master equation can be chosen which will produce the same first-moment dynamics in its MFA.

To illustrate the classes of terms that are lost in projecting from an actual stochastic process to the minimal representative that will generate the same first-moment dynamics in the MFA, we use the simplest non-trivial case of mutation between two types with conserved total number. The mutation rates for changes in the two directions are not assumed equal in this appendix. When they are set equal, we recover the very simplest minimal master equation, used to check the integration measure for fluctuations in section 4.3.4.1.

4.5.1 Synchronous changes by multiple agents and changes with equal frequency in both directions

The types are labeled $\{1, 2\}$. Their numbers, which index a state for the population, are denoted by $(\mathrm{n}_1, \mathrm{n}_2)$. We will choose a dynamic for which $\mathrm{n}_1 + \mathrm{n}_2 = N$ at all times. We consider the possibility that $k \geqslant 1$ individuals may switch type in a single event, and assign rates $p_k^{+}\mathrm{n}_1$ to such a switch for the change of k individuals from types 1 → 2 from an initial number n_1 and $p_k^- n_2$ for the change of k individuals from types 2 → 1 from an initial number n_2.[46]

The mutation rates for 1 → 2 and 2 → 1 will differ on average at general (n_1, n_2), becoming equal only at the steady state where detailed balance holds. In addition to the terms parametrized by $p_k^{\pm}$, we also consider type switching by k individuals with a rate $p_k^S(n_1, n_2)$ that is the same for changes in either direction. Such symmetric-rate switching does not change any population state on average, but does generally affect the population dispersion. The function $p_k^S(n_1, n_2)$ can be an arbitrary polynomial, except that we require $p_k^S(n_1, n_2)/(n_1 n_2)$ to be regular both when $n_1 \to 0$ and when $n_2 \to 0$. This condition ensures compatibility with evolutionary population processes reflected in the structure of the replicator equation (4.9), in which a type can never die at a rate that fails to vanish when all of its members are already gone. (Rates of the form $p^S \propto n_1 n_2 / N$ arise from the ϕ terms in (4.9) from the constant components of fitness.)

4.5.2 The generating master equation and its moments

The master equation for a distribution ρ_n on such populations is a sum over all events in which any allowed number k of agents change type. Its expression is cluttered but otherwise straightforward:

$$\begin{aligned}
\frac{d\rho_n}{dt} = & \sum_{k=1}^{N-n_1} \left[p_k^+ (n_1 + k) + p_k^S (n_1 + k, n_2 - k) \right] \rho_{n+(k,-k)} \\
& - \sum_{k=1}^{n_1} \left[p_k^+ n_1 + p_k^S (n_1, n_2) \right] \rho_n \\
& + \sum_{k=1}^{n_1} \left[p_k^- (n_2 + k) + p_k^S (n_1 - k, n_2 + k) \right] \rho_{n+(-k,k)} \\
& - \sum_{k=1}^{N-n_1} \left[p_k^- n_2 + p_k^S (n_1, n_2) \right] \rho_n .
\end{aligned} \tag{4.122}$$

To simplify the notation we will suppose that $N \gg k_{max} \equiv \max(\mathrm{supp}_k(p_k^{\pm}), \mathrm{supp}_k(p_k^S))$, and that the support of ρ_n, $\mathrm{supp}_n \rho_n \subset [k_{max}, N - k_{max}]$, so that sums over k or n may run over their entire ranges.

Now we consider what can be learned about the parameters in (4.122) from the dynamics of its moments. The first moment and also the first cumulant, $\bar{n}$ is given in (4.8). Computing its time dependence from $d\rho_n/dt$ in (4.122) is straightforward by a

[46] The choice of rates for k-individual switching events, which are linear in the initial number of individuals, is not appropriate for all processes: for instance, for k-particle chemical reactions it would be difficult to invent a model with such kinetics. One can probably think up an evolutionary story in which these rate equations would be motivated. We adopt this choice here to demonstrate a mathematical property: it produces first-moment equations which are closed and therefore exact in mean-field theory and yet which hide an entire hierarchy of higher-order moments that may be given different dynamics by changing their dependence on synchronized switching events involving different numbers of agents.

shift of indices. The result is expressed in terms of two lumped parameters. The first is a relaxation rate, with which we convert from physical time t to a dimensionless characteristic timescale τ, given by

$$\frac{d\tau}{dt} \equiv \sum_k k\left(p_k^- + p_k^+\right). \tag{4.123}$$

The second is the equilibrium at which the population reaches detailed balance, given for the coordinate $\bar{n}_1$ by

$$\bar{n}_1^{eq} \equiv N \frac{\sum_k k p_k^-}{\sum_k k\left(p_k^- + p_k^+\right)}, \tag{4.124}$$

or equivalently, by

$$\frac{2\bar{n}_1^{eq}}{N} - 1 = \frac{\sum_k k\left(p_k^- - p_k^+\right)}{\sum_k k\left(p_k^- + p_k^+\right)} \tag{4.125}$$

which we will use below. In terms of these, the first-moment dynamics is exactly linear and therefore closed. It is given by

$$\frac{d\bar{n}_1}{d\tau} = -(\bar{n}_1 - \bar{n}_1^{eq}). \tag{4.126}$$

Note that the $\{p_k^\pm\}$ enter only through the sums $\sum_k k p_k^\pm$ and that the symmetric-switching rate p_k^S does not enter at all.

The second cumulant (the variance) is represented without loss of generality (since the population has only one dynamical variable) by $\langle (n_1 - \bar{n}_1)^2 \rangle \equiv \sum_n (n_1 - \bar{n}_1)^2 \rho_n$. It likewise may be computed from (4.122) by a shift of indices. Two new lumped parameters enter, which we denote by $\beta_\pm^{(2)}$. They are given in terms of the rate parameters $p_k^\pm$ by

$$\beta_+^{(2)} \equiv \frac{\sum_k k^2\left(p_k^- + p_k^+\right)}{\sum_k k\left(p_k^- + p_k^+\right)}$$

$$\beta_-^{(2)}\left(\frac{2\bar{n}_1^{eq}}{N} - 1\right) \equiv \frac{\sum_k k^2\left(p_k^- - p_k^+\right)}{\sum_k k\left(p_k^- + p_k^+\right)} = \frac{\sum_k k^2\left(p_k^- - p_k^+\right)}{\sum_k k\left(p_k^- - p_k^+\right)}\left(\frac{\sum_k k\left(p_k^- - p_k^+\right)}{\sum_k k\left(p_k^- + p_k^+\right)}\right). \tag{4.127}$$

In the case that $p_k^{\pm} = 0$ for all $k > 1$, both of $\beta_{\pm}^{(2)} = 1$. Otherwise both are free except that they are regular and $\beta_{+}^{(2)}$ is strictly positive[47].

The time dependence of the variance is then given in terms of these lumped parameters by

$$\frac{\mathrm{d}}{\mathrm{d}\tau}\left\langle(\mathrm{n}_1 - \bar{n}_1)^2\right\rangle = -2\left\langle(\mathrm{n}_1 - \bar{n}_1)^2\right\rangle + \frac{N}{2}\left[\beta_{+}^{(2)} - \beta_{-}^{(2)}\left(\frac{2\bar{n}_1^{\mathrm{eq}}}{N} - 1\right)\left(\frac{2\bar{n}_1}{N} - 1\right)\right] + \frac{\mathrm{d}t}{\mathrm{d}\tau}\sum_k k^2\left\langle p_k^S\right\rangle. \tag{4.128}$$

The terms in the first line depend on the parameters from the dynamics of the mean and the new k^2-dependent ratios $\beta_{\pm}^{(2)}$. The term in the second line depends on the symmetric switching rates p_k^S which did not appear in the mean dynamics. Note that if p_k^S is higher than quadratic order in n, the second-moment dynamics cannot be solved without also solving the dynamics at higher orders.

If $p_k^S \propto \mathrm{n}_1\mathrm{n}_2/N$ at leading order—as it will be when produced by mean-fitness terms ϕ in the replicator equation (4.9)—the terms that serve as sources for the variance in (4.128) all scale as $O(N)$ or smaller. Therefore any relative variance $\langle(\mathrm{n}_1 - \bar{n}_1)^2\rangle/N^2$ towards which the system converges scales as $O(1/N)$, compared to the terms that appear in the dynamics for the relative mean $\mathrm{d}/\mathrm{d}\tau(\bar{n}_1/N)$ in (4.126), which scale as $O(N^0)$. This scaling and its extension to higher-order moments is the motivation behind replacing correlations $\langle \mathrm{n}_{i_1}\ldots\mathrm{n}_{i_m}\rangle$ with $\langle \mathrm{n}_{i_1}\rangle\ldots\langle \mathrm{n}_{i_m}\rangle$ in the MFA, in cases when the mean dynamics is not exact. The utility of the approximation therefore depends on the presence of distinctions larger than $O(1/N)$ for all degrees of freedom in $\mathrm{d}\bar{n}/\mathrm{d}t$ from which higher-order cumulants are dropped.

The reader with time to kill—an expression that we do not use idly [88]—is invited to compute the next few cumulants and to see that each cumulant at order m depends on the moment k^m under $p_k^{\pm}$ which is not fixed by $\mathrm{d}\tau/\mathrm{d}t$, $\bar{n}_1^{\mathrm{eq}}$, or any of the lower-order moments of k.

4.5.3 MFA and a minimal process

Since no higher-order moments appear in the mean dynamics (4.126), the exact solution is already identical to its MFA. The two rate parameters that appear in this equation could be denoted by

$$\begin{aligned} \bar{p}^{+} &\equiv \sum_k k p_k^{+} \\ \bar{p}^{-} &\equiv \sum_k k p_k^{-}. \end{aligned} \tag{4.129}$$

[47] Note from (4.127) that $\beta_{-}^{(2)}\,(2\bar{n}_1^{\mathrm{eq}}/N - 1)$ may be non-zero even if $2\bar{n}_1^{\mathrm{eq}}/N \to 1$; we have grouped terms in which $(2\bar{n}_1^{\mathrm{eq}}/N - 1)$ may be continued through zero as a regular limit.

Only the antisymmetric combination of rates $(\tilde{p}^- n_2 - \tilde{p}^+ n_1)$ appears in $d\bar{n}_1/dt$. It happens in this simple model that we can identify $\tilde{p}^\pm$ individually because in addition to the dynamics we know the constraint $n_1 + n_2 \equiv N$, but this is only one constraint for a system that could have been given many more types. The inability to identify general symmetric terms is shown by the absence of any contribution from p_k^S.

To identify a minimal stochastic process by the procedures followed in the main text, we would allow only $k = 1$ switching events and assign rates $\tilde{p}^\pm$ from (4.129) to match the mean dynamics, giving the master equation

$$\frac{d\rho_n}{dt} = \tilde{p}^+\left[\left(n_1 + 1\right)\rho_{n+(1,-1)} - n_1\rho_n\right] + \tilde{p}^-\left[\left(n_2 + 1\right)\rho_{n+(-1,1)} - n_2\rho_n\right]. \quad (4.130)$$

Since no further symmetric parameters can be identified from the mean, counterparts to p_k^S are all set to zero.

The process (4.130), like the generating process (4.122), has an MFA that is exact and the two agree by construction. Equation (4.130) generates expressions for the variance and higher-order moments as in (4.128) with $\beta_\pm^{(2)} = 1$, which generally will not match those of the generating process. Because of the simplicity of the mean-field equation, the minimal process *is* closed, so that dynamical equations for all cumulants can be solved recursively working upward from the mean.

4.5.4 The source of mismatch in higher-order moments

A feature of this minimal example, which we note in the text but will not pursue in our example systems illustrating symmetry classes, is the possibility of type switching by $k > 1$ agents in single events. This is the feature that permits higher-order moments in the generating process to differ from those in a minimal model even if the MFAs of the two processes agree, not only to $O(1/N)$, but *exactly*. Such synchronized multi-agent events are performed by higher-order polynomials in the algebraic raising and lowering operators developed in section 4.3.2. These are what we have referred to as 'contact terms', following a standard terminology in quantum field theory where the same methods are used.

4.6 Appendix. Taylor's series expansion of the Liouville operators

Partial derivatives of the Liouville operator are required for a number of the explicit evaluations of fluctuation structure and fluctuation-induced corrections to average behavior in the following chapters. From the general form (4.69) for the Liouville operator, we can present these for all games used in the examples.

4.6.1 The derivative expansion in coherent-state fields

The stationary-point conditions (4.87, 4.88) are defined by vanishing of the first variational derivative of S. The complete algebraic forms for this derivative at general $(\phi^\dagger, \phi)$ are

$$\begin{aligned}\frac{\mathrm{d}\phi_i}{\mathrm{d}t} &= -\frac{\partial \mathcal{L}}{\partial \phi_i^\dagger} \\ &= \sum_{j=1}^{D}\left(\mu_{ij}\phi_j - \mu_{ji}\phi_i\right) + \sum_{j=1}^{D}\phi_i\phi_j\left(\phi_i^\dagger\lambda_{ij} - \phi_j^\dagger\lambda_{ji}\right) \\ &\quad - \phi_i\left\{\sum_{j=1}^{D}\left(\phi_j^\dagger - \phi_i^\dagger\right)\phi_j\lambda_{ij} + \frac{1}{N}\sum_{j,l=1}^{D}\left(\phi_l^\dagger - \phi_j^\dagger\right)\phi_l a_{ij}^S\phi_j^\dagger\phi_j\right\}\end{aligned} \tag{4.131}$$

$$\begin{aligned}\frac{\mathrm{d}\phi_i^\dagger}{\mathrm{d}t} &= \frac{\partial \mathcal{L}}{\partial \phi_i} \\ &= \sum_{j=1}^{D}\left(\phi_i^\dagger - \phi_j^\dagger\right)\mu_{ji} + \sum_{j=1}^{D}\left(\phi_i^\dagger - \phi_j^\dagger\right)\phi_j^\dagger\phi_j\lambda_{ji} \\ &\quad + \phi_i^\dagger\left\{\sum_{j=1}^{D}\left(\phi_j^\dagger - \phi_i^\dagger\right)\phi_j\lambda_{ij} + \frac{1}{N}\sum_{j,l=1}^{D}\left(\phi_l^\dagger - \phi_j^\dagger\right)\phi_l a_{ij}^S\phi_j^\dagger\phi_j\right\}.\end{aligned} \tag{4.132}$$

Because only difference terms $(\phi_i^\dagger - \phi_j^\dagger)$ appear in (4.132), it is immediate that $\phi^\dagger \equiv 1$ is always a consistent solution. The terms grouped in parentheses are those which cancel in the stationary-point equation for the total number, $\mathrm{d}(\phi_i^\dagger\phi_i)/\mathrm{d}t$, which corresponds to $\mathrm{d}\nu_i/\mathrm{d}t$.

The second partial with respect to ϕ, which is the quantity averaged to compute the stochastic effective action (7.3) and whose variation is averaged to compute the corrections to the mean-field equations of motion (4.121), is

$$\begin{aligned}\frac{\partial^2\mathcal{L}}{\partial\phi_m\partial\phi_n} &= \left(\phi_m^\dagger - \phi_n^\dagger\right)\left[\left(\phi_n^\dagger\frac{a_{nm}}{2} - \phi_m^\dagger\frac{a_{mn}}{2}\right) + \frac{1}{N}\sum_{k=1}^{D}\left(\phi_n^\dagger a_{nk}^S - \phi_m^\dagger a_{mk}^S\right)\left(\phi_k^\dagger\phi_k\right)\right] \\ &\quad + \frac{1}{N}\sum_{k=1}^{D}\left(\phi_k^\dagger\phi_k\right)\left[\left(\phi_m^\dagger - \phi_k^\dagger\right)a_{kn}^S\phi_n^\dagger + \left(\phi_n^\dagger - \phi_k^\dagger\right)a_{km}^S\phi_m^\dagger\right] \\ &\quad - \phi_m^\dagger\phi_n^\dagger a_{mn}^S\frac{1}{N}\sum_{k=1}^{D}\left(\phi_m^\dagger + \phi_n^\dagger - 2\phi_k^\dagger\right)\phi_k.\end{aligned} \tag{4.133}$$

Because we will not compute examples of fluctuation corrections to the response and correlation functions (4.96) and will consider these only about classical backgrounds, we evaluate the mixed partial that defines the diffusion kernel only in the background $\phi^\dagger \equiv 1$,

$$\begin{aligned}\frac{\partial^2\mathcal{L}}{\partial\phi_i^\dagger\partial\phi_m} &= -\mu_{im} - \delta_{im}\left(\sum_{j=1}^{D}a_{ij}\phi_j - \frac{1}{N}\sum_{j,k=1}^{D}\phi_j a_{jk}\phi_k\right) \\ &\quad + \phi_i\left[\frac{1}{N}\sum_{j=1}^{D}\left(2a_{mj}^S - a_{ij}^S\right)\phi_j - a_{im}\right].\end{aligned} \tag{4.134}$$

The trace of this kernel, which gives the sum of eigenvalues as a function of the background, evaluates to

$$\sum_{i=1}^{D} \frac{\partial^2 \mathcal{L}}{\partial \phi_i^\dagger \partial \phi_i} = -\sum_{i=1}^{D} \mu_{im} - \sum_{i,j=1}^{D} \left(a_{ij} + \delta_{ij} a_{ii}\right) \phi_j + \frac{D+1}{N} \sum_{i,j=1}^{D} \phi_i a_{ij} \phi_j. \quad (4.135)$$

The Hessian which is the noise source for coherent-state field fluctuations, appearing in equations (4.104 – 4.106) and also in the equivalent Langevin source term (4.114), is given (also at $\phi^\dagger \equiv 1$) by

$$\frac{\partial^2 \mathcal{L}}{\partial \phi_i^\dagger \partial \phi_l^\dagger} = -\delta_{il} \phi_l \sum_{j=1}^{D} \left(2a_{lj} + a_{jl}\right) \phi_j + \phi_i \phi_l \left[\frac{2}{N} \sum_{j=1}^{D} \left(a_{lj}^S + a_{lj}^S\right) \phi_j - a_{il}^S \right]. \quad (4.136)$$

Finally, the third partial responsible for the leading corrections (4.121) to the stationary-point conditions is

$$\frac{\partial^3 \mathcal{L}}{\partial \phi_i^\dagger \partial \phi_m \partial \phi_n} = \delta_{im} \left[\frac{1}{N} \sum_{j=1}^{D} \left(2a_{nj}^S - a_{mj}^S\right) \phi_j - a_{mn} \right] + \delta_{in} \left[\frac{1}{N} \sum_{j=1}^{D} \left(2a_{mj}^S - a_{nj}^S\right) \phi_j - a_{nm} \right]$$
$$+ \frac{\phi_i}{N} \left(2a_{mn}^S - a_{in}^S - a_{im}^S\right). \quad (4.137)$$

We evaluate this quantity for the example in chapter 7 at $\phi^\dagger \equiv 1$, because all expectations except those of $\langle \phi' \phi' \rangle$ at equal times will cancel [58].

4.6.2 The derivative expansion in action-angle coordinates

The stationary-point condition (4.89) for the number field ν, resulting from first variation of the normalized action $\hat{S}$, may be expressed variously in terms of the population-corrected fitness λ_{ij} of (4.82) or the average flow rates $\hat{p}_{ij}$ of (4.83) as

$$\begin{aligned} \frac{d\nu_i}{d\tau} &= -\frac{\partial \hat{\mathcal{L}}}{\partial \eta_i} \\ &= \sum_{j=1}^{D} e^{\eta_i - \eta_j} \left(\hat{\mu}_{ij} + \nu_i \lambda_{ij}\right) \nu_j - e^{\eta_j - \eta_i} \left(\hat{\mu}_{ji} + \nu_j \lambda_{ji}\right) \nu_i \\ &= \sum_{j=1}^{D} e^{\eta_i - \eta_j} \hat{p}_{ij} - e^{\eta_j - \eta_i} \hat{p}_{ji}. \end{aligned} \quad (4.138)$$

In particular, the last line, at $\eta \equiv 0$ (or indeed $\eta =$ any constant) yields the condition of *detailed balance* $\hat{p}_{lj} = \hat{p}_{jl}$ between all pairs jl, expressed in the text as the fixed point condition of vanishing of r_{jl} in (4.91).

The corresponding (4.90) for η, which is used to implement steepest descents of $\hat{S}$ in section 5.2.4, evaluates to

$$\frac{d\eta_i}{d\tau} = \frac{\partial \hat{\mathcal{L}}}{\partial \nu_i}$$
$$= \sum_{j=1}^{D} \left(1 - e^{\eta_j - \eta_i}\right)\left(\hat{\mu}_{ji} + \nu_j \lambda_{ji}\right) + \left(1 - e^{\eta_i - \eta_j}\right)\lambda_{ij}\nu_j + \sum_{j,l=1}^{D} a_{ij}^{S} \nu_j \left(1 - e^{\eta_j - \eta_l}\right)\nu_l. \tag{4.139}$$

Again we verify that $\eta \equiv 0$ is always a consistent solution. All remaining quantities will be evaluated at $\eta \equiv 0$, because they will only be applied here to classical stationary trajectories.

The noise source and inverse mass matrix (4.94) has a considerably simpler form in action-angle coordinates than its counterpart (4.136) in coherent-state fields,

$$\frac{\partial^2 \hat{\mathcal{L}}}{\partial \eta_i \partial \eta_j} = \left(\hat{p}_{ij} + \hat{p}_{ji}\right) - \delta_{ij} \sum_{k=1}^{D} \left(\hat{p}_{ik} + \hat{p}_{ki}\right). \tag{4.140}$$

The diffusion kernel is also simple and may be expressed in a manner that emphasizes the antisymmetry of its argument in i and j, corresponding to the antisymmetry of the deterministic replicator in (4.16),

$$\begin{aligned}\frac{\partial^2 \hat{\mathcal{L}}}{\partial \eta_i \partial \nu_m} = {} & \delta_{im} \sum_{j=1}^{D} \left[\hat{\mu}_{jm} + \nu_j \left(\lambda_{jm} - \lambda_{mj}\right)\right] \\ & - \left[\hat{\mu}_{im} + \nu_i \left(\lambda_{im} - \lambda_{mi}\right)\right] \\ & + \nu_i \sum_{j=1}^{D} \nu_j \left(a_{jm}^{S} - a_{im}^{S}\right).\end{aligned} \tag{4.141}$$

Equation (4.141) evaluates to the same quantity as its coherent-state counterpart (4.134), up to an overall factor of N reflecting the definitions of rescaled number fields and time.

Finally, the third partial derivative that appears in fluctuation corrections to the stationary-point equations, counterpart to (4.121) in the text, but for the number field ν, is

$$\begin{aligned}\frac{\partial^3 \hat{\mathcal{L}}}{\partial \eta_i \partial \nu_m \partial \nu_n} = {} & \left(\lambda_{nm} - \lambda_{mn}\right)\left(\delta_{mi} - \delta_{ni}\right) \\ & + \sum_{j=1}^{D} \nu_j \left[\delta_{mi}\left(a_{jn}^{S} - a_{mn}^{S}\right) + \delta_{ni}\left(a_{jm}^{S} - a_{nm}^{S}\right)\right] \\ & - \nu_i \left[a_{in}^{S} + a_{im}^{S} - 2a_{nm}^{S}\right].\end{aligned} \tag{4.142}$$

This is the form used numerically to evaluate the corrections to the stationary point in going from the naïve mean-field limit to the self-consistent rest points in figure 7.3.

4.6.3 Relations between field and action-angle variables

Only the diffusion kernels and noise source appear in the evaluation of fluctuations in section 4.4.2.1, and section 4.6.2 has shown that the diffusion kernels about classical backgrounds are the same in coherent-state and action-angle variables. Therefore, although the correlation function $\langle \nu\nu \rangle$ is formally fourth-order in $\phi^\dagger$ and ϕ, it differs only by the addition of noise-source terms. The exact relation between the Hessian forms (4.136) and (4.140) is

$$-\frac{\partial^2 \hat{\mathcal{L}}}{\partial \eta_i \partial \eta_j} = -\frac{1}{N^2}\frac{\partial^2 \mathcal{L}}{\partial \phi_i^\dagger \partial \phi_j^\dagger} - \left(\hat{\mu}_{ij}\nu_j + \hat{\mu}_{ji}\nu_i\right) + \nu_i \nu_j \sum_{k=1}^{D}\left(a_{ik}^S + a_{jk}^S - 2a_{ij}^S\right)\nu_k. \tag{4.143}$$

In most presentations the correlation function for coherent-state fields is used as a proxy for fluctuations in the number field [47, 58, 75]. The evaluations in (6.40) of chapter 6 show that this difference, near a critical point, is the same as the $O(\bar{R}^2)$ correction in passing from the uniform background to the small-radius limit cycle and is $O(1/N)$.

Bibliography

[1] Hofbauer J and Sigmund K 1998 *Evolutionary Games and Population Dynamics* (New York: Cambridge University Press)

[2] Hofbauer J and Sigmund K 2003 Evolutionary game dynamics *Bull. Am. Math. Soc.* **40** 479–519

[3] Nowak M A 2006 *Evolutionary Dynamics: Exploring The Equations of Life* (New York: Belknap)

[4] Wilson K G and Kogut J 1974 The renormalization group and the ε expansion *Phys. Rep.* **12** 75–199

[5] Coleman S 1985 *Aspects of Symmetry* (New York: Cambridge)

[6] Weiss P 1907 L'hypothèse du champ moléculaire et la propriété ferromagnétique *J. Phys. Theor. Appl.* **6** 661–90

[7] Hamilton W D 1964 The genetical evolution of social behavior: I *J. Theor. Biol.* **7** 1–16

[8] Hamilton W D 1964 The genetical evolution of social behavior: II *J. Theor. Biol.* **7** 17–52

[9] Gould S J 2002 *The Structure of Evolutionary Theory* (Cambridge, MA: Harvard University Press)

[10] Huang K 1987 *Statistical Mechanics* (New York: Wiley)

[11] Ellis R S 1985 *Entropy Large Deviations, and Statistical Mechanics* (New York: Springer)

[12] Jaynes E T 1957 Information theory and statistical mechanics *Phys. Rev.* **106** 620–30

[13] Jaynes E T 1957 Information theory and statistical mechanics: II *Phys. Rev.* **108** 171–90

[14] Jaynes E T 2003 *Probability Theory: The Logic of Science* (New York: Cambridge University Press)

[15] Simon H A 1962 The architecture of complexity *Proc. Am. Phil. Soc.* **106** 467–82

[16] Simon H A 1973 The organization of complex systems *Hierarchy Theory: The Challenge of Complex Systems* ed H H Pattee (New York: George Braziller) pp 3–27

[17] Gerhart J and Kirschner M 1997 *Cells, Embryos, and Evolution* (New York: Wiley)
[18] Ashby W R 1956 *An Introduction to Cybernetics* (London: Chapman and Hall)
[19] Conant R C and Ashby W R 1970 Every good regulator of a system must be a model of that system *Int. J. Syst. Sci.* **1** 89–97
[20] Ya Khintchine A 1937 A new derivation of a formula of Paul Lévy *Bull. Moscow State Univ.* **1** 1–5
[21] Ya Khintchine A 1937 Zur Theorie der unbeschränkt teilbaren Verteilungsgesetze *Mat. Sbornik* **44** 79–119
[22] Ken-iti Sato 2001 Basic results on Lévy processes *Lévy Processes: Theory and Applications* eds T Mikosch, O E Barndorff-Nielsen and S I Resnick (Boston: Birkh'auser) pp 3–37
[23] Coles S 2001 *An Introduction to Statistical Modeling of Extreme Values* (Heidelberg: Springer)
[24] Touchette H 2009 The large deviation approach to statistical mechanics *Phys. Rep.* **478** 1–69
[25] Smith E 2011 Large-deviation principles, stochastic effective actions, path entropies, and the structure and meaning of thermodynamic descriptions *Rep. Prog. Phys.* **74** 046601
[26] Fermi E 1956 *Thermodynamics* (New York: Dover)
[27] Benaim M and Weibull J W 2003 Deterministic approximation of stochastic evolution in games *Econometrica* **71** 873–903
[28] Champagnat N 2009 Large deviations for singular and degenerate diffusion models in adaptive evolution *Markov Proc. Relat. Fields* **15** 289–342
[29] t'Hooft G 1976 Computation of the quantum effects due to a four-dimensional pseudo-particle *Phys. Rev.* D **14** 3432–50
[30] Dykman M I, Mori E, Ross J and Hunt P M 1994 Large fluctuations and optimal paths in chemical kinetics *J. Chem. Phys.* **100** 5735–50
[31] Onsager L and Machlup S 1953 Fluctuations and irreversible processes *Phys. Rev.* **91** 1505
[32] Bertini L, De Sole A, Gabrielli D, Jona-Lasinio G and Landim C 2002 Macroscopic fluctuation theory for stationary non equilibrium states *J. Stat. Phys.* **107** 635–75
[33] Bertini L, De Sole A, Gabrielli D, Jona-Lasinio G and Landim C 2009 Towards a non-equilibrium thermodynamics: a self-contained macroscopic description of driven diffusive systems *J. Stat. Phys.* **135** 857–72
[34] Ma S-K 1976 *Modern Theory of Critical Phenomena* (New York: Perseus)
[35] Polchinski J G 1984 Renormalization group and effective lagrangians *Nucl. Phys.* B **231** 269–95
[36] Goldenfeld N 1992 *Lectures on Phase Transitions and the Renormalization Group* (Boulder, CO: Westview)
[37] Weinberg S 1996 *The quantum Theory of Fields* vol 2 (New York: Cambridge)
[38] Price G R 1972 Fisher's 'fundamental theorem' made clear. *Ann. Human Genetics* **36** 129–40
[39] Frank S A 1995 George Price's contributions to evolutionary genetics *J. Theor. Biol.* **175** 373–88
[40] Maynard Smith J and Price G R 1973 The logic of animal conflict *Nature* **246** 15–18
[41] Ewens W J 2004 *Mathematical Population Genetics* 2nd edn (Heidelberg: Springer)
[42] Nicolis G and Prigogine I 1971 Fluctuations in nonequilibrium systems *Proc. Natl Acad. Sci. USA* **68** 2102–7
[43] Kurtz T G 1981 *Approximation of Population Processes* Society for Industrial and Applied Mathematics, Philadelphia, PA, CMBS-NSF Regional Conference Series in Applied Mathematics vol 36

[44] Ethier S N and Kurtz T G 1986 *Markov Processes: Characterization and Convergence* (New York: Wiley)
[45] Feng J and Kurtz T G 2006 *Large Deviations for Stochastic Processes (Mathematical Surveys and Monographs* vol 131*)* (Providence, RI: American Mathematical Society)
[46] Eyink G L 1996 Action principle in nonequilibrium statistical dynamics *Phys. Rev.* E **54** 3419–35
[47] Mattis D C and Glasser M L 1998 The uses of quantum field theory in diffusion-limited reactions *Rev. Mod. Phys* **70** 979–1001
[48] Doi M 1976 Second quantization representation for classical many-particle system *J. Phys. A: Math. Gen.* **9** 1465–78
[49] Doi M 1976 Stochastic theory of diffusion-controlled reaction *J. Phys. A: Math. Gen.* **9** 1479
[50] Peliti L 1985 Path-integral approach to birth–death processes on a lattice *J. Physique* **46** 1469
[51] Peliti L 1986 Renormalization of fluctuation effects in $a + a \to a$ reaction *J. Phys. A: Math. Gen* **19** L365
[52] Matsubara T 1955 A new approach to quantum-statistical mechanics *Prog. Theor. Phys. (Kyoto)* **14** 351–78
[53] Schwinger J 1961 Brownian motion of a quantum oscillator *J. Math. Phys.* **2** 407–32
[54] Keldysh L V 1965 Diagram technique for nonequilibrium processes *Sov. Phys.—JETP* **20** 1018
[55] Freidlin M I and Wentzell A D 1998 *Random perturbations in dynamical systems* 2nd edn (New York: Springer)
[56] Smith E 2008 Quantum-classical correspondence principles for locally non-equilibrium driven systems *Phys. Rev.* E **77** 021109
[57] Martin P C, Siggia E D and Rose H A 1973 Statistical dynamics of classical systems *Phys. Rev.* A **8** 423–37
[58] Kamenev A Keldysh and doi–peliti techniques for out-of-equilibrium systems. pages arXiv: cond-mat/0109316v2, 2001. NATO School on Field Theory of Strongly Correlated Fermions and Bosons in Low-Dimensional Disordered Systems (Windsor, UK, August 2001).
[59] Carberry D M, Reid J C, Wang G M, Sevick E M, Searles D J and Evans D J 2004 Fluctuations and irreversibility: an experimental demonstration of a second-law-like theorem using a colloidal particle held in an optical trap *Phys. Rev. Lett.* **92** 140601
[60] Bodineau T and Derrida B 2004 Current fluctuations in non-equilibrium diffusive systems: an additivity principle *Phys. Rev. Lett.* **92** 180601
[61] Seifert U and Speck T 2010 Fluctuation-dissipation theorem in nonequilibrium steady states *Europhys. Lett.* **89** 10007
[62] van Kampen N G 2007 *Stochastic Processes in Physics and Chemistry* 3rd edn (Amsterdam: Elsevier)
[63] Boland R P, Galla T and McKane A J 2008 How limit cycles and quasi-cycles are related in systems with intrinsic noise *J. Phys. A: Math. Theor.* **41** 435003
[64] Boland R P, Galla T and McKane A J 2009 Limit cycles, complex Floquet multipliers, and intrinsic noise *Phys. Rev.* E **79** 051131
[65] Cardy J L 1987 Electron localisation in disordered systems and classical solutions in ginzburg-landau field theory *J. Phys. C: Solid State Phys.* **11** L321–8
[66] Graham R 1977 Path integral formulation of general diffusion processes *Z. Phys.* B **26** 281–90
[67] Graham R and Tél T 1984 Existence of a potential for dissipative dynamical systems *Phys. Rev. Lett.* **52** 9–12

[68] Jaynes E T 1980 The minimum entropy production principle *Annu. Rev. Phys. Chem.* **31** 579–601
[69] Ghosh K, Dill K A, Inamdar M M, Seitaridou E and Phillips R 2006 Teaching the principles of statistical dynamics *Am. J. Phys.* **74** 123–33
[70] Stock G, Chosh K and Dill K A 2009 Maximum caliber: a variational approach applied to two-state dynamics *J. Chem. Phys.* **128** 194192
[71] Wu D, Ghosh K, Inamdar M, Lee H J, Fraser S, Dill K and Phillips R 2009 Trajectory approach to two-state kinetics of single particles on sculpted energy landscapes *Phys. Rev. Lett.* **103** 050603
[72] Wilf H S 2006 *Generatingfunctionology* ed A K Peters 3rd edn (MA: Wellesley)
[73] Graham R and Tél T 1985 Weak-noise limit of Fokker–Planck models and nondifferential potentials for dissipative dynamical systems *Phys. Rev.* A **31** 1109–22
[74] Goldstein H, Poole C P and Safko J L 2001 *Classical Mechanics* 3rd edn (New York: Addison Wesley)
[75] Cardy J 1999 Field theory and non-equilibrium statistical mechanics http://www-thphys.physics.ox.ac.uk/users/JohnCardy/home.html
[76] Pilgram S, Jordan A N, Sukhorukov E V and Büttiker M 2003 Stochastic path integral formulation of full counting statistics *Phys. Rev. Lett.* **90** 206801
[77] Sinitsyn N A and Nemenman I 2007 Universal geometric theory of mesoscopic stochastic pumps and reversible ratchets *Phys. Rev. Lett.* **99** 220408
[78] Sinitsyn N A and Ohkubo J 2008 Hannay angle and geometric phase shifts under adiabatic parameter changes in classical dissipative systems *J. Phys. A: Math. Theor.* **41** 262002
[79] Sinitsyn N A 2009 The stochastic pump effect and geometric phases in dissipative and stochastic systems *J. Phys. A: Math. Theor.* **42** 193001
[80] Sinitsyn N A, Hengartner N and Nemenman I 2009 Adiabatic coarse-graining and simulations of stochastic biochemical networks *Proc. Natl Acad. Sci. USA* **106** 10546–51
[81] Landauer R 1962 Fluctuations in bistable tunnel diode circuits *J. Appl. Phys.* **33** 2209–16
[82] Maier R S and Stein D L 1993 Escape problem for irreversible systems *Phys. Rev.* E **48** 931–8
[83] Cover T M and Thomas J A 1991 *Elements of Information Theory* (New York: Wiley)
[84] Gluckheimer J and Holmes P 1988 Structurally stable heteroclinic cycles *Math. Proc. Cambridge Phil. Soc.* **103** 189–92
[85] Gluckheimer J and Holmes P 2002 *Nonlinear Oscillations, Dynamical Systems, and Bifurcations of Vector Fields (Applied Mathematical Sciences* vol 42*)* (Berlin: Springer)
[86] Maier R S and Stein D L 1993 Effect of focusing and caustics on exit phenomena in systems lacking detailed balance *Phys. Rev. Lett.* **71** 1783–6
[87] Maier R S and Stein D L 1996 Oscillatory behavior of the rate of escape through an unstable limit cycle *Phys. Rev. Lett.* **77** 4860–3
[88] Smith E, Farmer J D, Gillemot L and Krishnamurthy S 2003 Statistical theory of the continuous double auction *Quant Finance* **3** 481–514

Chapter 5

Discrete symmetries and emergent multiscale dynamics

Population models that break exchange symmetries among types by forming stationary non-symmetric population states create the simplest robust form of multiscale evolutionary dynamics. A coordination game played by three types of agents—asymmetric left (L) and right (R) types, and a symmetric middle type—illustrates this phenomenon. At large mutation rates, the population is symmetrically distributed over types, but at small mutation rates, the population spends almost all its time in one of two asymmetric states dominated by either L or R types. Individual dynamics in the short term is limited to small fluctuations about the average, no matter what symmetry the population state has. In the long term, in cases where asymmetric population states form, individual dynamics takes on an additional freedom from rare events of state switching by the population. This model illustrates the use of semiclassical trajectories, introduced in section 4.4.1, in the large-deviations formulae from chapter 4 to compute exponentially rare fluctuation events, both their rates and non-trivial escape trajectories. The coordination game also provides an illustration of Hamilton's rule for group selection through an elementary and transparent derivation.

5.1 Simple discrete symmetry breaking and the exponential separation of timescales

The form of symmetry breaking in population models that most closely resembles discrete phase transitions in equilibrium thermodynamics is breaking of a discrete permutation symmetry among agent types to a fixed number of stationary rest points. A symmetry of the underlying fitness function is broken when the population settles on one of the rest points, but the eligible rest points must transform under a representation of the hidden symmetry. When the population is in one of its asymmetric states, the distribution of agents remains within small fluctuations of a fixed

doi:10.1088/978-0-7503-1137-3ch5

composition determined by the mutation–selection balance. The variance of typical fluctuations in the fractional agent distribution is $O(1/N)$. Fluctuations by fractions of $O(N^0)$ are exponentially rare in N. These comprise the rare large excursions of the population from its average composition. A sub-set of these excursions lead to escapes of the population from the basin of attraction of its most recently occupied rest point, following which it may settle into a different rest point. The symmetry of the population rest points implies that the capacity for rare large excursions persists in populations of any size. However, the characteristic timescale for these, which are the only residual dynamics at the population scale, increases exponentially in N relative to the scale of individual moves. For large N, such separations of scale are costly to capture in simulations, but they are typical of the separation between molecular and macroscopic timescales in organisms [1, 2], or between individual and group dynamics (where N may not be as large) in social systems.

We illustrate this class of phenomena with a minimal model of a coordination game with three types, which breaks an exchange symmetry between two of the types that may be visualized as a reflection. In the classical replicator equation, the breaking of a single stable population state to a pair of symmetric rest points takes the form of a pitchfork bifurcation. In the stochastic process it is a non-equilibrium second-order phase transition. The part of the large-deviations expression that controls the escapes between basins is $S^{(0)}$ in (4.86) and we show how to compute the most likely trajectory as well as the rate of escapes.

It is a feature of the stochastic model with linear frequency-dependent selection that a coordination game played by only two agent types will not break reflection symmetry; a third type is needed. In this example the third type cannot benefit by coordination with either of the two asymmetric types. A natural gene description thus arises which partitions the types capable of coordinating from the type that cannot. The ability to form asymmetric states leads to a form of relatedness among those who can coordinate, in a derivation of Hamilton's rule from the Price equation (2.9) following the exposition by Steven Frank [3]. The relaxation to the long-run dynamics of populations that spend almost all their time at the asymmetric rest points with intermittent switching provides an example of group selection of the coordinators over the non-coordinators. The existence of two coordinated states rather than one is essential in this construction, because it isolates relatedness as the source of the fitness advantage.

Finally, we show how discrete symmetry breaking in evolutionary games has a representation in terms of potential functions similar to the energy potentials of equilibrium thermodynamics.

5.1.1 Breaking of reflection symmetry in a coordination game

Consider a population in which agents may have types Left (L), Right (R) and Middle (M). We write the instantaneous population state as a column vector

$$\mathrm{n} \equiv \begin{bmatrix} \mathrm{n_L} \\ \mathrm{n_R} \\ \mathrm{n_M} \end{bmatrix}. \tag{5.1}$$

With respect to this vector, the fitness is given by the matrix

$$[a] = \bar{a}\begin{bmatrix}1\\1\\1\end{bmatrix}\begin{bmatrix}1 & 1 & 1\end{bmatrix} + \begin{bmatrix}a & -a & 0\\-a & a & 0\\0 & 0 & 0\end{bmatrix}. \tag{5.2}$$

This payoff matrix is uniquely defined by the reflection symmetry between L and R, equal payoff to M from all types and the existence of a uniform-population Nash equilibrium in all dynamical regimes in the infinite-population limit. The parameters $\bar{a}$ and a are otherwise arbitrary and the qualitative behavior of the model is associated with broad parameter ranges.

The mean-field replicator equation with payoff and mutation, first introduced in (2.15), is

$$\frac{\mathrm{d}\bar{n}_i}{\mathrm{d}t} = \bar{n}_i\left(f_i - \phi\right) + \sum_{j=1}^{3}\mu_{ij}\bar{n}_j. \tag{5.3}$$

In the solutions below, we will use the fitness function (2.16) that excludes terms describing self-play. These lend corrections at $O(1/N)$ to the naïve infinite-population mean-field limit that was our guide in choosing the structure of the game. (Recall also from (2.6) that $\mu_{ij} = 1 - D\delta_{ij}$, where the number of types $D = 3$ here.) The parameter combinations that control mean-field dynamics in the leading large-N limit are aN and D. A 'classical' population dynamics with well-defined mean behavior in the limit of zero fluctuation corresponds to the equivalence class of models with aN and D held fixed as $N \to \infty$.

The long-term classical solutions to the coordination game are rest points, where all $\mathrm{d}\bar{n}_i/\mathrm{d}t = 0$. At all of these rest points except the critical point for bifurcations (which has parameter-measure zero), the gradients $\partial/\partial\bar{n}_j(\mathrm{d}\bar{n}_i/\mathrm{d}t) \neq 0$, $\forall\, i, j$ at all N. Therefore convergence to the rest points is exponential in time. Since fluctuations have leading order $O(1/N)$, non-vanishing mean regression of fluctuations away from the rest point ensures that models of this form converge to their mean-field approximations (MFAs) in the large-N limit.

For sufficiently weak coordination payoff aN (weak selection) in relation to the (D-dependent but N-independent) randomizing effects of mutation, the model has a single, L/R-symmetric evolutionary stable state (ESS) [4] that approaches the uniform population composition at small aN/D. With increase of aN through a critical value, the system undergoes a pitchfork bifurcation. A symmetric population near the uniform distribution remains a rest point but becomes a saddle point, while a pair of left and right rest points emerge as the ESSs, which favor populations of mostly L or mostly R. As $1/aN \to 0$ the population comes to reside asymptotically in either an all L or an all R state. This pair of asymmetric ESSs forms the *stochastically stable set* of [5, 6].

The mean-field phase diagram of the model is most easily obtained by expressing the replicator equations in terms of the L/R antisymmetric variable $(\bar{n}_{\mathrm{R}} - \bar{n}_{\mathrm{L}})/N$ and the symmetric combination $(\bar{n}_{\mathrm{R}} + \bar{n}_{\mathrm{L}})/N = 1 - \bar{n}_{\mathrm{M}}/N$.

The equilibrium equation for the symmetric component is

$$\left[D + aN\left(\frac{\bar{n}_{\mathrm{R}} - \bar{n}_{\mathrm{L}}}{N}\right)^2\right]\frac{n_{\mathrm{M}}}{N} - 1 = a\left(1 - \frac{n_{\mathrm{M}}}{N}\right)\frac{n_{\mathrm{M}}}{N}. \tag{5.4}$$

(Here we have grouped terms that govern the leading scaling in the large-population limit on the left-hand side and corrections of $O(1/N)$ on the right-hand side. Recall that the primary scaling variable is aN, so relative to this scaling we regard $a \sim 1/N$.) The equilibrium condition for the antisymmetric component has two branches: either $(\bar{n}_{\mathrm{R}} - \bar{n}_{\mathrm{L}})/N = 0$ in symmetric phases, or

$$aN\left[\left(1 - \frac{n_{\mathrm{M}}}{N}\right) - \left(\frac{\bar{n}_{\mathrm{R}} - \bar{n}_{\mathrm{L}}}{N}\right)^2\right] - D = a\frac{n_{\mathrm{M}}}{N} \tag{5.5}$$

in asymmetric phases, if (5.5) admits non-negative solutions for $(\bar{n}_{\mathrm{R}} - \bar{n}_{\mathrm{L}})^2$. Canceling the terms in $(\bar{n}_{\mathrm{R}} - \bar{n}_{\mathrm{L}})^2$ between (5.4) and (5.5), we obtain the equilibrium condition for $\bar{n}_{\mathrm{M}}/N$ in asymmetric phases, as

$$aN\left(1 - \frac{n_{\mathrm{M}}}{N}\right)\frac{n_{\mathrm{M}}}{N} - 1 = a\frac{n_{\mathrm{M}}}{N}. \tag{5.6}$$

The solution to (5.4), giving the symmetric equilibrium point, is

$$\frac{n_{\mathrm{M}}}{N} = \frac{1}{2}\left(\frac{D - a}{a}\right)\left\{\sqrt{1 + \frac{4a}{(D - a)^2}} - 1\right\}. \tag{5.7}$$

At large N, this approaches $n_{\mathrm{M}}/N \approx (1/D)[1 + a(D - 1)/D^2]$. (We will see that this is the same as $n_{\mathrm{M}}/N \approx 1 + 1/N_{\mathrm{crit}}$ for the critical value of N.) The solution to (5.6) for the asymmetric phases is

$$\frac{n_{\mathrm{M}}}{N} = \frac{1}{2}\left(\frac{N - 1}{N}\right)\left\{1 - \sqrt{1 - \frac{4N}{a(N - 1)^2}}\right\}. \tag{5.8}$$

The critical point N_{crit} for any a is the value of N for which the asymmetric phases coalesce and the solutions (5.7) and (5.8) are equal. At leading order in $1/N$, the solution is approximately $aN_{\mathrm{crit}} \approx D^2/(D - 1) = 9/2$. Substituting (5.8) back into (5.5) then gives the value of $(\bar{n}_{\mathrm{R}} - \bar{n}_{\mathrm{L}})^2/N^2$.

Figure 5.1 shows dynamical trajectories solving (5.3) at $aN = 8$ and $N \to \infty$. (The actual parameters we will use in simulations are $a = 0.5$ and $N = 16$.) The horizontal axis is $(\bar{n}_{\mathrm{R}} - \bar{n}_{\mathrm{L}})/N$ and the vertical axis is $\bar{n}_{\mathrm{M}}/N$, normalized to orthonormal coordinates. (We provide algebraic forms for the corresponding basis elements in $(\bar{n}_{\mathrm{L}}/N, \bar{n}_{\mathrm{R}}/N, \bar{n}_{\mathrm{M}}/N)$ coordinates in section 5.4.1.) Since for this model $aN_{\mathrm{crit}} = 4.5$, (5.3) has asymmetric stable solutions. In this regime, the deterministic solutions from arbitrary initial conditions flow rapidly toward an arc running through the uniform distribution and the ESS, shown as the light red background band in the figure.

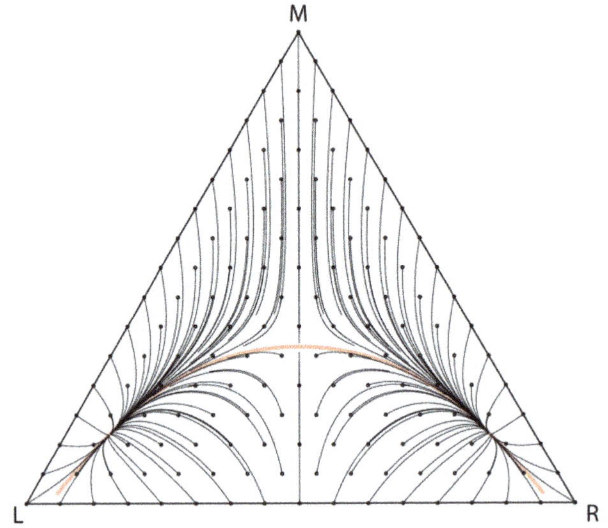

Figure 5.1. Discrete symmetry breaking in the normal-form coordination game (5.2), at $aN = 8$ and $N \to \infty$. The simplex of constant N is shown, with the populations of fixed types (L,R,M) indicated at the vertices. Dots represent arbitrary initial values of $\bar{n}$ in the MFA, and lines are the solutions of the mean-field evolutionary game equation (5.3) from those initial conditions. The light red band is the contour of slowest relaxation, which passes through the three fixed points.

Diffusion is then slowest along this arc, with both departure from the saddle point as well as the approach to the stable points being exponential in time with a rate constant linear in N.

5.1.2 The large-deviations scaling limit and multiscale dynamics

The pitchfork bifurcation in the coordination game—the point at which one stable solution splits into a pair of stable solutions and an unstable solution—is dominated near the critical point by large fluctuations. They become possible because one direction of the gradient $\partial/\partial\bar{n}_j(\mathrm{d}\bar{n}_i/\mathrm{d}t)$ develops a zero eigenvalue; fluctuations in that direction do not regress even in the large-N limit. The qualitative behaviors of fluctuations about the symmetric phase in the coordination game and in the rock–paper–scissors game with a Hopf bifurcation, developed in chapter 6, are similar. (For the coordination game this similarity applies in the bifurcated phase as well.) Therefore we mention only a few qualitative features of interest in section 5.3 and defer a more detailed treatment of Gaussian-order fluctuations to chapter 6.

In this section we will focus on a feature that no deterministic replicator can capture and which even expansions of fluctuation effects in a power series in $1/N$ cannot systematically approximate. This is the escape from either ESS to the unstable state, permitting subsequent re-exploration of the configuration space.

Such rare fluctuations remain possible in populations of arbitrarily large N because of the symmetry between L and R states in this game. Precisely because a L/R symmetry of the payoff function (5.2) is hidden by convergence to either individual ESS its presence must be expressed in the strict equivalence between the two mirror-image states. Because either ESS itself has no time dependence, macroscopic aggregate dynamics is limited to events of escape from one ESS by rare sequences of cumulative single-agent changes that evade correction by selection and subsequent (ordinary) evolution toward the other ESS. A problem of central interest for systems with broken discrete symmetry is the computation of the most likely trajectory for escapes and the probability with which these events occur.

This problem comprises two distinct large-deviations formulae. First, escapes are rare among the set of fluctuations and are governed by aggregate-event probabilities of the form (4.1). However, in *conditional* probabilities, conditioned on a trajectory's being in more than one ESS within some finite time interval, escape events make the leading contribution[1]. Within these conditional distributions, a single trajectory (unique except for uncertainty in when it occurs) dominates the probability and deviations from it are exponentially rare, resulting in a second large-deviations formula in which the rate function is a functional of histories. Even for relatively simple systems the dominant escape trajectory and probability can be somewhat difficult to compute and the difficulty grows rapidly with increasing complexity of the evolutionary game dynamics. Because successful escapes are rare, they are also costly to simulate. Therefore systematic methods of computation that give both trajectories and their probabilities within a unified framework, which the Doi–Peliti construction provides, are very useful.

Whole-population escapes are also the events that return a density initially concentrated about a single ESS—which break symmetry in the initial conditions—back to the ergodic symmetric distribution at late times.

Escape events will be described by an application of the 'non-classical' stationary trajectories solving the equations of motion (4.87)–(4.90) from chapter 4. Their probabilities are governed by non-zero values of the classical action $S^{(0)}$ of (4.86). We begin in section 5.2 with the computation and interpretation of the stationary paths of the Doi–Peliti action and their associated Hamiltonian dynamical system. We return in section 5.3 to a discussion of the role of weak and strong fluctuations in relation to escapes. In section 5.4 we use the property that escapes are (*ex post*) infrequent with finite, asymptotically N-independent support in time, to form an ensemble description of population states in the coordination game and to illustrate the emergence of Hamilton's rule and the interpretation of aggregate-population dynamics as an emergent dynamics of coalitions.

5.2 Hamiltonians and instanton methods for systems with multiple discrete equilibria

The computation of instanton trajectories and probabilities for irreversible stochastic processes has both important similarities to, and important differences from, the corresponding computation for the phenomena of barrier penetration in equilibrium

[1] The next sub-leading contribution again comes from small Gaussian-order fluctuations, though now about backgrounds that include occasional large excursions.

thermal or quantum systems. These concern the role of saddle points in relation to stable rest points and the asymmetry of probabilities for relaxation and escape. They also lead to a 'kinematic' description of the stochastic process which is different from its canonical description. It is the kinematic description that provides the most direct comparison of potentials and symmetries to the equilibrium problem.

We begin by reviewing the familiar role of potentials in equilibrium symmetry breaking and discuss the different roles that potentials play for non-equilibrium stochastic processes. We then consider the structure of the dynamical system defined by the Liouville operator and finally the way this can be used to efficiently compute trajectories and probabilities for escape events.

5.2.1 The role of potentials in representing broken symmetries

The symmetries of potential functions were understood to govern the properties of motion in mechanics with the work of Lagrange and Hamilton [7]. The idea of an *effective potential* defined phenomenologically from symmetries was introduced by Landau [8, 9] as a general model for the Gibbs free energies responsible for phase transitions. Over time it became understood that the energy scale dependence of the effective potential came from absorbing fluctuation effects into average parameters [10]. The *free energy landscape* of thermodynamics is just such an effective potential [11], in which the entropy component comprises the fluctuation contributions. For this reason, it has become common to convey the picture of symmetry breaking in equilibrium systems in terms of symmetries of an effective potential.

Spontaneous symmetry breaking by pitchfork bifurcation in equilibrium thermodynamics results from effective potentials of the form shown in figure 5.2. The curvature of the wells at their minima determines the strength with which fluctuations regress toward the mean population state and thus the fluctuation variance. The depth of the wells determines the large-deviations rate function $\hat{S}^{(0)}$ in expansions of the form (4.86) for escapes from one well to the other. The energy dependence of fixed points makes a pitchfork because the center is a fixed point in all phases; in the

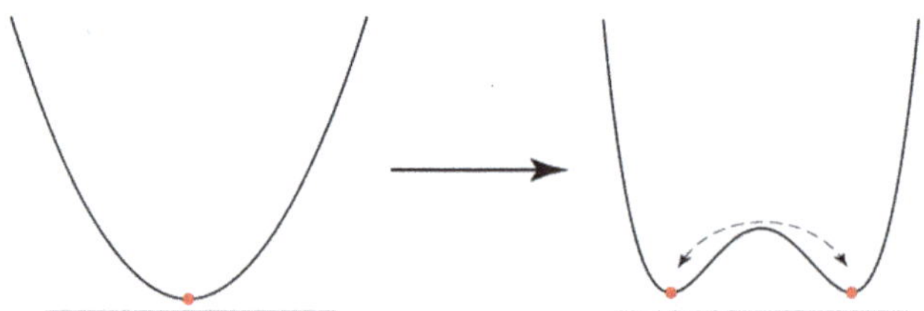

Figure 5.2. In equilibrium, the minima of a potential determine the rest points of dynamics (red dots), the gradients of the potential in neighborhoods of its minima determine rates of mean regression and the barrier of the potential between wells determines the leading exponential dependence of escape rates. Phase transitions are described by changes of the potential from the convex form shown on the left to the non-convex form on the right, as an interaction strength is increased. As the wells of an equilibrium potential deepen, the fluctuations in population states are suppressed [12] and the dynamics becomes increasingly well described by a two-state, continuous-time Markov process, suggested by the dashed line with arrows.

symmetric phase it is stable, while in the symmetry-broken phase it becomes unstable and a pair of uncentered stable fixed points emerges.

Since the depth and curvatures of the two wells of $N\hat{S}^{(0)}$ are both linear in N, regression rates increase linearly with N and the escape times increase exponentially in N. However, as noted in section 5.1.2, the underlying reflection symmetry ensures that the well depths are equal and so the ratio of occupation frequencies in the two wells does *not* grow exponentially with N,[2] allowing population-scale dynamics at any N.[3]

5.2.1.1 Uses of potentials in non-equilibrium stochastic processes

The term 'potential' can be applied in at least four ways to stochastic evolutionary games. Only one of these preserves the feature of the potential energy in mechanics, that its local minima are fixed points of the dynamical system, which transform under a representation of the symmetries of population states. We will briefly summarize here the meanings of the four uses and their relations to each other and then return to construct the kinematic potential that characterizes symmetry breaking in the following section.

The most common usage refers to *potential games* [15], for which the vectors of population change (2.15) may be written as the gradient of a scalar 'potential' function of the state $\bar{n}$. Non-equilibrium stochastic processes do not generally admit such potential descriptions, so the potential games are a restricted class. The results presented here do not depend on any of the simplifications provided by potential games, so we do not assume these restrictions.

Two generally applicable notions of potential were introduced through the work of Freidlin and Wentzell [16], which apply also to the Doi–Peliti construction used here. They correspond formally to the mechanical Hamiltonian and the thermodynamic potentials in equilibrium systems and they always reflect the symmetries of population states. We will introduce them starting with the thermodynamic analog, in which the mechanical analogue appears as one of the terms.

A quantity known as the *quasipotential* [17–21] arises in Friedlin–Wentzell theory as the rate function in a large-deviation principle [22, 23] for fluctuation probabilities at the macroscale. It is so-named because it is a non-equilibrium generalization of the thermodynamic potentials such as the entropy or free energies[4]. The quasipotential is constructed as a functional Legendre transform of the cumulant-generating

[2] For the symmetric models in this chapter, the ratio of occupation frequencies in the two wells is just unity. More general models can preserve degenerate minima but create asymmetric basins of attraction. The resulting occupation frequencies would not grow or shrink exponentially at large N, but could take values other than unity.

[3] The absence of such a protecting symmetry was used in [5, 6, 13, 14], along with a selection-strength relaxation protocol analogous to simulated annealing in [14], to provide a further equilibrium refinement beyond the ESS. This equilibrium refinement works because, generically, all but the lowest effective minimum becomes depopulated. The effective minimum, however, depends on non-local properties of the basin of attraction and therefore requires the inclusion of fluctuation corrections to compute reliably.

[4] A slight variant on the quasipotential is constructed by Young [6], which counts events of improbable movement rather than accumulating their log-likelihoods.

functional in the Doi–Peliti construction—often it will be approximated to leading order by the action functional S from (4.68), evaluated along a particular trajectory.

Within that action functional, the Liouville operator fills the role of a Hamiltonian in the construction of section 4.3. This Liouville–Hamiltonian is like an energy function at the classical level; it is conserved in systems without explicit time dependence, and provides the symplectic structure on the space of stationary solutions. More generally, the Liouville function of fields appears in Gibbs exponential factors determining the probabilities of histories, in the functional integral (4.64). However, due to important differences that arise from irreversibility, this Hamiltonian generally does not carry the interpretation of the mechanical potential responsible for trapping and escapes from basins of attraction[5] in thermodynamics.

It is also important not to mistake the formal correspondence of the Liouville operator to a Hamiltonian for an implied interpretation in terms of true mechanical energy. The conjugate momentum in the Freidlin–Wentzell Hamiltonian system has an interpretation similar to the source terms in a generating function that measure the improbability of deviations from the classical behavior. This formally defined canonical momentum is therefore related to the inference of most likely paths leading to fluctuations and is more like a chemical potential than like a mechanical momentum[6].

The last case is our 'kinematic' use of the term potential, introduced in section 4.4.1.2. For games with discrete symmetry that may be hidden by strong selection, the Doi–Peliti dynamical system represents the population state as a particle moving in an energy potential with an additional velocity-dependent force [11]. The fixed points of the energy potential (having zero velocity) are also the rest points of the dynamical system and so are rest points of the evolutionary dynamics. For systems with point-group symmetry, the local minima of the kinematic potential furnish a representation of the broken symmetries. The kinematic potential defined in (4.93) is shown in figure 5.3.

We now consider the consequences of conservation laws implied by symmetries of the Doi–Peliti field theory for an action with a potential of this form.

5.2.2 Conservation laws and non-classical stationary points

The expression of the large-deviation properties of complex stochastic processes in terms of a deterministic Hamiltonian greatly simplifies calculations of complex rare events. However, the heteroclinic network of a general dynamical system may still be complicated to extract. The heteroclinic orbits are saddle points, so that numerical integration along them (with real η and real ν) is generally unstable. Therefore it is useful to further restrict the space of possible solutions using the conservation laws of the Hamiltonian system. Doing so leads to a topological

[5] For a discussion of equilibrium potentials and escapes, see [24], chapter 7.

[6] For a systematic description of the meaning of position and momentum coordinates in Freidlin–Wentzell theory with respect to inference, see [11].

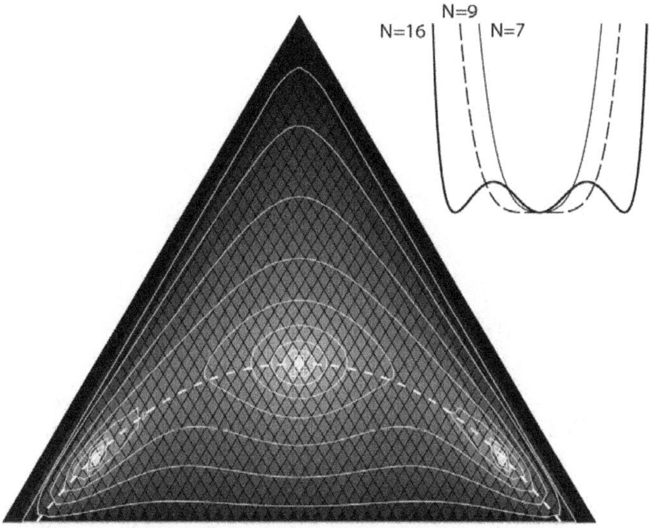

Figure 5.3. The 'kinematic' potential $V(\nu)$ from (4.93), for the coordination game at the parameters shown in figure 5.1. The greyscale and contours are uniform in log $V(\nu)$. The dashed white line is drawn to follow the saddle of $V(\nu)$ and closely approximates the direction of slowest flow in the diffusive field of figure 5.1. It also approximates the most probable trajectory for escapes. (Inset) Linear plot of the potential along the dashed white contour at three values of N: $N = 7$ stable uniform rest point, $N = 9$ the classical critical value for phase transition and $N = 16$ the ordered phase of figure 5.1. The symmetry breaking of the potential resembles that shown in figure 5.2 for equilibrium systems, except that in the ordered regime both the saddle point and the stable rest points are zeros.

characterization of the heteroclinic network and provides efficient steepest-descent algorithms to identify the orbits.

For Markovian systems of the kind we have constructed, two continuous symmetries always exist, which imply the existence of two conserved quantities by *Noether's theorem* [7]. First, in the Liouville operator (4.92) only differences $\eta_j - \eta_l$ arise, implying that global shifts of all η_i by a constant is a symmetry of the Hamiltonian. These shifts affect the diagonal component of η, $(1/D)\sum_{i=1}^{D}\eta_i$, which appears in (4.84) for S only in the term $\sum_{i=1}^{D}\eta_i \mathrm{d}\nu_i/\mathrm{d}\tau$. The conjugate Noether charge, conserved along all stationary paths, is precisely the total number of agents[7]

$$\frac{\mathrm{d}}{\mathrm{d}\tau}\left(\sum_{i=1}^{D}\nu_i\right) = 0. \tag{5.9}$$

[7] Because this symmetry was built in from the beginning, we have factored out the scale parameter N to define the field ν. Equation (5.9) verifies that this rescaling step is consistent with the dynamics. In a pedantic treatment we could have worked in the original fields n, in which case the conserved quantity would have been $\sum_i n_i$. It is important to understand that the conservation law (5.9) reflects a non-trivial aspect of the dynamics and is not the same as tautological conservation of the sum of normalized population fractions built into the deterministic replicator equation.

The other global symmetry is any shift of τ by a constant, reflecting the stationarity of the transition probabilities in the Markov process. The conservation law implied by this symmetry is of the Liouville operator itself,

$$\frac{\mathrm{d}\hat{\mathcal{L}}}{\mathrm{d}\tau} = 0, \tag{5.10}$$

which follows also from the equations of motion (4.89, 4.90) as[8]

$$\frac{\mathrm{d}\bar{\hat{\mathcal{L}}}}{\mathrm{d}\tau} = \sum_{i=1}^{D}\left(\overline{\frac{\partial\hat{\mathcal{L}}}{\partial\nu_i}}\frac{\mathrm{d}\bar{\nu}_i}{\mathrm{d}\tau} + \overline{\frac{\partial\hat{\mathcal{L}}}{\partial\eta_i}}\frac{\mathrm{d}\bar{\eta}_i}{\mathrm{d}\tau}\right). \tag{5.11}$$

We may use these conservation laws to restrict the manifold of possible stationary paths as follows: We have observed that all classical stationary paths satisfy $\eta \equiv 0 \Rightarrow \hat{\mathcal{L}} = 0$ and that these paths terminate in the stable set of the replicator dynamic. It is far from apparent that all solutions to the stationary-path conditions actually have $\hat{\mathcal{L}} = 0$, but that result may be shown as follows. Although the set of stationary paths consistent with late-time equilibrium is, strictly speaking, singular in the ν simplex, actual trajectories involving escapes merely pass *arbitrarily near* the stable set, for arbitrarily long time intervals [24]. Thus they are formally continuous with the classical stationary paths, so the conservation relation (5.11) implies, starting from the classical solutions, that

$$\hat{\mathcal{L}} = 0 \tag{5.12}$$

at all times.

For population processes in which the Liouville operator takes the form (4.85), the topology of the $\hat{\mathcal{L}} = 0$ set can be seen by itself to determine the start and end points of trajectories that are eligible to describe escapes. To see this, recall first the equivalent representation (4.92) for $\hat{\mathcal{L}}$ from section 4.4.1.2. The second line, giving the quadratic-order approximation in η, provides intuition for the parabolic curvature of the landscape of $\hat{\mathcal{L}}$ which is also a property of the exact form due to the convexity of the cosh function. The set of solutions $\hat{\mathcal{L}} = 0$ at any $\bar{\nu}$ is topologically a $(D - 2)$ sphere (here, a circle) in η, centered approximately at η_0. All stationary paths, whether classical or non-classical, lie within a fiber bundle of $(D - 2)$ spheres on the base space $\sum_i \nu_i \equiv 1$.

The velocity field η_0 reflects the imbalance between influx and efflux rates r_{ij}, introduced in (4.91), of any agent type. $\eta = 0$ solutions have flowfields $\mathrm{d}\bar{\nu}/\mathrm{d}t$ driven only by these fluxes. Escape solutions, where $\mathrm{d}\bar{\nu}/\mathrm{d}t$ at a given $\bar{\nu}$ approximately

[8] We present both conservation laws as properties of the deterministic paths of the Hamiltonian dynamical system. In general, when such paths arise as stationary points within a functional integral, they reflect a larger symmetry associated with shifts in the dummy variables of integration. (These dummy variables are the fields $(\phi^\dagger\phi)$ or (η, ν).) The expression of the symmetry by shifts of variables of integration, in the correlation functions of the theory, are known as its *Ward identities* [25].

reverses the classical direction, require $\eta \approx 2\eta_0$, expressing the fact that a generating function which samples only a small bundle of trajectories on the shoulder of the main distribution dominates the probability for escapes. Any escape trajectories must, however, have both segments with $\eta = 0$ and segments with $\eta \approx 2\eta_0$, by symmetry of (4.92) under a change of sign in the argument of the second cosh function that depends on η. For flips between steady states that punctuate otherwise normal periods of classical stasis, the shift in the value of η must occur in both directions. The only places these two opposite extrema are joined are at the rest points, where $\eta_0 = 0$ and the fiber bundle pinches off. Therefore all escape trajectories thread through these points. We will use this property of the $\hat{\mathcal{L}} = 0$ manifold in section 5.2.4 to efficiently converge to the heteroclinic orbits by gradient descent[9].

5.2.3 Clarifying the main structures and dynamical interpretations using a quadratic approximation to the Liouville operator

It is easiest to understand the consequences of conservation laws for escape trajectories, the role of the kinematic potential and also a variety of other basic structural characteristics of Doi–Peliti actions, by treating the quadratic expansion in $\eta - \eta_0$ to the Liouville operator of the form (4.92) in section 4.4.1.2 as an exact form. This amounts to replacing the actual model by an approximate but still internally consistent model, in which fluctuations in η may always be removed by Gaussian integration.

For such quadratic forms, we may remove the explicit index notation and use a condensed inner product notation for vectors and matrices. We will let $\eta^T m^{-1} \eta$ stand for $\sum_{i,j=1}^{D} \eta_i m_{ij}^{-1} \eta_j$, $\eta \cdot \dot{\nu}$ stand for $\eta^T \dot{\nu}$, and we will use overdot to denote $\mathrm{d}/\mathrm{d}\tau$. With these conventions, the starting form for the action $\hat{S}$ in (4.84) is written

$$\hat{S} = \int \mathrm{d}\tau \left(\eta \cdot \dot{\nu} + \hat{\mathcal{L}} \right). \tag{5.13}$$

Treating the second-order expansion of (4.92) as exact, the Liouville functional $\hat{\mathcal{L}}$ in (5.13) becomes

$$\hat{\mathcal{L}} = -\frac{1}{2}(\eta - \eta_0)^T m^{-1} (\eta - \eta_0) + V(\nu). \tag{5.14}$$

[9] The complex set-up procedure for the Doi–Peliti construction, reviewed in chapter 4, can obscure the fact that the construction of the stationary-path conditions is in the end fairly straightforward. Dykman [26] illustrates such a construction with relatively low overhead for models of stochastic chemical reaction networks. In one-dimensional systems the entire solution is therefore elementary, but in higher dimensions the solution for the stationary paths remains a potentially difficult problem even if the equations are arrived at simply. For these more complicated cases the gradient-descent method in section 5.2.4 provides an efficient alternative to direct numerical integration.

Although this form is quadratic in η, we permit it to retain arbitrary dependence on ν. For the quadratic action, requiring $\hat{\mathcal{L}}(\eta \equiv 0, \nu) = 0$ replaces the approximation (4.93) with an equality:

$$V(\nu) = \frac{1}{2}\eta_0^T m^{-1}\eta_0. \tag{5.15}$$

Recall η_0 and m^{-1} are defined by (4.93) and (4.94). The potential $V(\nu)$ is bounded below by zero and has local minima at rest points given by $\eta_0 \equiv 0$. This is the 'trapping' potential that corresponds to the ordinary kinematic potential in equilibrium thermodynamics. When $(\eta - \eta_0)$ is integrated along a convergent contour, the lower bound on V will also be responsible for the non-negative spectrum of the Liouville operator and convergence of the density ρ_n to a stationary distribution.

From the equilibrium treatment of instantons [24], we expect that escapes will be described by the heteroclinic orbits of a configuration ν moving on potential $-V(\nu)$. To see how this description arises, we expand $\hat{S}$ as a bilinear form using three different groupings of terms:

$$\begin{aligned}\hat{S} &= \int d\tau\left[\eta \cdot \dot{\nu} - \frac{1}{2}(\eta - \eta_0)^T m^{-1}(\eta - \eta_0) + V(\nu)\right] \\ &= \int d\tau\left[-\frac{1}{2}\left(\eta - \eta_0 - m_\perp\dot{\nu}\right)^T m^{-1}\left(\eta - \eta_0 - m_\perp\dot{\nu}\right) + \frac{1}{2}\dot{\nu}^T m_\perp\dot{\nu} + V(\nu) + \eta_0 \cdot \dot{\nu}\right] \\ &= \int d\tau\frac{1}{2}\left[-\left(\eta - \eta_0 - m_\perp\dot{\nu}\right)^T m^{-1}\left(\eta - \eta_0 - m_\perp\dot{\nu}\right) + \left(\dot{\nu} + \eta_0 m^{-1}\right)^T m_\perp\left(\dot{\nu} + m^{-1}\eta_0\right)\right].\end{aligned} \tag{5.16}$$

In order to invert the matrix m^{-1} in (5.16), it has first been necessary to remove the zero-eigenvalue associated with shifts of all η_i by a constant, mentioned in section 5.2.2. The projection of m^{-1} onto the $(D - 1)$-dimensional hyperplane $\sum_{i=1}^{D}\eta_i = 0$ is positive-definite, and $m_\perp$ is its matrix inverse.

Using the forms (5.16), we can quickly summarize the various equivalent dynamical representations the Doi–Peliti action can take, as well as justifying the main steps in evaluation of the path integral (4.66). The first line in (5.16) is the standard Lagrange–Hamilton definition of the action with canonical momentum η and a Hamiltonian equal to the Liouville operator, which has non-canonical dependence on η and potential $-V(\nu)$.

It is clear that integrating $\eta - \eta_0 - m_\perp\dot{\nu}$ along a real contour in the second line of (5.16) produces a divergence and that $\eta - \eta_0 - m_\perp\dot{\nu}$ must be rotated to an imaginary contour to define the Gaussian integral by steepest descent. After such an integration, the remaining terms on the second line give the *Lagrangian* for potential $-V(\nu)$. This Lagrangian corresponds to the so-called *Euclidean action* in the equilibrium method of instantons [24]. For an equilibrium barrier-crossing problem, η_0 would equal zero. The real-valued offset it creates for η and the final expression $\eta_0 \cdot \dot{\nu}$

that it adds in the second line of (5.16) will lead to different probabilities for relaxation and escape trajectories, despite the fact that the trajectories will be nearly time-reverses of one another, as we will see below.

It is necessary that escape probabilities have these features in order to be consistent with the difference between the potentials for equilibrium escapes shown in figure 5.2 versus non-equilibrium escapes shown in figure 5.3. A velocity-dependent force must exist relative to the 'kinematic' momentum variables in order that the classical diffusion solution be unpenalized, but this same force must then double the probability penalty for trajectories that reverse the classical solution.

The third line in (5.16), after integration over $\eta - \eta_0 - m_\perp \dot{\nu}$, gives the stochastic action first derived by Onsager and Machlup [27]. It is the small-fluctuation approximation to the large-deviations rate function for histories [11].[10]

The stationary-path condition for η from the second line of (5.16),

$$m_\perp \dot{\nu} = \eta - \eta_0, \tag{5.17}$$

verifies that $\eta - \eta_0$ is the kinematic momentum, and $m_\perp$ the mass matrix relating velocities $\dot{\nu}$ to momenta. In particular, the classical stationary-path trajectories (for which $\eta \equiv 0$) express the mean-field replicator dynamic through the relation

$$m_\perp \dot{\nu}_{\mathrm{Cl}} = -\eta_0. \tag{5.18}$$

Thus we recognize $-m^{-1}\eta_0$ as the tangent vector to the classical flowfield at each ν.

To see that the term $\eta_0 \cdot \dot{\nu}$ in (5.16)—a formal analog to a velocity-dependent energy term that arises from the macroscopic irreversibility of diffusion equations—does not qualitatively alter the kinematic interpretation of barrier crossing in the potential $-V(\nu)$, we may evaluate the stationary-point condition for ν from the third line of (5.16) to obtain

$$\frac{\mathrm{d}}{\mathrm{d}\tau}(m_\perp \dot{\nu}) = \frac{\partial}{\partial \nu} V(\nu) + \eta_0 \frac{\partial m^{-1}}{\partial \nu}\left(m_\perp \dot{\nu} + \frac{\eta_0}{2} \right). \tag{5.19}$$

The leading term in (5.19) is the acceleration due to the gradient of $-V(\nu)$ and the second term is small whenever m^{-1} is not a strong function of ν. An exact evaluation of m^{-1} is plotted in figure 5.4, showing weak to moderate ν dependence and anisotropy. The correction proportional to $\partial m^{-1}/\partial \nu$ contains a viscosity linear in $\dot{\nu}$ as well as a force term not captured by the potential, due to the momentum offset η_0.

[10] To understand why a quadratic expansion in η corresponds to a small-fluctuation limit, recall the construction of the Langevin stochastic differential equation in section 4.4.2.2. That construction was performed in coherent-state fields, but its counterpart in action-angle variables has the same structure. The small-fluctuation limit for the Langevin field is the limit in which the Hessian $\partial^2 \hat{\mathcal{L}} / \partial \eta^2$ has small eigenvalues. This is the limit in which the expansion of $\mathcal{L}$ to quadratic order in η is a good approximation over a large integration range in η. Hence the approximation of the integral over η as imposing a functional δ-function between the field ν and its corresponding Langevin field, which would be exact for a Liouvillian that was exactly quadratic in η, is a good approximation for the actual $\mathcal{L}$.

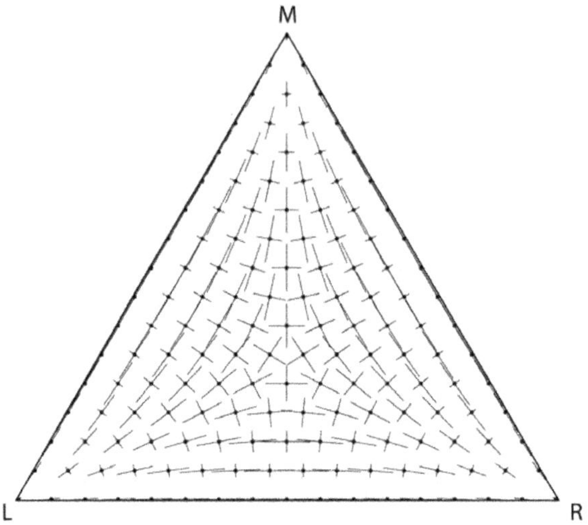

Figure 5.4. In-plane projection of the fluctuation kernel $\partial^2\hat{\mathcal{L}}/\partial\eta^2$ for the parameters that produce the flowfield shown in figure 5.1 for the coordination game. The major and minor principle axes of fluctuation (variances) for η, at a regular sample of ν values, are plotted as crosses centered at ν (and divided by 20 from actual magnitudes to aid viewing). The fluctuation spectrum is only weakly ν-dependent and only moderately anisotropic at any ν.

We note that the combination of the conservation law $\hat{\mathcal{L}} = 0$, the equation of motion (5.18),[11] and the quadratic expansion (5.14) for $\hat{\mathcal{L}}$, implies that the velocity fields along all stationary paths must satisfy

$$\dot{\nu}^T m_\perp \dot{\nu} = (\eta - \eta_0)^T m^{-1} (\eta - \eta_0) = \eta_0^T m^{-1} \eta_0. \tag{5.20}$$

As claimed in section 5.2.2, the solutions to this equation for $\dot{\nu}$ are $(D - 2)$ spheres, which pinch off at the rest points where $\eta_0 = 0$.

5.2.4 A steepest-descent algorithm to extract non-classical stationary points

Local minima of the action can be found directly with a method of steepest descent, as we now demonstrate. The key observation is that locally consistent momentum

[11] A word of explanation on our use of conservation laws and equations of motion in this and later sections, and on the relation of semi-classical stationary-path solutions to the exact evaluation of correlation functions: $\mathcal{L}$ is the field-valued expression of the generator of time translations. Starting from any correlation function written as the expectation of a function $\mathcal{A}$ of fields in the path integral (4.66), infinitesimal time displacement produces a sum of correlation functions from the expectation of $\partial\mathcal{A}/\partial t - \mathcal{A}\mathcal{L}$ which must equal zero. Similarly, a shift of the dummy variables of integration—$(\phi, \phi^\dagger)$ or (ν, η)—produces sums of correlations from $\partial\mathcal{A}/\partial\phi - \mathcal{A}\delta S/\delta\phi$, etc, which must likewise equal zero. In the bare path integral (the observable $\mathcal{A} \equiv 1$), these are simply the expectations of $\mathcal{L}$ and the equations of motion, which may be combined to produce field expressions such as (5.19). When we write these conditions as functions of definite field configurations we make the mean-field assumption that corrections from higher-order cumulants are suppressed by powers of $1/N$, which for this model is the case. Since we have dropped these corrections implicitly, we also omit explicit corrections of $O(1/N)$ in the equations of motion, effectively taking the stationary paths in the $N \to \infty$ limit as the reference values about which corrections are made at all large N.

values, which are needed to evaluate the stationary-point equations (4.89, 4.90), can be uniquely inferred for all points along any smooth trajectory in ν, from the equation of motion (5.18) and the condition $\hat{\mathcal{L}} = 0$. If we know the endpoints of a trajectory, we may vary ν-paths pinned at those endpoints, minimizing the difference between the path-derivative for the stationary-point momentum and the derivative implied by the equation (4.90).

Consider a general variation about an arbitrary, τ-dependent, pair of functions η and ν. Ignoring total derivatives, the resulting variation in S will be

$$\delta S = N \int \mathrm{d}\tau \left[\left(\frac{\partial \hat{\mathcal{L}}}{\partial \nu} - \dot{\eta} \right) \cdot \delta \nu + \delta \eta \cdot \left(\dot{\nu} + \frac{\partial \hat{\mathcal{L}}}{\partial \eta} \right) \right]. \tag{5.21}$$

Now define a *trajectory* as a smooth curve of points $\{\nu\}$ and let $\mathrm{d}\nu$ at any point be a vector tangent to the curve at that point. The normalization of the tangent vector may be arbitrary, but its orientation is not. For each point ν in the trajectory, choose a corresponding value for η, denoted $\eta(\nu)$, such that

$$\hat{\mathcal{L}}(\nu, \eta(\nu)) = 0, \qquad \text{and}$$
$$\mathrm{d}\nu = -\mathrm{d}\tau \left. \frac{\partial \hat{\mathcal{L}}}{\partial \eta} \right|_{\nu, \eta(\nu)}, \tag{5.22}$$

for some positive constant $\mathrm{d}\tau$. As long as the surface $\hat{\mathcal{L}} = 0$ is convex, $\eta(\nu)$ and $\mathrm{d}\tau$ will be uniquely specified. $\mathrm{d}\tau$ in (5.22) is the time interval appropriate to make

$$\frac{\mathrm{d}\nu}{\mathrm{d}\tau} = - \left. \frac{\partial \hat{\mathcal{L}}}{\partial \eta} \right|_{\nu, \eta(\nu)} \tag{5.23}$$

a velocity field. The integral of (5.23), $\nu(\tau)$, is a well-defined time-dependent trajectory, satisfying the ν stationary-path condition (4.138). Similarly, for the function $\eta(\nu)$, a time derivative $\mathrm{d}\eta(\nu)/\mathrm{d}\tau$ along the trajectory is implied as well. In general it will *not* satisfy the η equation of motion (4.139), since the foregoing construction can be carried out for any smooth curve in ν-space.

Now we reconsider the variation (5.21) of S, among the restricted class of functions that admit general $\delta\nu$ and $\delta\eta$, but which start from paths satisfying equations (5.22, 5.23). The second term in the integral vanishes by (5.23). The remaining gradient $\partial \hat{\mathcal{L}} / \partial \nu \equiv \dot{\eta}_{\mathrm{EOM}}$ is the velocity of the η-field along a *true* stationary point instantaneously tangent to the trajectory $\{\nu\}$, by (4.139). The variation of S then becomes

$$\delta S = N \int \mathrm{d}\tau \left[\left(\dot{\eta}_{\mathrm{EOM}} - \frac{\mathrm{d}\eta(\nu)}{\mathrm{d}\tau} \right) \cdot \delta\nu \right]. \tag{5.24}$$

Note that, starting from $(\nu, \eta(\nu))$, the variations $\dot{\eta}_{\mathrm{EOM}} \mathrm{d}\tau$ and $\mathrm{d}\eta(\nu)$ (along with $\mathrm{d}\nu$), both leave their paths in the surface $\hat{\mathcal{L}} = 0$ at $\nu + \mathrm{d}\nu$. Therefore the difference

$\dot{\eta}_{\rm EOM}{\rm d}\tau - {\rm d}\eta(\nu)$ must satisfy $\left(\dot{\eta}_{\rm EOM}{\rm d}\tau - {\rm d}\eta(\nu)\right) \cdot \partial\hat{\mathcal{L}}/\partial\eta = 0$ at $\nu + {\rm d}\nu$, or equivalently, from (5.23)

$$\left(\dot{\eta}_{\rm EOM} - \frac{{\rm d}\eta(\nu)}{{\rm d}\tau}\right) \cdot \frac{{\rm d}\nu}{{\rm d}\tau} = 0. \tag{5.25}$$

In other words, the functional variation of the action about this class of curves yields a vector field instantaneously perpendicular to the trajectory at each point. Thus the gradient in (5.24) may be used in a Newton's method to converge on the non-classical stationary paths.

Figure 5.5 shows a portion of the ν-simplex from the flowfield of figure 5.1 for the coordination game, with the classical diffusion paths in the background. Overlaid on these are segments of actual solutions to the stationary-path equations of motion, starting from arcs that begin in the stable rest point and terminate in the saddle point. The arcs are chosen to lie above and below the contour of slowest approach in the classical diffusion solution, which serves as a proxy for the saddle path of the potential $-V(\nu)$. As expected from the kinematic approximation (5.19), these trajectories diverge from the saddle path between the two rest points. The actual escape

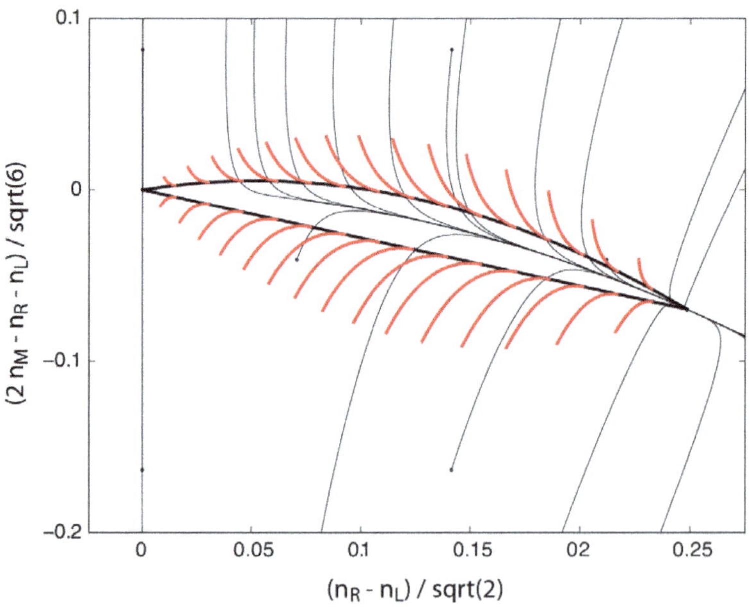

Figure 5.5. Solutions to the non-classical stationary point conditions (4.89, 4.90) for the coordination game with the payoffs of (5.2). Symmetry is weakly broken ($\bar{a} = 1$, $aN = 5$). Flowfields of the classical solution from figure 5.1 are shown as light lines in the background for reference and the path of slowest diffusion is indicated by the confluence of these lines. Two ν trajectories are tested (heavy black lines), one above and one below the classical direction of slowest flow, which approximates the saddle contour of the potential $-V(\nu)$ in figure 5.3. Exact solutions initially tangent to these trajectories (red) curve away from the saddle, as expected if $-V(\nu)$ defines the relevant potential for kinematics.

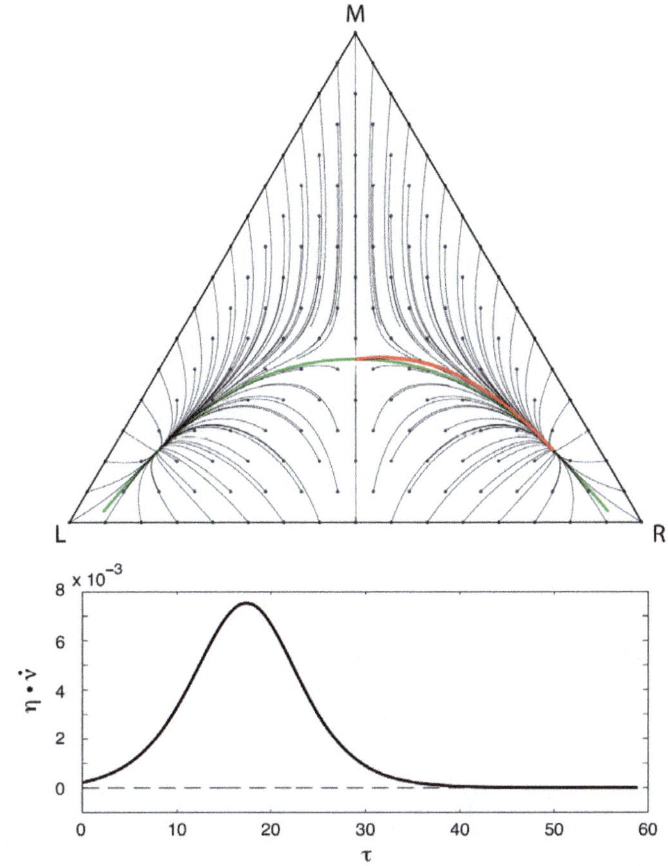

Figure 5.6. The result of the steepest-descent method giving the most probable trajectory for escape from the stable ESS in the coordination game defined by (5.2). The red curve, moving from the *R*-dominated ESS to the uniform population, is the escape trajectory. It differs slightly from the contour of slowest flow (green line, from figure 5.1, traversed outward from the uniform population), which lies along the contour of convergence of the classical stationary paths (thin black). Plotted below is the numerical value of $\eta \cdot \dot{\nu}$ along the escape contour. Abscissa is rescaled time τ and the integral of $\int d\tau \, \eta \cdot \dot{\nu} = \hat{S}$, which here evaluates to 0.1123. We have shown in [2, 28] that this analytic approximation for large-deviations rate functions agrees well with numerical simulations (which become costly for rare events).

trajectory, obtained by the method of steepest descent described here, is shown in figure 5.6.

To estimate the actions associated with reverse-direction paths, we use the condition $\hat{\mathcal{L}} = 0$, both explicitly in (5.13) and through equations (5.14, 5.17) to remove τ as an integration variable, leaving

$$\dot{\nu}^T m_\perp \dot{\nu} \approx \frac{\sqrt{d\nu^T m_\perp d\nu}}{d\tau} \sqrt{2V(\nu)} \, . \tag{5.26}$$

Then S of (5.13) may be approximated

$$\begin{aligned} S &= N\int d\tau\, \eta\cdot\dot{\nu} \\ &= N\int d\tau\big((\eta-\eta_0)\cdot\dot{\nu}+\eta_0\cdot\dot{\nu}\big) \\ &\approx N\int d\tau\big(\dot{\nu}^T m_\perp \dot{\nu} - \dot{\nu}_{\mathrm{Cl}}^T m_\perp \dot{\nu}\big) \\ &\approx N\int \sqrt{d\nu^T m_\perp d\nu}\,\sqrt{2V(\nu)}\left(1-\frac{\dot{\nu}_{\mathrm{Cl}}^T m_\perp \dot{\nu}}{\dot{\nu}^T m_\perp \dot{\nu}}\right), \end{aligned} \tag{5.27}$$

in which the approximations correspond to the second-order expansion of $\hat{\mathcal{L}}$ in η, used in the previous sub-section.

$\sqrt{d\nu^T m_\perp d\nu}$ is a line element along the trajectory ν and, by (5.20), the term in the fraction on the fourth line of (5.27) is a direction cosine between the tangent fields $\dot{\nu}$ and $\dot{\nu}_{\mathrm{Cl}}$ in the inner product defined by $m_\perp$. For the classical stationary paths, this fraction equals unity by definition, giving the required $S \equiv 0$. To the extent that the non-classical solutions approximately *reverse* $\dot{\nu}_{\mathrm{Cl}}$, the direction cosine is ≈ -1, and the action evaluates to the positive quantity

$$S \approx 2N\int \sqrt{d\nu^T m_\perp d\nu}\,\sqrt{2V(\nu)}. \tag{5.28}$$

Equation (5.28) is twice the Wentzel–Kramers–Brillouin (WKB) value that would have been expected for equilibrium barrier crossing in the same potential. We see that the last term $\eta_0\cdot\dot{\nu}$ in the second line of (5.27) has only weakly affected the stationary trajectories, but provides the offset to the value $S^{(0)}$ that permits free classical diffusion, doubly penalizing reverse-direction diffusion responsible for escape from the stable fixed points.

A numerical evaluation of the first line of (5.27)—no longer using the quadratic approximation—along the escape trajectory is shown in figure 5.6. It shows a small difference between the escape trajectory (red) and the time-reverse of the diffusion solution (green), but the two are very similar and the constructions in this section provide a good approximation to the actual escape probability.

5.3 Qualitative behavior of fluctuations in the coordination game

Fluctuations about the rest points are approximately Gaussian with variance $\sim O(1/N)$ far from the critical point of the pitchfork bifurcation and the covariance matrix is roughly the inverse of the matrix of relaxation rates indicated by the classical flow-field (see figure 5.1). At the bifurcation (using N as the control parameter) the variance diverges as $\sim 1/(N - N_{\mathrm{crit}})$, similarly to the Curie–Weiss model. We have demonstrated elsewhere [28], for a model with gross phenomenology similar to that of the coordination game, the close agreement of the analytic calculation of Gaussian-order fluctuations with simulations from a Gillespie algorithm. Therefore we do not present simulation results of fluctuations for the coordination game here. However, simulations are presented for fluctuations about the stationary population state for the rock–paper–scissors game in chapter 6. Although the nature of its symmetry breaking is different, and the fluctuations differ accordingly as *histories*, the fluctuating states sampled are similar for the two games about the symmetric population background.

5.3.1 Analytic structure of the fluctuation decomposition

From the scaling of fluctuations $\sim O(1/N)$ away from the critical point, we see that $1/N$ is the small parameter for expansions of higher-order non-linear corrections to the action in Gaussian moments, collectively termed *perturbative corrections*. It is important that, in the power-series expansion of the action in (4.86), with $S^{(0)} \propto N$, the factor $e^{-S^{(0)}}$ has an *essential singularity* in the small parameter $1/N$—that is, it does not converge to its power series in $1/N$, in any radius about $1/N \to 0$.[12]

The stationary-point expansion in the symmetry-broken phase, about the entire set of essential singularities corresponding to escape trajectories, is necessary in order that the remaining perturbation series be convergent [29]. The inclusion of a continuum of escape trajectories compensates for zero-modes of $S^{(2)}$ that it is necessary to re-arrange by hand in the Taylor's series expansion of S, a construction presented in section 4.4.2.4.

The strong analytical distinction between essential singularities and perturbative corrections that can be expanded as powers in $1/N$, enforced in this way of arranging the calculations, becomes somewhat misleading near the critical point, where escapes and the normal background of fluctuations merge into a single distribution. The degree to which the two kinds of excursions separate defines the difference between regimes of weak and of strong selection.

5.3.2 Weak and strong selection

Characteristic time series for agent numbers n_L/N and n_R/N, from samples of stochastic simulations of the model (5.2), are shown in figures 5.7 and 5.8.

[12] In quantum field theoretic presentations of background fields and fluctuations, it is conventional to identify a small parameter $g \ll 1$ which is the expansion parameter for corrections computed from moments of Gaussian-order fluctuations. The polynomial order of a fluctuation correction is related to the power of g with which it appears in the equations of motion or stable-set conditions. Here we may identify $g \equiv 1/N$. Since $S = N\hat{S}$, and likewise we have $S^{(0)} \equiv N\hat{S}^{(0)}$, we may write $e^{-S^{(0)}} = e^{-(1/g)\hat{S}^{(0)}}$. Because this expression does not converge to its Taylor's series expanded about $g = 0$, no polynomial order of incorporation of Gaussian fluctuations can converge to a complete representation of the functional integral if instanton terms are not also included in the sum over backgrounds.

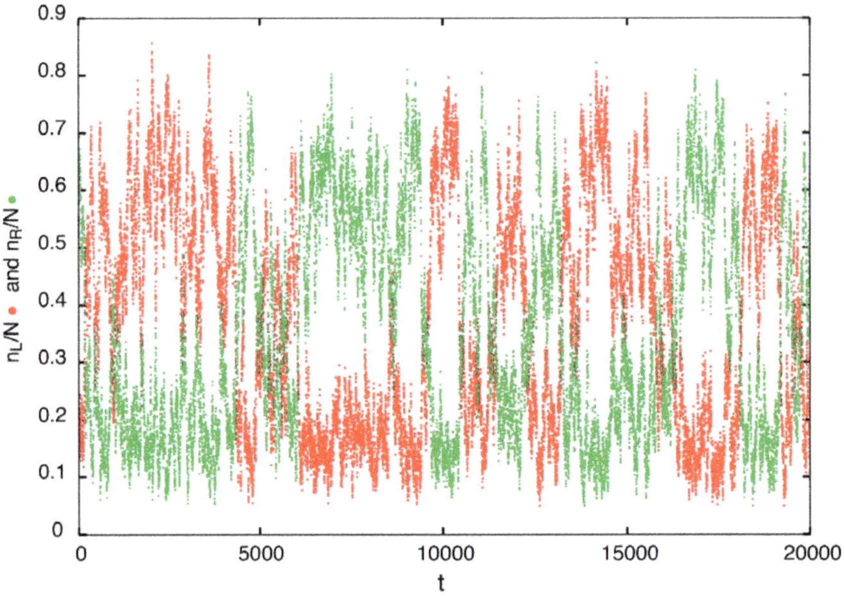

Figure 5.7. Characteristic time series of two number components n_L and n_R for weakly broken symmetry, in a simulation of the stochastic game model (5.2). The average occupancies are asymmetric between domains of high-occupancy and low-occupancy, because the numbers n_L and n_R project differently onto the ESSs in which they are dominant and in which they are the minority. Fluctuations are also asymmetric between high- and low-occupancy, because the projections onto the saddle path of slowest mean regression differs.

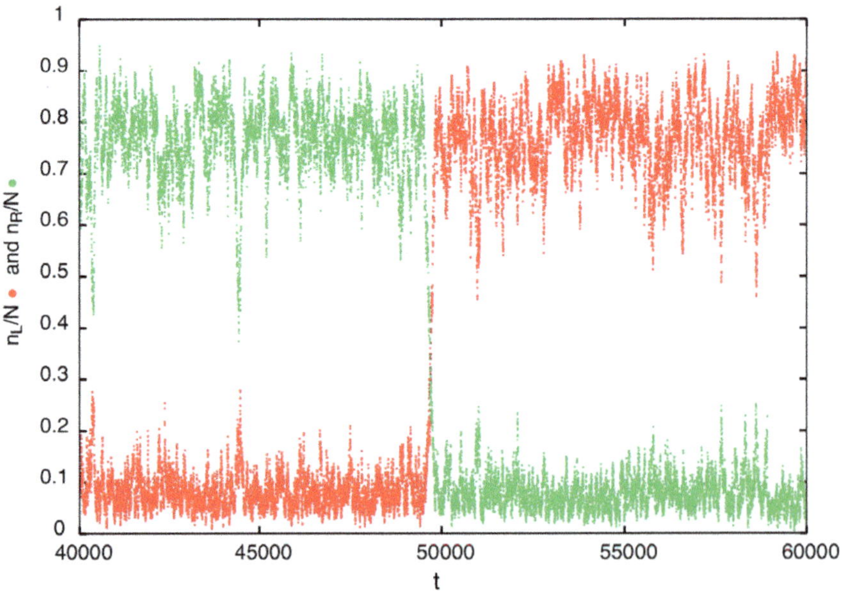

Figure 5.8. Characteristic time series of number components n_L and n_R for strongly broken symmetry, in comparison to figure 5.7. Population variance about either rest point is now much smaller than the separation between the two ESSs, causing the large excursions that lead to escapes to become exponentially rare with N.

These simulations illustrate the transition from the weak-selection regime $aN \gtrsim 4.5$ to the strong-selection regime $aN \gg 4.5$.

Figure 5.7 shows the behavior of $\mathrm{n_L}/N$ and $\mathrm{n_R}/N$ in a neighborhood of the bifurcation point. In this regime, fluctuations involving large fractions of the population are common and transitions between the basins of attraction of the two ESSs are frequent.

Figure 5.8 shows corresponding time series, over the same number of generations as figure 5.7, for a regime of strong selection. The typical residence time in either basin of attraction is now much longer than the time window shown, so this window was selected to include one such transition. The magnitude of fluctuations has decreased due to strong mean regression, as indicated by the curvatures of the kinematic potential in figure 5.3. More importantly the frequency of large excursions has been suppressed relative to that of smaller ones, leading to rare but sharp transitions between basins of attraction. The distribution over individual types is thus tightly entrained by the mean population state, and this distribution changes as a 'coalition'.

5.3.2.1 Symmetry in the short-term and long-term ensembles

Figure 5.9 shows the effect of the transition between weak and strong selection, on the time-averaged description of the population state. A moving window was used to time-average $(\mathrm{n_L} - \mathrm{n_R})/N$, corresponding to the T-generation average of (5.29). We denote this average as $\langle \mathrm{n_L} - \mathrm{n_R} \rangle_T / N$. The ensemble average of this quantity over window positions is then plotted in absolute value versus $1/N$. In the realm of weak selection, any window long enough to suppress fluctuations within a basin of attraction has high probability to include a transition and the ensemble mean of $|\langle \mathrm{n_L} - \mathrm{n_R} \rangle_T| / N$ falls rapidly to zero. Thus even short time averages fail to reflect the analytic solution for the rest points (derived in section 5.4.1 below), shown in the solid line. The point of departure of the time-averaged population statistic from the analytic result depends sensitively on window length in this weak-selection regime.

In the regime of strong selection, population states are persistent, permitting long windows to average over Poisson fluctuations about the population mean while rarely including transitions. In this regime, a time window shorter than the typical persistence time readily identifies group states as dynamical variables in the two-state Markov process. The persistence time, which grows exponentially with population size in the strong-selection regime, sets the rate of convergence to the ergodic distribution over L/R population states. The group-level description is not sensitively dependent on window length, because a large separation of timescales has opened between Poisson fluctuations whose frequencies are polynomial in $1/N$ and escapes whose frequencies are exponential in $-N$.

Any game in which large N concentrates individual behavior almost always around the population average, but in which the population average retains dynamics on the macroscale, admits a multilevel description in which either the players or the populations may be the elementary units. The symmetry-broken coordination game, with its two population states, provides a particularly simple

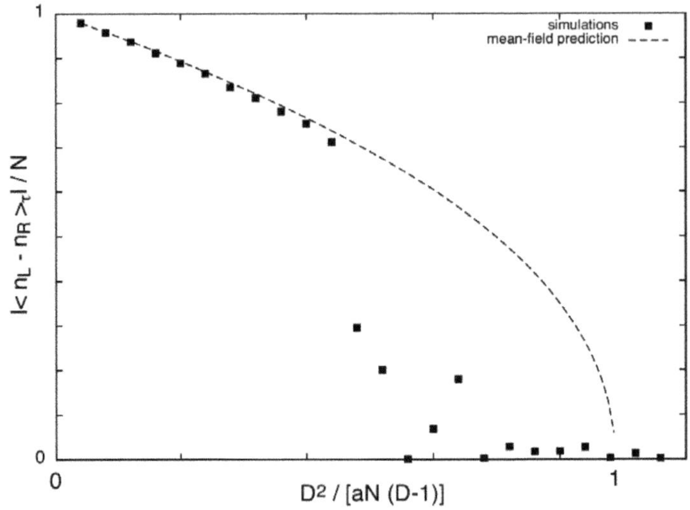

Figure 5.9. Short-term breaking and long-term restoration of symmetry. The MFA for asymmetry is plotted against the time-averaged population composition in samples. Average population asymmetry $|\langle n_L - n_R \rangle_\tau|$, over windows of length $\tau/\Delta t$ generations, is plotted versus $D^2/[aN(D-1)]$, which takes value 1 at the mean-field critical point. When state changes have rate $\gtrsim 1/\tau$, the time average $|\langle n_L - n_R \rangle_\tau|$ (symbols) approximates the symmetric, ergodic distribution, departing from the analytic formula (dashed line) for mean asymmetry—even though the typical population state is asymmetric as shown in figure 5.7. When state changes have rate $\ll 1/\tau$, $|\langle n_L - n_R \rangle_\tau|$ captures the time dependence of a two-state stochastic process (symbols follow the line). The sample average begins to deviate from the analytic form when the typical residence time is comparable to the window timescale. For weak symmetry breaking as in figure 5.7, changes in τ of order unity result in visible shifts of the transition point on the diagram. For strong symmetry breaking as in figure 5.8, changes of τ by factors $\gg 1$ are required to shift the transition point, corresponding to the exponential dependence of the residence times on N.

demonstration of Hamilton's rule and an interpretation in terms of the emergence of coalitional behavior from non-cooperative interactions.

5.4 Symmetry breaking in the Price equation and Hamilton's rule

Hamilton's rule for group selection follows as a general result from the Price equation, whenever the variance within typical group states is smaller than the total variance shown by the population as a whole. The origin of the ensemble over group states does not matter; they may be contemporaneous groups in an island model, or temporally segregated group states of a single population. The transitions in the coordination game (as we will show below) with broken symmetry are typically rapid compared to the time spent in either asymmetric state. Therefore population states at successive times form a natural ensemble of group states, whose frequencies are the same as those sampled independently from an ensemble of model instances. Here we derive the form of Hamilton's rule for the coordination game, following the exposition of Steven Frank [3].

We may write the Price equation for the change in population state $\bar{n}$ in a long-time average from times 0 to T, as the sum of terms of the form (2.14) at single generations,

$$\begin{aligned}
\sum_{\mathrm{n}}(\rho_{\mathrm{n},T} - \rho_{\mathrm{n},0})\mathrm{n}_J &= \sum_{t=0}^{T-\Delta t}\sum_{\mathrm{n}}\left(\rho'_{\mathrm{n},t} - \rho_{\mathrm{n},t}\right)\mathrm{n}_J \\
&= \sum_{t=0}^{T-\Delta t}\sum_{\mathrm{n}}\rho_{\mathrm{n},t}\left(\sum_i \left\langle \mathrm{n}'_{J|i}\right\rangle_{\mathrm{n}} - \mathrm{n}_J\right) \\
&= \sum_{t=0}^{T-\Delta t}\sum_{\mathrm{n}}\rho_{\mathrm{n},t}\sum_i \mathrm{n}_i\left\{\delta_{Ji}\left\langle \mathrm{w}_i - 1\right\rangle_{\mathrm{n}} + \sum_j \delta_{Jj}\left\langle \frac{\mathrm{n}'_{j|i}}{\mathrm{n}_i} - \delta_{ji}\mathrm{w}_i\right\rangle_{\mathrm{n}}\right\} \\
&= TN\left\{\mathrm{Cov}_\rho\left(\delta^J, \mathrm{w}\right) + \mathrm{E}_\rho\left(\mathrm{w}\cdot\Delta\delta^J\right)\right\}.
\end{aligned} \tag{5.29}$$

Because the population process is Markovian, the update is conditionally independent of t given any state n. The fourth line is a covariance in the time-dependent distribution $\rho_{\bar{n},t}$.

With respect to the same time-dependent distribution, we may define mean values over the whole interval

$$\bar{n}_J \equiv \frac{\Delta t}{T}\sum_{t=0}^{T-\Delta t}\sum_{\mathrm{n}}\rho_{\mathrm{n},t}\mathrm{n}_J. \tag{5.30}$$

About the collection of these mean values, the variance of the observable δ^J is then

$$V_J \equiv \frac{\Delta t}{T}\sum_{t=0}^{T-\Delta t}\sum_{\mathrm{n}}\rho_{\mathrm{n},t}\frac{1}{N}\sum_i \mathrm{n}_i\left(\delta_{Ji} - \frac{\bar{n}_J}{N}\right)^2. \tag{5.31}$$

If the instantaneous population distributions $\rho_{\mathrm{n},t}$ flip stochastically and symmetrically between rest points, the average $\bar{n}_J$ may be small while the instantaneous averages of the indicator functions δ_{Ji} are large. Thus the relatedness from switching of population states can dominate the transient relatedness from fluctuations about the instantaneous average.

Now we perform a regression for w_i at each time in the same variables as those of (2.17), but we require that the set of (fixed) coefficients a_{ij} minimize the squared error in the joint population/time covariance, rather than in the single-time population covariance of (2.18).[13] The time-averaged generalization of the regression (2.18) then becomes

$$\frac{\mathrm{w}_i - 1}{\Delta t} = \sum_{kj}\left(\delta_{ik} - \frac{\mathrm{n}_k}{N}\right)(\alpha_{kj} + a_{kj}\mathrm{n}_j) + O\left(\frac{1}{N}\right) + \epsilon_i. \tag{5.32}$$

[13] To reduce clutter in the next several equations, we will collect terms arising from the $-\delta_{ij}$ factor that eliminates self-play into an overall summand at $O(1/N)$.

We now wish to decompose the covariance term in the long-time Price equation (5.29). Following [3], we define a 'cost' component C_J for each type, normalized by the variance (5.31), as

$$\begin{aligned} -V_J C_J &= \frac{\bar{n}_J}{N}\frac{\Delta t}{T}\sum_{t=0}^{T-\Delta t}\sum_{\mathrm{n}}\rho_{\mathrm{n},t}\sum_j\left(\delta_{Jj}-\frac{\mathrm{n}_j}{N}\right)\left(\alpha_j+\sum_k a_{jk}\mathrm{n}_k\right) \\ &\equiv \frac{\bar{n}_J}{N}\left\langle\sum_j\left(\delta_{Jj}-\frac{\mathrm{n}_j}{N}\right)\left(\alpha_j+\sum_k a_{jk}\mathrm{n}_k\right)\right\rangle_\rho \\ &= \frac{\bar{n}_J}{N}\left[\left(\alpha_J+\sum_k a_{Jk}\bar{n}_k\right)-\frac{1}{N}\sum_j\left(\alpha_j\bar{n}_j+\sum_k a_{jk}\left\langle \mathrm{n}_j\mathrm{n}_k\right\rangle_\rho\right)\right]. \end{aligned} \tag{5.33}$$

The cost measures the fitness contribution from the long-term average population state $\bar{n}$, and in the last line of (5.33) we have separated the MFA from covariance corrections. In the regime of strong selection, the latter will be small.

In the case where multiple regressions against several predictor types k are used, the 'benefit' term B_{Jk} in Hamilton's rule is a matrix indexed by 'benefactor' k and 'beneficiary' J, with components

$$\begin{aligned} B_{Jk} &\equiv \sum_j\left[\left(\delta_{jJ}-\frac{\bar{n}_j}{N}\right)a_{jk}-a_{kj}\frac{\bar{n}_j}{N}\right]-\frac{\alpha_k}{N} \\ &= a_{Jk}-\frac{\alpha_k}{N}-\sum_j(a_{jk}+a_{kj})\frac{\bar{n}_j}{N}. \end{aligned} \tag{5.34}$$

Here 'benefit' stands for the difference between the actual payoff coefficients per interaction and the average in the mean population state $\bar{n}$.

The relatedness r_{Jk} between benefactor and beneficiary is also indexed by J and k and is written as a set of regression coefficients against the variance of the beneficiary, so

$$\begin{aligned} V_J r_{Jk} &\equiv \frac{\Delta t}{T}\sum_{t=0}^{T-\Delta t}\sum_{\mathrm{n}}\rho_{\mathrm{n},t}\frac{1}{N}\sum_i \mathrm{n}_i\left(\delta_{Ji}-\frac{\bar{n}_J}{N}\right)(\mathrm{n}_k-\bar{n}_k) \\ &= \frac{\Delta t}{T}\sum_{t=0}^{T-\Delta t}\sum_{\mathrm{n}}\rho_{\mathrm{n},t}\frac{1}{N}(\mathrm{n}_J-\bar{n}_J)(\mathrm{n}_k-\bar{n}_k). \end{aligned} \tag{5.35}$$

We see from the second line of (5.35) that relatedness may be mediated primarily by population state switching in the realm of strong selection.

Finally we may write the covariance term from (5.29) as

$$\mathrm{Cov}_\rho\left(\delta^J, \mathrm{w}\right) = V_J\left(\sum_k B_{Jk} r_{Jk} - C_J\right) - \frac{1}{N}\left\langle (\mathrm{n}_J - \bar{n}_J) \sum_{jk}\left(\mathrm{n}_j - \bar{n}_j\right)(\mathrm{n}_k - \bar{n}_k) \right\rangle_\rho . \tag{5.36}$$

The least-squares condition for the parameter α_J ensures vanishing of $\mathrm{Cov}_\rho(\delta^J, \varepsilon)$. In the large-population limit, the third-order variation in the average $\langle\ \rangle_\rho$ of (5.36) will be small and asymptotically ignorable compared to terms kept in the first line.

5.4.1 Relatedness and effective fitness for the coordination game in the regime of broken symmetry

We now compute explicit forms for C_J, B_{Jk} and r_{Jk}, and show how the long-term average over population states coarse-grains the type space together with the distribution. The two 'types' that are left invariant by reflection, even in the domain of broken symmetry, are the union of L and R types, versus the M type. The fluctuating population state appears within this coarse-graining as a fluctuating phenotype within the $(\mathrm{n}_L + \mathrm{n}_R)$-population. Both variants of the fluctuating phenotype share the feature of being capable of coordination which confers benefit when they coordinate on the same phenotype, relative to the M type, who are not capable of coordinating.

5.4.1.1 Emergence of a 'genic' distinction between coordinating and non-coordinating types in the case of pitchfork bifurcation

The coarse-graining from three types into two, preserving reflection symmetry, results in a partition which is equivalent to a 'gene' whose alleles distinguish types that are sensitive to population state from those that are not. The alleles take on an effective fitness difference in long-term regressions, generated entirely via relatedness in intermittently switching population states.

5.4.1.2 Simplex coordinates and rest points of the replicator dynamic with mutation

We begin with notation for a basis that separates the constrained and dynamical degrees of freedom for the population fractions of the three types left, right, and middle. Denote by $(\hat{e}_0, \hat{e}_1, \hat{e}_2)$ the three basis vectors

$$\hat{e}_0 \equiv \frac{1}{\sqrt{D}}\begin{bmatrix}1\\1\\1\end{bmatrix}; \qquad \hat{e}_1 \equiv \frac{1}{\sqrt{D-1}}\begin{bmatrix}1\\-1\\0\end{bmatrix}; \qquad \hat{e}_2 \equiv \frac{1}{\sqrt{D(D-1)}}\begin{bmatrix}-1\\-1\\2\end{bmatrix}, \tag{5.37}$$

the last two of which form an orthonormal basis in the simplex $\sum_i \mathrm{n}_i/N \equiv 1$.

A general population state with the coordinates of (5.1) may be expanded in two simplex coordinates β and γ as

$$\frac{n}{N} = \frac{1}{\sqrt{D}}\hat{e}_0 \pm \beta\hat{e}_1 + \gamma\hat{e}_2. \tag{5.38}$$

With this definition for the basis vectors,

$$\begin{aligned}\frac{n_{L/R}}{N} &= \frac{1}{D} \pm \frac{\beta}{\sqrt{D-1}} - \frac{\gamma}{\sqrt{D(D-1)}} \\ \frac{n_M}{N} &= \frac{1}{D} + \gamma\sqrt{\frac{D-1}{D}}.\end{aligned} \tag{5.39}$$

The difference in the occupation of the L and R states, which is also the order parameter of this problem, is given in terms of β by

$$\frac{n_L - n_R}{N} = \frac{2\beta}{\sqrt{D-1}} \tag{5.40}$$

γ gives the excess in the combined occupation of the L and R states over the fraction $2/D$ and is also proportional to the fitness accruing to either L or R types as a consequence of the relatedness that characterizes either, even as the population shifts between its stable states (as we will see in the following section).

The non-constant part of the payoff matrix in (5.2) is written in the simplex basis vectors as

$$[a] = a(D-1)\hat{e}_1\hat{e}_1^T. \tag{5.41}$$

The isotropic mutation matrix (2.6) used for all the game examples is written in the simplex basis as

$$[\mu] = -D\left(\hat{e}_1\hat{e}_1^T + \hat{e}_2\hat{e}_2^T\right). \tag{5.42}$$

In a time average long enough that the population state has flipped many times between the left and right ordered forms, the time-averaged population number for each type $J \in \{L, R, M\}$ becomes

$$\frac{\bar{n}_J}{N} = \left(\frac{1}{\sqrt{D}}\hat{e}_0 + \bar{\gamma}\hat{e}_2\right)_J. \tag{5.43}$$

A regression on this long-time average would resolve only the distinction between type M and the union of types L and R, corresponding to indicator variables σ_M and σ_{L+R} which partitioned the population into two sets. We could characterize this 'genic' description by presence/absence at a single locus of an allele for 'ability to coordinate', irrespective of which move (L or R) the players coordinate *on*. In this coarse-grained description, only $\bar{\gamma}$ is directly estimated from the average population state, so we regard the typical population configuration as a random variable converging in probability on one of the two values $\pm|\beta(\bar{\gamma})|$, where the solutions to the equation

$$\beta^2(\bar{\gamma}) = \frac{-\bar{\gamma}}{\sqrt{D(D-1)}} + \frac{1}{D} - \frac{D^2}{Na(D-1)} \tag{5.44}$$

will be proportional to the typical relatedness among individuals in the long-time average.

In the decomposition of section 5.4, the expression (5.33) for 'cost'—the regression of fitness on individual type J in the context of the *long-time average* population composition—evaluates to

$$\begin{aligned} V_J C_J &= a(D-1)\beta^2(\bar{\gamma})\bar{n}_J \\ &= Na(D-1)\beta^2(\bar{\gamma})\left(\frac{1}{\sqrt{D}}\hat{e}_0 + \bar{\gamma}\hat{e}_2\right)_J. \end{aligned} \tag{5.45}$$

The expression (5.34) for the benefit conferred by any type J on any type k—the regression of k's fitness given encounters with type J in the *instantaneous* population—evaluates to

$$B_{Jk} = a(D-1)\left(\hat{e}_1\right)_J\left(\hat{e}_1\right)_k. \tag{5.46}$$

Finally, the relatedness (5.35) between types J and k determining their frequency of encounter—the regression of the frequency of type J in the environment on the indicator variable for type k—evaluates to

$$V_J r_{Jk} = N\beta^2(\bar{\gamma})\left(\hat{e}_1\right)_J\left(\hat{e}_1\right)_k. \tag{5.47}$$

Whereas the cost (5.45) contains a term decreasing linearly in $\bar{\gamma}$, the product of benefit and relatedness projects into the same sub-space spanned by $\hat{e}_0$ and $\hat{e}_2$ with constant coefficients, as

$$\begin{aligned} \sum_k B_{Jk} V_J r_{Jk} &= Na(D-1)\beta^2(\bar{\gamma})\left[\left(\hat{e}_1\right)_J\right]^2 \\ &= Na(D-1)\beta^2(\bar{\gamma})\left(\frac{1}{\sqrt{D}}\hat{e}_0 + \frac{1}{\sqrt{D(D-1)}}\hat{e}_2\right)_J. \end{aligned} \tag{5.48}$$

5.4.1.3 *Absorbing relatedness from time segregation into effective fitness in a time-averaged model for two populations*

We may now see how non-zero average relatedness leads to time evolution in the 'genic' description in terms of ± 'alleles for coordination' and how we may absorb relatedness terms into an effective fitness parameter in the regression on slow timescales. We suppose that the weak average selective advantage of types L + R leads to a slow shift out of type M that can be described in a coarse-grained theory over long times, while the strong selection between L and R in most population configurations leads to internal fluctuations of the L + R population that we absorb into the coefficient of relatedness. The Price equation (5.29) becomes an equation of motion for $\bar{\gamma}$ written as

$$\begin{aligned}
\mathrm{Cov}_\rho\left(\vec{\delta}^J, w\right) + \mathrm{E}_\rho\left(w\Delta\vec{\delta}^J\right) &= \frac{\mathrm{d}\bar{\gamma}}{\mathrm{d}t}(\hat{e}_2)_J \\
&= V_J\left(\sum_k B_{Jk} r_{Jk} - C_J\right) + \frac{1}{N}\sum_k \mu_{Jk}\bar{n}_k \\
&= -\left[Na(D-1)\beta^2(\bar{\gamma})\left(\frac{1}{\sqrt{D(D-1)}} + \bar{\gamma}\right) + D\bar{\gamma}\right](\hat{e}_2)_J.
\end{aligned} \tag{5.49}$$

(As required, terms proportional to $\hat{e}_0$ in the direct (5.45) and population-mediated (5.48) fitness have canceled, leaving motion only within the simplex.)

Recognizing that

$$\begin{aligned}
\bar{\gamma} &= \sqrt{\frac{D}{D-1}}\left(\frac{n_\mathrm{M}}{N} - \frac{1}{D}\right) \\
&= -\sqrt{\frac{D}{D-1}}\left(\frac{n_\mathrm{L} + n_\mathrm{R}}{N} - \frac{D-1}{D}\right),
\end{aligned} \tag{5.50}$$

(5.49) is, as required, an evolutionary game equation for a game between two types, M and L + R. The quantity expressed explicitly in (5.49) is $\bar{n}_\mathrm{M}$ (in parentheses in the third line), which we may interpret as a combination of an intrinsic fitness $f_\mathrm{M} \equiv 0$ minus a population-averaged fitness $\phi \equiv (\bar{n}_L + \bar{n}_\mathrm{R})f_{\mathrm{L+R}}/N$. The effective fitness for the joint population L + R is then given by

$$\begin{aligned}
f_{\mathrm{L+R}} &= \frac{N\phi}{\bar{n}_\mathrm{L} + \bar{n}_\mathrm{R}} \\
&= a(D-1)\frac{N^2\beta^2(\bar{\gamma})}{\bar{n}_\mathrm{L} + \bar{n}_\mathrm{R}} \\
&= a(D-1)\frac{\left\langle(n_\mathrm{L} - n_\mathrm{R})^2\right\rangle_\rho}{4(\bar{n}_\mathrm{L} + \bar{n}_\mathrm{R})}.
\end{aligned} \tag{5.51}$$

5.4.2 The transition from relatedness to the emergence of coalitional behavior from non-cooperative interactions

The frequency with which sub-sets of a population deviate from the mean population behavior (in these examples) becomes exponentially rare in the regime of strong selection. When the evolutionary game equation has a unique solution, the suppression of deviations describes concentration around an adaptive optimum. When broken symmetry requires multiple population states and autonomous dynamics, we may view the population as a *coalition* and the convergence of individual behaviors to the population mean as a mechanism for the emergence of coalitional behavior from non-cooperative behavior.

The ability of coalitions to entrain the strategy choices of individuals is the basis of *cooperative solution concepts* in strategic game theory [30, 31]. Usually the mechanisms that define coalitions, such as side payments to share game payoffs, are assumed to be exogenous and given. Symmetry breaking provides one mechanism to endogenize the formation of coalitional behavior, so we provide an explicit map from the shared benefit of coordinated group states to a form of side payment.

5.4.2.1 Scaling of within- and between-population variance and the emergence of effective side payments

For a Poisson process, $\langle (n_{\mathrm{L}} - n_{\mathrm{R}})^2 \rangle_\rho / (\bar{n}_{\mathrm{L}} + \bar{n}_{\mathrm{R}})$ in (5.51) would typically scale as N^0. This scaling is characteristic both of the symmetric state when selection is too weak to lead to bifurcation and of the population fluctuations about either of the asymmetric states *individually* in the ordered regime. If, rather than studying games with fixed constitutive parameters at varying N, we wish to study games with fixed fluctuation variances, an invariant $\bar{\gamma}$ in a population with large N can be achieved by scaling $a \sim 1/N$ so that fitness remains in a fixed proportion to mutation[14]. With this scaling, the contribution to excess fitness for the L + R population from Poisson fluctuations is therefore $O(1/N)$, comparable to terms we have ignored.

With scaling for $a \sim 1/N$, in contrast to the disordered regime, in the ordered regime produced when $Na(D - 1) > D^2$—so that related sub-populations at $(n_{\mathrm{L}} - n_{\mathrm{R}})/N \sim \pm 2\beta(\bar{\gamma})/\sqrt{D - 1}$ become typical—the variance $\langle (n_{\mathrm{L}} - n_{\mathrm{R}})^2 \rangle_\rho / (\bar{n}_{\mathrm{L}} + \bar{n}_{\mathrm{R}})$ scales as N^1 and L + R types have an asymptotically finite excess fitness in the infinite-population limit. In the limit $Na(D - 1)/D^2 \gg 1$, where the within-group variance remains $O(1)$ as N grows large, it becomes sensible to regard *the group* as the dynamical actor, whose strategy is the $\pm$ sign for β in (5.39). As the benefits to relatedness are apportioned equally to all group members, they constitute a sort of side payment, leading to an interpretation of the group as acting in coalitional form relative to agents of type M, who are effectively independent.

A feature of the pitchfork bifurcation model that makes the notion of 'emergent' coalitions natural is the fact that $\bar{\beta}$ scales as $\sqrt{\bar{\gamma}}$ near the bifurcation point. That is, the mean regression leading to exponential suppression of deviations from the instantaneous L- or R-dominated population state, may be made arbitrarily stronger

[14] The retention of fluctuations in large populations can arise if the populations have spatial structure, so that individuals experience only local interactions of fixed size in total populations of any size. Scaling the constitutive fitness parameters as $\sim 1/N$ in well-mixed models is a proxy for this limitation that does not involve the full complexity of modeling spatially extended systems. This scaling is conventionally adopted, for roughly this reason, in models such as Curie–Weiss [32].

than the fitness difference $\propto \bar{\gamma}$ that distinguishes the actors *within* the L + R coalition from the M types, which make up a distinct (single-member) coalition.

In two respects, however, the particular mechanism of symmetry breaking may miss concepts of importance in cooperative game theory. Symmetry breaking will not generally lead to rest points for all possible combinations of coalitional action, so that concepts such as the core of a cooperative game may not be fully represented[15]. Likewise, since coalition formation is given in cooperative game theory, it has no characteristic timescale or inertia and switching of the population is effectively free between any coalitional states. The long timescales for stability of population states that accompanies suppression of deviations may fail to capture this flexibility. An important problem in systems ranging from neuronal memory [1] to social dynamics [33] is how slow-timescale variables may emerge which confer order on short-term individual behavior, but which may also rapidly switch in response to particular sets of local events. Understanding the origin of variables which can be alternately rigid and responsive requires a more complex class of models than the examples developed here.

Bibliography

[1] Bialek W 2001 Stability and noise in biochemical switches *Advances in Neural Information Processing* eds T K Leen, T G Dietterich and V Tresp (Cambridge: MIT Press)

[2] Krishnamurthy S, Smith E, Krakauer D C and Fontana W 2007 The stochastic behavior of a molecular switching circuit with feedback *Biol. Direct* **2** 13

[3] Frank S A 1997 The Price equation, Fisher's fundamental theorem, kin selection, and causal analysis *Evolution* **51** 1712–29

[4] Maynard Smith J and Price G R 1973 The logic of animal conflict *Nature* **246** 15–18

[5] Foster D and Young P 1990 Stochastic evolutionary game dynamics *Theor. Population Biol.* **38** 219–32

[6] Young H P 1993 The evolution of conventions *Econometrica* **61** 57–84

[7] Goldstein H, Poole C P and Safko J L 2001 *Classical Mechanics* 3rd edn (New York: Addison Wesley)

[8] Landau L D 1937 *Zh. Eksp. Teor. Fiz.* **7** 19–32

[9] Ma S-K 1976 *Modern Theory of Critical Phenomena* (New York: Perseus)

[10] Wilson K G and Kogut J 1974 The renormalization group and the ε expansion *Phys. Rep.* **12** 75–199

[11] Smith E 2011 Large-deviation principles, stochastic effective actions, path entropies, and the structure and meaning of thermodynamic descriptions *Rep. Prog. Phys.* **74** 046601

[12] Benaim M and Weibull J W 2003 Deterministic approximation of stochastic evolution in games *Econometrica* **71** 873–903

[13] Canning D 1992 Average behavior in learning models *J. Econ. Theory* **57** 442–72

[14] Kandori M, Mailath G J and Rob R 1993 Learning, mutation, and long run equilibria in games *Econometrica* **61** 29–56

[15] The core, like the feasible set in repeated game theory, emphasizes maximum strategic flexibility. If restrictive mechanisms are desired which enable the formation of only a sub-set of coalitions, symmetry breaking may be applicable.

[15] Hofbauer J and Sigmund K 1998 *Evolutionary Games and Population Dynamics* (New York: Cambridge University Press)
[16] Freidlin M I and Wentzell A D 1998 *Random Perturbations in Dynamical Systems* 2nd edn (New York: Springer)
[17] Graham R and Tél T 1984 Existence of a potential for dissipative dynamical systems *Phys. Rev. Lett.* **52** 9–12
[18] Graham R and Tél T 1985 Weak-noise limit of Fokker–Planck models and nondifferential potentials for dissipative dynamical systems *Phys. Rev.* A **31** 1109–22
[19] Maier R S and Stein D L 1993 Escape problem for irreversible systems *Phys. Rev.* E **48** 931–8
[20] Maier R S and Stein D L 1993 Effect of focusing and caustics on exit phenomena in systems lacking detailed balance *Phys. Rev. Lett.* **71** 1783–6
[21] Maier R S and Stein D L 1997 Asymptotic exit location distributions in the stochastic exit problem *SIAM J. Appl. Math.* **57** 752
[22] Ellis R S 1985 *Entropy, Large Deviations, and Statistical Mechanics* (New York: Springer)
[23] Touchette H 2009 The large deviation approach to statistical mechanics *Phys. Rep.* **478** 1–69
[24] Coleman S 1985 *Aspects of Symmetry* (New York: Cambridge)
[25] Weinberg S 1995 *The Quantum Theory of Fields* vol 1 (New York: Cambridge)
[26] Dykman M I, Mori E, Ross J and Hunt P M 1994 Large fluctuations and optimal paths in chemical kinetics *J. Chem. Phys.* **100** 5735–50
[27] Onsager L and Machlup S 1953 Fluctuations and irreversible processes *Phys. Rev.* **91** 1505
[28] Smith E, Krishnamurthy S, Fontana W and Krakauer D C 2011 Non-equilibrium phase transitions in biomolecular signal transduction *Phys. Rev.* E **84** 051917, PMID 22181454
[29] Cardy J L 1987 Electron localisation in disordered systems and classical solutions in ginzburg–landau field theory *J. Phys. C: Solid State Phys.* **11** L321–8
[30] Shubik M 1984 *Game Theory in the Social Sciences: Concepts and Solutions* (Cambridge, MA: MIT Press)
[31] Fudenberg D and Tirole J 1991 *Game Theory* (Cambridge, MA: MIT Press)
[32] Kochmański M, Paszkiewicz T and Wolski S 2013 Curie–Weiss magnet—a simple model of phase transition *Eur. J. Phys.* **34** 1555
[33] deDeo S, Krakauer D C and Flack J 2010 Inductive game theory and the dynamics of animal conflict *PLoS Comp. Biol.* **6** e1000782

IOP Publishing

Symmetry and Collective Fluctuations in Evolutionary Games

Eric Smith and Supriya Krishnamurthy

Chapter 6

Limit cycles and noisy clocks

Evolutionary dynamics can exhibit a new form of classical symmetry breaking possible in non-equilibrium stochastic processes but outside the scope of equilibrium systems. This is the breaking of unique rest points into limit cycles with a continuous phase symmetry, in systems that have only discrete symmetries among the agent types. The classic rock–paper–scissors (RPS) game exhibits this form of symmetry breaking. Such symmetry breaking is possible because in non-equilibrium systems, ordered phases are extended-time trajectories and time translation is a symmetry in addition to the discrete symmetries among agent types. Like symmetries among agent types, time-translation invariance can be hidden in ordered background states that result from dynamics, introducing phenomena associated with continuous symmetry in systems where the type spaces themselves have no such symmetries. Broken time-translation symmetry results in residual population dynamics which is a continuously occurring random walk relative to the uniform phase advance along the limit cycle. This form of group-level motion contrasts with the rare escape events by populations that break discrete symmetries. This model illustrates the calculation of the fluctuation spectrum from chapter 4 in cases when it contains a zero eigenvalue reflecting the presence of the hidden symmetry. The proof that a random walk exists, which is not subject to mean regression, even if all orders of fluctuation corrections are taken into account, is the stochastic version of Goldstone's theorem.

6.1 Simple continuous symmetry breaking and a new role for time in non-equilibrium processes

The sets of ordered population states formed by classical symmetry breaking may be either discrete or continuous and distinct population-level dynamics arises in the two cases. In equilibrium phase transitions, in order for a continuous set of ordered population states to exist, the set of transformations that take one state into another must first have existed as symmetries of the underlying configuration space. In non-equilibrium systems, however, a new form of continuous symmetry breaking becomes not only possible but common. A population process having a finite set of agent types, and only discrete exchange symmetries among them, may still form limit cycles (or even more complex attractors). Along these the population states transform under a continuous symmetry even though no such symmetry is a feature of the agent type space. The source of a continuous symmetry group for population states is the symmetry of *time translation* in the underlying dynamics.

To understand how time translation generates transformations among ordered phases, we must first recognize that in non-equilibrium Markov processes, an ordered phase corresponds to an entire *history* rather than to an instantaneous state. Non-equilibrium systems in which discrete symmetry breaking to rest points resembles equilibrium symmetry breaking make it easy to miss this change of concept, since their ordered histories are just extended-time trajectories at fixed points in type-space. (However, the extended-time character of the history re-asserts itself if we compute probabilities of fluctuations [1–3].) For limit cycles or other dynamically non-trivial backgrounds, recognizing the entire history as the ordered state becomes essential even at the classical level, in order to identify the symmetries relevant to the phase transition. Under shifts of time, limit cycles are extended-time trajectories that are transformed as wholes, as the phase along the cycle is advanced or retarded.

We illustrate the way that time-translation symmetry can be spontaneously broken by dynamics the same as any other symmetry and the distinctive consequences that result from this kind of symmetry breaking, in the example of the classic game of RPS played by three agent types[1]. In the classical replicator equation, the breaking of a single stable population state to a limit cycle takes the form of a supercritical Hopf bifurcation. The part of the large-deviations expression that controls this new kind of symmetry breaking is the second-order fluctuation kernel $S^{(2)}$ in (4.86). The Green's function for diffusive relaxation takes on a zero eigenvalue, whose eigenvector is locally tangent to the limit cycle, reflecting the equivalence of all phases for the cycle as a consequence of the hidden symmetry.

In the classical replicator analysis, the existence of a limit cycle seems unproblematic. In the stochastic analysis, however, where fluctuation effects can feed back to change average dynamics, we are forced to ask whether these corrections might destroy the equivalence of ordered configurations corresponding to different phases around the cycle. After all, the agent types have only a discrete, cyclic permutation symmetry and the cycle itself can be strongly non-linear in the configuration space and in its rate of advance. It is the recognition that the limit cycle is an extended-time trajectory, protected by the underlying time-translation symmetry, which allows us to prove that the fluctuation corrections do not destroy its degeneracy, even though the explicit calculation of these fluctuations to all orders is beyond the scope of our analysis.

[1] We do not take up in this monograph another interesting and more subtle question, which is what determines whether the continuous symmetry of time translation, or only one of the discrete agent-exchange symmetries, will be broken in a particular evolutionary game. That question falls within the domain of bifurcation theory in dynamical systems, which is a large area outside the scope of our topic.

The proof is a stochastic process version of the celebrated result known as Goldstone's theorem in statistical mechanics and field theory. In equilibrium systems, Goldstone's theorem ensures the existence of massless particles, sound waves, or other so-called 'gapless' excitations in the face of stochastic or quantum corrections. In population processes, it ensures the existence of phase noise around the limit cycle as a form of population dynamics that is not subject to mean regression and so persists in populations of arbitrarily large size. The noisy walk around the limit cycle is the multilevel population dynamics for continuous-symmetry breaking, which replaces the rare state-switching dynamics in systems with discrete symmetry breaking. We compute the relation between the diffusivity around the cycle and the rate of mean regression toward the cycle, as an application of the non-equilibrium fluctuation-dissipation theorem developed in equations (4.104)–(4.106) of section 4.4.2.1.

6.1.1 Continuous degeneracy of the order parameter in a game with a discrete type space

The simplest game with a supercritical Hopf bifurcation is the totally-symmetric RPS game [4].[2]

Types are indexed (R,P,S), and the population state is denoted

$$\mathrm{n} \equiv \begin{bmatrix} \mathrm{n_R} \\ \mathrm{n_P} \\ \mathrm{n_S} \end{bmatrix}. \tag{6.1}$$

The normal-form payoff matrix assuming random matching is [7]

$$[a] = \bar{a} \begin{bmatrix} 1 \\ 1 \\ 1 \end{bmatrix} \begin{bmatrix} 1 & 1 & 1 \end{bmatrix} + \begin{bmatrix} & -a & b \\ b & & -a \\ -a & b & \end{bmatrix}. \tag{6.2}$$

This form is uniquely specified, up to the magnitudes of a and b, by antisymmetry of payoffs and invariance under cyclic permutation of the agent types. As in our other simple game models, choosing the number of agent types $D = 3$ provides the simplest non-trivial example and lends itself to illustration with two-dimensional figures. However, all the equations in this chapter extend to larger numbers D of types with cyclic permutation symmetry in the payoff matrix.

The RPS model, like the coordination game, converges to its mean-field limit in all respects except for the accumulation of noise along the limit cycle in the broken-symmetry phase. The bifurcation, within mean-field approximation (MFA), is derived by introducing a radius variable

$$\bar{r}^2 \equiv \frac{\bar{n}_R^2 + \bar{n}_P^2 + \bar{n}_S^2}{N^2} - \frac{1}{D}. \tag{6.3}$$

[2] Similar conclusions apply, however, to a wide variety of stochastic processes with limit cycles and many of these have been developed in reaction–diffusion theory [5, 6].

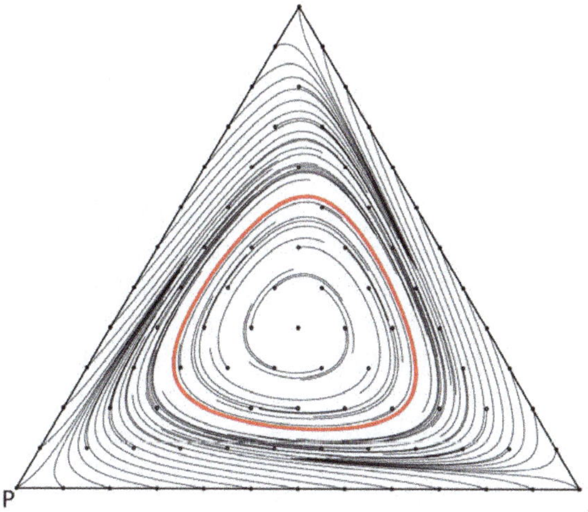

Figure 6.1. Mean-field flow lines for the RPS game (6.2) with the selection parameter $(a - b)N = 24$. The critical population size for this model is $2D^2 = 18$. Dots represent initial mean states $\bar{n}$, and thin black lines are solutions to (2.15) from those initial states. The red line is the limit cycle, to which all solutions converge.

Since the population vector is constrained by $0 \leqslant \bar{n}_R, \bar{n}_P, \bar{n}_S \leqslant N$ and $\bar{n}_R + \bar{n}_P + \bar{n}_S = N$, $\bar{r}^2 = 0$ corresponds to the uniform population $\bar{n}_R = \bar{n}_P = \bar{n}_S = N/D$ and only positive values of $\bar{r}^2$ permit dynamics. Evaluating (2.15) for log r^2 gives

$$\frac{\mathrm{d} \log \bar{r}^2}{\mathrm{d}t} = (a - b)N\left[\frac{1}{D} - \frac{2D}{(a - b)N} + O(r) - \bar{r}^2\right]. \tag{6.4}$$

The term denoted $O(r)$ is oscillatory in the angular coordinate on the simplex and does not accumulate over time, as revealed by the simple Floquet analysis in section 6.3.2 below.

From (6.4), $a < b + 2D^2/N$ gives a unique, stable, static equilibrium at $r = 0$ (the uniformly mixed population). When $a > b + 2D^2/N$, the uniform population becomes unstable and all solutions to (2.15) converge to a limit cycle, as shown in figure 6.1. Near the bifurcation, the cycle is approximately circular, with a mean value given by

$$\bar{r}^2 \approx \frac{1}{D} - \frac{2D}{(a - b)N}. \tag{6.5}$$

Symmetric mutation among the D types, which on its own would drive the population toward the uniform distribution, is responsible for the stability of the limit cycle along an interior trajectory. Without mutation the limit cycle would accumulate to the simplex boundary.

6.2 Gaussian-order response and correlation functions

The robust and dynamically important fluctuation effects in this game may all be qualitatively understood and quantitatively approximated to leading order in $1/N$, from the Gaussian kernel $S^{(2)}$ for fluctuations, introduced in section 4.4.2. The behavior of fluctuations in both the symmetric and symmetry-breaking phases is of interest, along with the quantitative continuation of the spectrum across the transition between phases. We will also illustrate calculation of the Gaussian kernel in both coherent-state and action-angle variables, to show the relation between the two constructions.

We begin with fluctuations about the uniform background, computed in coherent-state variables. We then explain what it means for the limit-cycle order parameter in the broken-symmetry phase to be an extended-time trajectory, show how this leads to differences between non-equilibrium and equilibrium systems and compute the solution for the cycle in the MFA. About this background of the deterministic cycle, we then return and compute the fluctuation spectrum along the radial and tangential directions in action-angle variables, using a weak non-linearity expansion to simplify the limit-cycle coordinates. We show the form of the fluctuation-dissipation theorem that relates mean regression in the radial direction to the rate of accumulation of Brownian noise along the phase of the cycle.

6.2.1 Fluctuations about uniform backgrounds in coherent-state fields

To demonstrate the effect of the terms in the action from section 4.4.2 which control fluctuations, we begin with the expansion about the uniform-population background, where these are constant matrices. We return in section 6.3.1 to the problem of fluctuation expansion about a time-dependent background.

General algebraic forms for the coherent-state expansion are given in appendix 4.6.1. We will keep the number of types D explicit, even though in the examples $D = 3$, to distinguish it from numerical factors related to the order of derivatives and to indicate the scaling of magnitudes with D.

The uniform background satisfies $\bar{\phi}_m = N/D$ for each m. In this background, the diffusion kernel (4.134) evaluates to

$$\overline{\frac{\partial^2 \mathcal{L}}{\partial \phi_i^\dagger \partial \phi_m}} = N\left(\frac{D}{N} + \frac{b-a}{2D}\right)\left(\delta_{im} - \frac{1}{D}\right) - a^A_{im}. \tag{6.6}$$

Continuity in the sum of the eigenvalues about more general backgrounds will also be of interest. From (4.135) and the definition (6.3) of r^2 on the population simplex, we obtain

$$\begin{aligned}\sum_{i=1}^{D} \overline{\frac{\partial^2 \mathcal{L}}{\partial \phi_i^\dagger \partial \phi_i}} &= (D-1)\left[D + N\left(\frac{b-a}{2D}\right)\right] + (D+1)N\left(\frac{a-b}{2}\right)r^2 \\ &\to -2\left|D + N\left(\frac{b-a}{2D}\right)\right|,\end{aligned} \tag{6.7}$$

where in the last line we have used the expression (6.5) for r^2 in the phase with a limit cycle. We will justify this approximation in (6.19) from the Floquet analysis in section 6.3.2 below.

Returning to the uniform background, the noise source corresponding to (6.6), which is the Hessian (4.136), becomes

$$-\overline{\frac{\partial^2 \mathcal{L}}{\partial \phi_i^\dagger \partial \phi_j^\dagger}} = N^2\left[\frac{3}{D}\left(\bar{a} + \frac{b-a}{2}\right) - \frac{2(b-a)}{D^2}\right]\left(\delta_{ij} - \frac{1}{D}\right). \tag{6.8}$$

The matrix in both equations (which has the same form as the mutation matrix up to a scale factor) is a projector into the transverse simplex $\sum_{m=1}^{D}(\bar{\phi}_m + \phi_m') = N$. In this simplex, we may diagonalize the diffusive Laplacian by writing the configuration fields ϕ in a suitable basis, similarly to the diagonalization performed for the pitchfork bifurcation in (5.37). Defining

$$v \equiv \frac{1}{2\sqrt{3}}\begin{bmatrix} 2 \\ -1 \\ -1 \end{bmatrix} + \frac{\mathrm{i}}{2}\begin{bmatrix} 0 \\ 1 \\ -1 \end{bmatrix}, \tag{6.9}$$

and v^* to be its complex conjugate, the projection matrix in (6.8) may be written

$$\begin{bmatrix} 1 & & \\ & 1 & \\ & & 1 \end{bmatrix} - \frac{1}{D}\begin{bmatrix} 1 \\ 1 \\ 1 \end{bmatrix}\begin{bmatrix} 1 & 1 & 1 \end{bmatrix} = vv^\dagger + v^* v^T. \tag{6.10}$$

The diffusive Laplacian then has eigenvalue

$$\frac{\partial^2 \mathcal{L}}{\partial \phi^\dagger \partial \phi} v = N\left[\left(\frac{D}{N} + \frac{b-a}{2D}\right) + \mathrm{i}\frac{(b+a)}{2\sqrt{D}}\right]v, \tag{6.11}$$

for v, and the eigenvalue for v^* is the complex conjugate.

About a uniform background both the eigenvalues and eigenvectors are time-independent, so the time-ordered exponential integrals in the Green's functions (4.99, 4.100) reduce to simple scalar exponents in $(t - t')$. These are readily integrated in the formula (4.105) for M_t, in which we can take $t'' \to -\infty$ because all noise is damped. The result is simply to divide the noise source (6.8) by the sum of the eigenvalue in (6.11) and its complex conjugate (one factor from D^R and one from D^A), yielding the expression for the correlation function of ϕ'

$$\left\langle \phi_i' \phi_l' \right\rangle_t = N\frac{\left[3\left(\bar{a} + \frac{b-a}{2}\right) - \frac{2(b-a)}{D}\right]}{\left| a - b - \frac{2D^2}{N} \right|}\left(\delta_{il} - \frac{1}{D}\right). \tag{6.12}$$

If we expand the instantaneous radial coordinate in the RPS simplex about its mean value as $r \equiv \bar{r} + r'$, with $\bar{r}$ satisfying (6.5), we may estimate the fluctuation magnitude for r' in the symmetric phase where $\bar{r} \equiv 0$,

$$\left\langle (r')^2 \right\rangle = \frac{2D}{N} \frac{\bar{a} - \dfrac{a-b}{3} + \dfrac{1}{N}}{\left| a - b - \dfrac{2D^2}{N} \right|}. \tag{6.13}$$

Equation (6.13) provides half of the comparison of the analytic calculation for fluctuations against numerical simulation, shown in figure 6.4. The comparison in the broken-symmetry phase will be more complicated because of the limit-cycle background, to which we return shortly.

6.2.2 Symmetries governing the Hopf bifurcation act on a space of histories

The graph of mean-field solutions to the evolutionary game equation for RPS, shown in figure 6.1, makes it difficult to see from symmetry alone why the ordered population state should follow a limit cycle. Why should fluctuations not cause the cycle to get 'stuck' in one of the corners so that the ordered states would respect the point-group symmetry of the type space, breaking to three trigonally symmetric fixed points the way the coordination game broke to two reflection-symmetric fixed points?[3]

Figure 6.2 shows the relevant symmetries of the problem by embedding population states in a three-dimensional space of both types and time. As developed at length in [3], the elementary entities to which non-equilibrium stochastic processes assign probabilities are not single-time configurations, but entire *histories* extended over time. For example, a fixed steady state becomes not merely a point in type-space, but a vertical trajectory in which only the type coordinates are fixed. In figure 6.2, the history representing the limit cycle is a spiral path that winds around the symmetric fixed point and which may be periodically identified in time at integer multiples of the limit-cycle period. The symmetry that transforms different ordered histories is translation in *time* (vertical displacement in the graph). If any particular ordered history is displaced in time by continuous amounts ranging from zero to a full cycle period, the resulting one-parameter family of spiral histories traces out a sleeve in the type-space/time prism. This non-trivial action of time translation is in contrast to what happens to rest points, which are vertical lines that simply transform into themselves under any time displacement. Because the ordered histories

[3] The corresponding difficulty is called the *hierarchy problem* in the quantum field theory of elementary particles and also has counterparts in condensed matter. Its statement is that any property that solutions to classical equations of motion may have, if it is not required to exist by symmetries of the problem, will generically be lost when fluctuation corrections are added to the classical solution. The typical property of interest is masslessness of elementary particles. If particles are not required to have zero mass by some symmetry, not only will they generically have non-zero masses, the masses should be characteristic of the strongest forces with which the particles interact and these are often large. The circumventing of the hierarchy problem is the reason all theories of massless particles are based on either gauge or global symmetries.

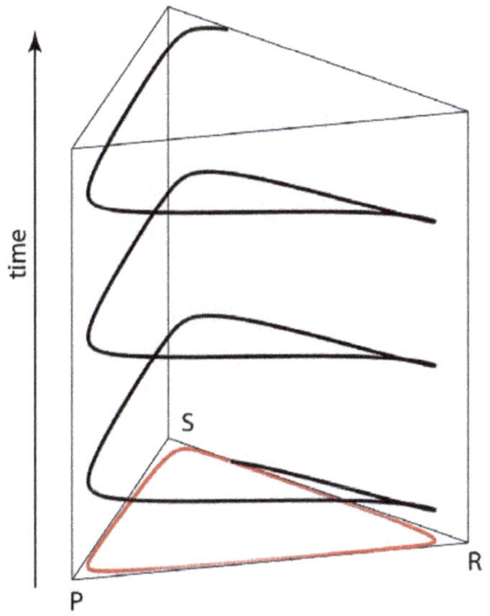

Figure 6.2. The order parameter in a space of histories, corresponding to a limit cycle, is an extended-time path. The projection of the limit cycle onto the space of types is shown in red in the base plane of the graph. Time translation of the spiral trajectory over one cycle period would fill out a non-cylindrical but threefold-symmetric 'sleeve' within the prism shown. This sleeve, which contains the full one-parameter family of ordered backgrounds, is time-independent, reflecting the symmetry hidden by any single background. For a limit cycle near the boundary of the RPS simplex, the cycle slows at the vertices, as shown by vertical inclination of the trajectory. However, because any trajectory winds around the sleeve, while fluctuation effects can change the rate of advance, they cannot collapse the trajectories down to fixed points without changing the topological feature of this winding number.

couple types and time, this translation maps to a rotation in the phase of the projected limit cycle[4].

The sleeve containing all possible phases of the limit cycle *is* transformed into itself by time translation. This is the expression of the fact that the group of ordered states as a whole respect the symmetry that is hidden by any one of them. The *winding number* of any ordered history around the cylindrical sleeve is a topological feature that cannot be continuously deformed into the zero-winding behavior of a set of rest points whose trajectories are mere vertical lines. This topological property is what protects the limit cycle from being a fine-tuned artifact of the classical equations of motion, which might be violated by fluctuation effects. Fluctuation corrections may alter the local speed of advance, smoothly distorting the embedding of the spiral trajectory in time, but they cannot break it into three vertical lines that hide only the point-group symmetry of the type space through anything short of inducing a full topology-changing phase transition.

[4] To appreciate that it is the time translation which is fundamental, note that the time translation acts *uniformly* on the spiraling history. If we wish to project this into an equivalent map on the limit cycle in the base space, the map is not a uniform advance along the line element of the limit cycle, but one that 'stretches' or 'compresses' different segments to reflect the different rates of advance along the cycle.

6.3 Stochastic Goldstone's theorem and noisy clocks

Recall from section 4.4.2.3 that for any stationary path of S, if $\phi^{\dagger}(t)$, $\phi(t)$ is a solution to $S^{(1)} = 0$, then $\phi^{\dagger}(t - \delta t)$, $\phi(t - \delta t)$ must also be a solution for any finite δt. The leading term in an expansion of the stationary-path equations (4.87, 4.88) in δt yields the stochastic version of a famous result from field theory known as *Goldstone's theorem* [8]. The mathematical statement of Goldstone's theorem, in terms of representations of symmetry, is that *the generator of the symmetry* hidden by the background (in this case, time translation, or $\mathrm{d}/\mathrm{d}t$), acting on that background, must produce a fluctuation with zero eigenvalue in $S^{(2)}$.

For RPS the theorem states that the cumulant expansion in the distribution for fluctuations has zero-eigenvalue modes corresponding to the direction of symmetry around the limit cycle. We derive only low-order approximations to this cumulant expansion, to produce quantitative estimates for fluctuations transverse to and Brownian motion along the cycle, which we compare to simulations in the next section. It is well understood that low-order approximations to cumulant expansions are not by themselves reliable, especially for exact cancellations such as zero eigenvalues. The hidden symmetry of time translation implies, however, that these zero eigenvalues exist at *all* orders of approximation and even when the approximate expansion in $1/N$ fails to converge, as long as the topological winding number of the average ordered history persists.

The analogy of the limit cycle in RPS to a continuous circle of degenerate ordered states that leads to Goldstone's theorem for equilibrium thermodynamics is limited. Ordinarily time-translation symmetry is not broken by the solutions of equilibrium thermodynamics[5]. Therefore, in order for a continuous symmetry to emerge among spatial ordered states, the potential would need to have the form shown in figure 6.3. The type space of RPS has no such rotational symmetry and the kinematic potential $V(\nu)$ constructed according to the prescription of section 4.4.1.2 shows a single zero, at the unstable rest point for the uniform population. Hence, a grayscale plot for the RPS game analogous to the potential plot for the coordination game in figure 5.3 would show no distinctive contour or other feature at the position of the cycle.

Since the limit-cycle background is a classical solution to the evolutionary game equations of motion, it has $\bar{\phi}^{\dagger} = 0$ and $\overline{\partial^2 \mathcal{L}/\partial\phi_i\partial\phi_j} \equiv 0$ as well. Therefore the form of (4.115) applicable to the time derivative of the limit cycle is

$$0 = \sum_{j=1}^{D} \left(\delta_{ij} \frac{\mathrm{d}}{\mathrm{d}t} + \overline{\frac{\partial^2 \mathcal{L}}{\partial \phi_i^{\dagger} \partial \phi_j}} \right) \frac{\mathrm{d}\bar{\phi}_j}{\mathrm{d}t}. \tag{6.14}$$

[5] Ordinarily equilibrium thermodynamics is *defined* through physical context, as a theory of time-independent states. However, the mathematics of equilibrium may readily be extended by analytic continuation to the range of thermodynamically reversible dynamics and systems in this domain may break time-translation symmetry [9, 10]. The paired state of superconductors may also be said to break time translation in the collective phase of the pair wave function, though this is not an observable under most conditions [11].

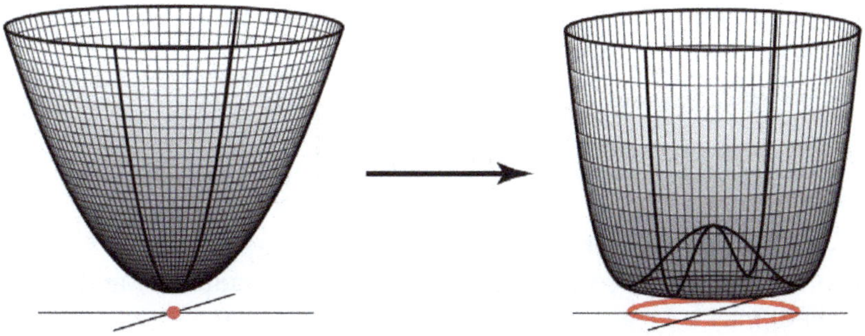

Figure 6.3. Spontaneous symmetry breaking in a two-dimensional equilibrium system. The exact degeneracy of a continuous, one-dimensional loop of solutions (red) requires a rotational symmetry in the underlying potential. Cross-sections of the potential are otherwise equivalent to the one-dimensional case of figure 5.2. The exact degeneracy around the loop in the equilibrium potential ensures that arbitrarily small energies can reach any minimum, independent of the steepness of the mean regression in the radial direction. These most accessible deformations are known as the *Goldstone modes* in the ordered state.

The equivalent form in action-angle variables in the de-scaled time coordinate τ is

$$0 = \sum_{j=1}^{D} \left(\delta_{ij} \frac{\mathrm{d}}{\mathrm{d}\tau} + \overline{\frac{\partial^2 \hat{\mathcal{L}}}{\partial \eta_i \partial \nu_j}} \right) \frac{\mathrm{d}\bar{\nu}_j}{\mathrm{d}\tau}. \tag{6.15}$$

When we transform the vector ν to polar coordinates (r, θ) in the type-simplex, the angular position along the limit cycle will advance as $\mathrm{d}\bar{\theta}/\mathrm{d}\tau \approx$ constant, so (6.15) implies that $\mathrm{d}\bar{\nu}/\mathrm{d}\tau$ is a zero-mode of $\overline{\partial^2 \hat{\mathcal{L}}/\partial \eta_i \partial \nu_j}$ within the same approximation. A zero mode of the matrix that governs mean regression is a collective fluctuation subject to free diffusion.

6.3.1 Frenet coordinates on the limit cycle in the RPS game

A *Frenet coordinate system* on a limit cycle is one whose principle axes are instantaneously tangent and normal to the cycle [5, 6]. These principle axes provide a convenient separation between mean-regressing noise normal to the cycle and Brownian motion tangent to it. For RPS near the bifurcation, the limit cycle is nearly circular and the Frenet coordinates approximate polar coordinates. A *Floquet* analysis integrates properties around cycles, to arrive at the long-time repeated dynamical states or convergence toward them.

Here we compute the leading dependence of the radial coordinate on angle around the cycle and verify that the transformation from polar to Frenet coordinates approaches the identity. By showing that the deviations from circularity are harmonic at leading order, we justify the omission of oscillatory terms in the mean-field radius estimates in section 6.1.1.

Let the normalized radius r introduced in (6.3) and the angle θ in the simplex be polar coordinates for ν. For definiteness, choose $\theta = 0$ to correspond to the axis $\mathrm{n_R} = \mathrm{n_P}$. Finally, perform a similar transformation on η to coordinates (h, φ). Thus,

$$\begin{bmatrix} \nu_1 \\ \nu_2 \end{bmatrix} \equiv \begin{bmatrix} r \cos \theta \\ r \sin \theta \end{bmatrix}; \quad \begin{bmatrix} \eta_1 \\ \eta_2 \end{bmatrix} \equiv \begin{bmatrix} h \cos \varphi \\ h \sin \varphi \end{bmatrix}. \tag{6.16}$$

The only combinations of these fields that actually appear in $S^{(2)}$ refer η to the direction of ν, not to the rectilinear frame, so define radial (r) and tangential (t) components of η as

$$\begin{bmatrix} \eta_{\mathrm{r}} \\ \eta_{\mathrm{t}} \end{bmatrix} \equiv \begin{bmatrix} h \cos(\varphi - \theta) \\ h \sin(\varphi - \theta) \end{bmatrix}. \tag{6.17}$$

For small r^2, we will approximate the noise kernel (4.140) by its value in the uniform background for simplicity, because it is not a strong function of ν near the center of the simplex, as was shown in figure 5.4. In polar coordinates, the action to *all orders* in ν and to second order in η then takes a simple form corresponding to the expansion (5.14) used to study stationary points,

$$S \approx N \int \mathrm{d}\tau \begin{bmatrix} \eta_{\mathrm{r}} & \eta_{\mathrm{t}} \end{bmatrix} \begin{bmatrix} \dfrac{\mathrm{d}r}{\mathrm{d}\tau} + \dfrac{\partial \hat{\mathcal{L}}}{\partial \eta_{\mathrm{r}}} \\ r\dfrac{\mathrm{d}\theta}{\mathrm{d}\tau} + \dfrac{\partial \hat{\mathcal{L}}}{\partial \eta_{\mathrm{t}}} \end{bmatrix} + \frac{1}{2} \begin{bmatrix} \eta_{\mathrm{r}} & \eta_{\mathrm{t}} \end{bmatrix} \frac{\partial^2 \hat{\mathcal{L}}}{\partial \eta^2} \begin{bmatrix} \eta_{\mathrm{r}} \\ \eta_{\mathrm{t}} \end{bmatrix}, \tag{6.18}$$

in which $\partial^2 \hat{\mathcal{L}} / \partial \eta^2$ is computed with respect to the $(\eta_{\mathrm{r}}, \eta_{\mathrm{t}})$ basis.

The stationary point conditions are most easily decomposed in terms of a set of scaling parameters which are functions of the payoffs. The reference scale for radius, $\bar{R}$, will be the value suggested by (6.5),

$$\bar{R}^2 \equiv \frac{1}{D} - \frac{2D}{N(a - b)}. \tag{6.19}$$

The mean rate of advance of the phase θ is

$$\omega = \frac{a + b}{2\sqrt{3}}. \tag{6.20}$$

A normalized radius coordinate will be denoted

$$\rho \equiv \frac{r}{\bar{R}}. \tag{6.21}$$

(The notation here is distinguished from ρ_n used for the original probability density.) Three further combinations of the game parameters appear in the exact stationary point equations:

$$
\begin{aligned}
A_1 &\equiv \bar{R}^2 \frac{a-b}{2\omega}, \\
2A_2^2 &\equiv \bar{R}^2 + A_1^2, \\
\tan\alpha &\equiv \frac{1}{3}\frac{a-b}{2\omega}.
\end{aligned}
\tag{6.22}
$$

The small expansion parameter will be $\bar{R}$, in terms of which $A_2 \sim O(\bar{R})$, $A_1 \sim O(\bar{R}^2)$.

Then the stationary point condition for ν in (ρ, θ) coordinates, from the action (6.18), becomes

$$
\begin{aligned}
\frac{d\rho}{\omega\, d\tau} &= -\frac{1}{\bar{R}\omega}\frac{\partial \hat{\mathcal{L}}}{\partial \eta_r} = \rho\left[A_2\rho \sin(3\theta + \alpha) + A_1\left(1 - \rho^2\right)\right] \\
\frac{\rho\, d\theta}{\omega\, d\tau} &= -\frac{1}{\bar{R}\omega}\frac{\partial \hat{\mathcal{L}}}{\partial \eta_t} = -\rho\left[1 - \eta_2\rho\cos(3\theta + \alpha)\right]
\end{aligned}
\tag{6.23}
$$

The limit-cycle trajectories are not perfect circles, as shown in figure 6.1 and the angle of inclination to a pure radial vector due to the variation of ρ is

$$
\begin{aligned}
\frac{\partial \hat{\mathcal{L}} / \partial \eta_r}{\partial \hat{\mathcal{L}} / \partial \eta_t} = \frac{\dot{\rho}}{\rho\dot{\theta}} &= -\frac{A_2\rho\sin(3\theta + \alpha) + A_1\left(1 - \rho^2\right)}{1 - \eta_2\rho\cos(3\theta + \alpha)} \\
&\equiv \tan\xi \\
&\equiv \frac{d\log\rho}{d\theta}.
\end{aligned}
\tag{6.24}
$$

These polar coordinates lead to a simple small-parameter expansion if we solve (6.24) for the limit-cycle trajectory, to give

$$
\bar{\rho}\left(\bar{\theta}\right) = 1 + \frac{A_2}{3}\cos\left(3\bar{\theta} + \alpha\right) + O\left(\bar{R}^2\right) \tag{6.25}
$$

At leading order $r = \bar{r}$ around the cycle and the deviation at $O(\bar{r}^2)$ is oscillatory with period three. This term cancels around one cycle, giving the approximate radial convergence equation (6.4).

6.3.2 Gaussian-order fluctuations in the Frenet frame

We may now carry out the derivation parallel to that in section 6.2.1, above the critical point where the stationary solutions for RPS converge on a limit cycle of non-zero radius. We continue to work in action-angle variables, drawing on algebraic forms provided in appendix 4.6.2.[6]

We begin with the action form for $S^{(2)}$ given in (4.109) with fluctuating fields (η', ν'). In the limit-cycle background, the diffusion kernel $\overline{\partial^2 \hat{\mathcal{L}} / \partial\eta\partial\nu}$ is time-dependent and does not generally have any zero entries, which makes direct evaluation of time-ordered exponentials complicated. Yet we know from Goldstone's theorem that the time-derivative of the limit cycle must be a zero mode and we know by symmetry that near the critical point it is approximately a circle. Therefore in the polar coordinates of the previous section, the zero eigenvalue may be made explicit and the resulting lower-triangular form for the kernel makes evaluation of 2×2 time-ordered exponentials elementary. This construction is a simple version of the transformation to a Frenet frame used in [5].

Because the limit cycle is not exactly a circle, we begin with a general invertible transformation V from the rectilinear basis (η^T, ν) to variables $\eta^T V$ and $V^{-1}\nu$. We may always choose V so that

$$\frac{\mathrm{d}V}{\mathrm{d}\tau} + \overline{\frac{\partial^2 \hat{\mathcal{L}}}{\partial\eta\partial\nu}} V = V\Lambda, \tag{6.26}$$

for some Λ which is lower triangular, by letting the second column of V be the time derivative of the limit-cycle trajectory. For now, however, assume only that V and Λ are smooth, i.e., $V\Lambda V^{-1}$ has a continuous first derivative so that in the continuum limit the retarded response function becomes

$$\left\langle \nu'_\tau \eta'_{\tau'} \right\rangle = \theta\left(\tau - \tau'\right) \frac{1}{N} V_\tau \mathcal{T}\left(\mathrm{e}^{-\int_{\tau'}^{\tau} \mathrm{d}u \Lambda_u}\right) V_{\tau'}^{-1}. \tag{6.27}$$

Under the same basis transformation, the correlation function (4.106) at equal times becomes

$$\begin{aligned} \langle \nu'_\tau \nu'_\tau \rangle = {} & V_\tau \mathcal{T}\left(\mathrm{e}^{-\int_{\tau'}^{\tau} \mathrm{d}u \Lambda_u}\right) V_{\tau'}^{-1} \langle \nu'_{\tau'} \nu'_{\tau'} \rangle V_{\tau'}^{-1\dagger} \mathcal{T}^{-1}\left(\mathrm{e}^{-\int_{\tau'}^{\tau} \mathrm{d}u \Lambda_u^*}\right) V_\tau^\dagger \\ & - \frac{1}{N} V_\tau \int_{\tau'}^{\tau} \mathrm{d}z \mathcal{T}\left(\mathrm{e}^{-\int_z^{\tau} \mathrm{d}u \Lambda_u}\right) V_z^{-1} \left.\overline{\frac{\partial^2 \hat{\mathcal{L}}}{\partial\eta^2}}\right|_z V_z^{-1\dagger} \mathcal{T}^{-1}\left(\mathrm{e}^{-\int_z^{\tau} \mathrm{d}u \Lambda_u^*}\right) V_\tau^\dagger. \end{aligned} \tag{6.28}$$

We now construct V through a series of successive approximations, beginning with the transformation (6.16) to static polar coordinates, which may be performed directly within the action.

[6] We note, as an aside, that the correlation function in action-angle variables differs from that in coherent state variables only by corrections of order $1/N$ relative to the leading magnitudes. These arise, as they must, from expectations $\langle \tilde{\phi}\phi' \rangle^2$, which the action-angle correlator includes but the coherent-state correlator does not. The closed-form relation between the noise sources for these two forms, for general payoff matrices, is provided in appendix 4.6.3.

In appendix 6.4 we solve for the matrices V and the fluctuation kernel as an expansion in small r^2 and plug the solutions into the formula (6.28) for the $\nu'\nu'$ correlation function. The result, still using polar coordinates, is that

$$\left\langle \begin{bmatrix} r' \\ \bar{r}\theta' \end{bmatrix} \begin{bmatrix} r' & \bar{r}\theta' \end{bmatrix} \right\rangle_t \rightarrow \begin{bmatrix} 0 & \\ & 1 \end{bmatrix} \left\langle \begin{bmatrix} r' \\ \bar{r}\theta' \end{bmatrix} \begin{bmatrix} r' & \bar{r}\theta' \end{bmatrix} \right\rangle_{t'} \begin{bmatrix} 0 & \\ & 1 \end{bmatrix}$$
$$+ \left(\bar{a} - \frac{a-b}{3} + \frac{1}{N} \right) \begin{bmatrix} \frac{D}{2N} \frac{1}{\left| a - b - \frac{2D^2}{N} \right|} & \\ & t - t' \end{bmatrix}$$
$$+ O\left(\bar{R}\right). \tag{6.29}$$

Figure 6.4 compares time averages of simulation results for the squared radius $\langle r^2 \rangle$ to these expressions. In the symmetric phase, fluctuations dominate the average and agree closely with the Gaussian-order estimate of (6.13). In the broken-symmetry phase, $\langle r^2 \rangle$ is dominated by the limit cycle itself, which we estimate from numerical simulations of the solutions to the mean-field dynamics (2.15).

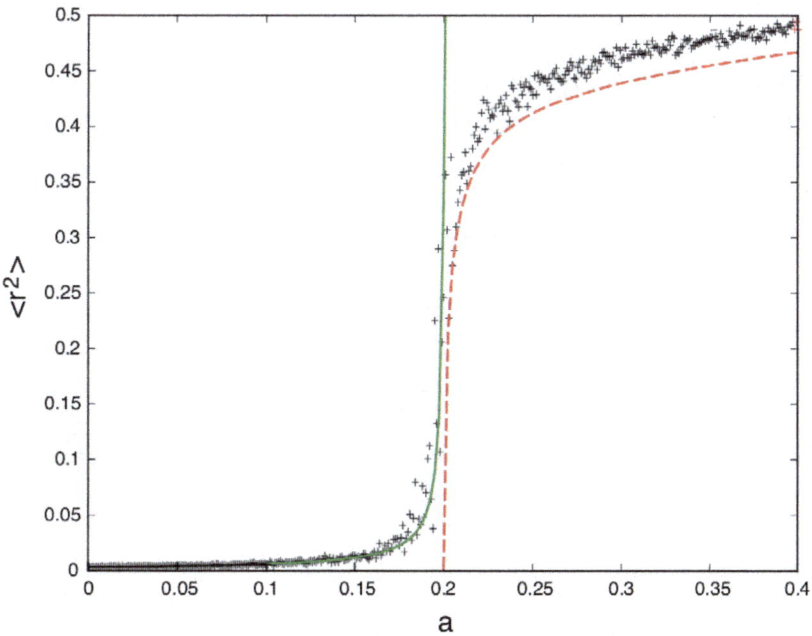

Figure 6.4. Comparison of fluctuation results from analytic estimates and simulations. Time-averaged squared radii $\langle r^2 \rangle$ (crosses), sampled at all a values, are well defined without regard to the circular approximation to the Frenet frame. Equation (6.13) for radial variance in the unbroken phase (green solid) agrees closely. In the broken phase, where most limit cycles are close to the periphery and the Gaussian fluctuation approximation (6.29) is poor, time averages $\langle r^2 \rangle$ over an integer number of limit cycles are computed numerically (red dashed). These underestimate stochastic $\langle r^2 \rangle$ by omission of variance terms. Here $b = 0.2$ and $N = 10000$.

Certain features of this model lead to a rapid transition away from the critical point and into the strongly non-Gaussian regime of fluctuations near the boundary of the simplex, making quantitative comparison to the radial and tangential estimates of (6.29) difficult. If we insert (6.5) for the mean $\bar{r}^2$ on the limit cycle into (6.29) for $\langle (r')^2 \rangle$, we obtain the result that for large N,

$$\left\langle (r')^2 \right\rangle \approx \frac{\bar{a}}{\left(2D\bar{r}\right)^2} \tag{6.30}$$

near the critical point, asymptotically independently of N. For the parameters in figure 6.3 and $\bar{a} \sim 1$ (needed so that $\bar{a} \gtrsim a$), $\bar{r}$ only becomes large enough for $\sqrt{\langle (r')^2 \rangle} \ll 1$ near the simplex boundary. At interior positions of the limit cycle, large fluctuations fill most of the simplex. This feature, together with the coordinate singularity at $r = 0$, makes numerical assignment of fluctuations to the Frenet frame ambiguous. The Gaussian approximation for fluctuations is therefore corrected by boundary terms until the limit cycle enters the strongly non-linear regime near the boundary. A thorough treatment of the oscillatory diffusion constants of very similar models in this non-linear regime is given in [5, 6], so we do not duplicate that analysis here.

6.4 Appendix. Rotating backgrounds, polar coordinates and accumulating Brownian noise

In this appendix we construct the Frenet transformation V of section 6.3.2 through a series of successive approximations. The first of these is the conversion (6.16) to polar coordinates, which may be performed directly in the action, and for which the notation of a time-dependent V is not yet needed.

The second-order action (4.109) in polar coordinates becomes

$$S^{(2)} = N \int d\tau [\eta_r \quad \eta_t] \left(\frac{d}{d\tau} - \overline{\begin{bmatrix} \frac{\partial \dot{\rho}}{\partial \rho} & \frac{\partial \dot{\rho}}{\rho \partial \theta} \\ \frac{\rho \partial \dot{\theta}}{\partial \rho} & \frac{\partial \dot{\theta}}{\partial \theta} + \frac{\dot{\rho}}{\rho} \end{bmatrix}} \right) \begin{bmatrix} r' \\ \bar{r}\theta' \end{bmatrix} + \frac{1}{2} [\eta_r \quad \eta_t] \overline{\frac{\partial^2 \hat{\mathcal{L}}}{\partial \eta^2}} \begin{bmatrix} \eta_r \\ \eta_t \end{bmatrix}. \tag{6.31}$$

Having removed most of the frame dependence of the limit cycle with this transformation, we may now return to the systematic approximation of V and Λ as a small parameter expansion. Equation (6.26) becomes

$$\left(\frac{d}{d\tau} - \overline{\begin{bmatrix} \frac{\partial \dot{\rho}}{\partial \rho} & \frac{\partial \dot{\rho}}{\rho \partial \theta} \\ \frac{\rho \partial \dot{\theta}}{\partial \rho} & \frac{\partial \dot{\theta}}{\partial \theta} + \frac{\dot{\rho}}{\rho} \end{bmatrix}} \right) V = V\Lambda. \tag{6.32}$$

If we take the time-derivative of the trajectory $(\bar{\rho}, \bar{\theta})$ to be the second column of V, and choose the first column simply to be orthogonal, then

$$V = -\frac{1}{\omega}\begin{bmatrix} \rho\dot{\theta} & \dot{\rho} \\ -\dot{\rho} & \rho\dot{\theta} \end{bmatrix} = \frac{1}{\omega}\sqrt{\dot{\rho}^2 + \left(\rho\dot{\theta}\right)^2}\begin{bmatrix} \cos\xi & \sin\xi \\ -\sin\xi & \cos\xi \end{bmatrix}, \tag{6.33}$$

with ξ defined in (6.24). Solving (6.32) then gives

$$\Lambda = \begin{bmatrix} \Lambda^{(11)} & 0 \\ \Lambda^{(21)} & 0 \end{bmatrix}, \tag{6.34}$$

in which

$$\begin{aligned} -\Lambda^{(11)} &= \omega\left[4A_2\bar{\rho}\sin\left(3\bar{\theta} + \alpha - 2\xi\right) - 2A_1\bar{\rho}^2\cos(2\xi)\right] \\ -\Lambda^{(21)} &= \omega\left[4A_2\bar{\rho}\cos\left(3\bar{\theta} + \alpha - 2\xi\right) - 2A_1\bar{\rho}^2\sin(2\xi)\right]. \end{aligned} \tag{6.35}$$

Time-ordered exponentials of 2×2 lower-triangular matrices reduce to elementary scalar integrals, as

$$\mathcal{T}\left(e^{-\int_{\tau 1}^{\tau 2} du\Lambda_u}\right) = \begin{bmatrix} e^{-\int_{\tau 1}^{\tau 2} du\Lambda_u^{(11)}} & 0 \\ -\int_{\tau 1}^{\tau 2} du\Lambda_u^{(21)} e^{-\int_{\tau 1}^{u} du'\Lambda_{u'}^{(11)}} & 1 \end{bmatrix}. \tag{6.36}$$

V and Λ may be constructed in this way for any stationary solution. Using the solution (6.25) for $\bar{\rho}(\bar{\theta})$, the leading order in V is then

$$V = \begin{bmatrix} 1 & \\ & 1 \end{bmatrix} + O(\bar{R}). \tag{6.37}$$

Thus the polar coordinates turn out to be adequate by themselves. The time-ordered exponential (6.36) similarly simplifies to

$$\mathcal{T}\left(e^{-\int_{\tau 1}^{\tau 2} du\Lambda_u}\right) = \begin{bmatrix} e^{-2A_1\omega(\tau 1 - \tau 2)} & 0 \\ 0 & 1 \end{bmatrix} + O(\bar{R}). \tag{6.38}$$

Now it remains only to compute the noise kernel. Evaluation of (4.140) about a uniform background gives

$$\begin{aligned} -\frac{\partial^2\hat{\mathcal{L}}}{\partial\eta_i\partial\eta_j} &= -\frac{1}{N^2}\frac{\mathcal{L}}{\partial\phi_i^\dagger\partial\phi_j^\dagger} + \left[\frac{2}{N} + \frac{b-a}{D^2}\right]\left(\delta_{ij} - \frac{1}{D}\right) \\ &= \left[\frac{3}{D}\left(\bar{a} + \frac{b-a}{2}\right) + \frac{2}{N} + \frac{a-b}{D^2}\right]\left(\delta_{ij} - \frac{1}{D}\right). \end{aligned} \tag{6.39}$$

The leading constant correction in small $\bar{R}^2$ can readily be included, leaving out oscillatory terms as they were left out of V in (6.37),

$$
\begin{aligned}
-\frac{\partial^2 \hat{\mathcal{L}}}{\partial \eta_i \partial \eta_j}\Bigg|_{\bar{R}^2 \neq 0} &= -\frac{\partial^2 \hat{\mathcal{L}}}{\partial \eta_i \partial \eta_j}\Bigg|_{\bar{R}^2 = 0} + \frac{(a-b)}{2D}\bar{R}^2\left(\delta_{ij} - \frac{1}{D}\right) \\
&= -\frac{1}{N^2}\frac{\mathcal{L}}{\partial \phi_i^\dagger \partial \phi_j^\dagger}\Bigg|_{\phi_i = N/D} + \left[\frac{1}{N} + \frac{b-a}{2D^2}\right]\left(\delta_{ij} - \frac{1}{D}\right) \\
&= \left[\frac{3}{D}\left(\bar{a} + \frac{b-a}{2}\right) + \frac{1}{N} - \frac{3(b-a)}{2D^2}\right]\left(\delta_{ij} - \frac{1}{D}\right),
\end{aligned}
\tag{6.40}
$$

but this result differs from (6.39) only by terms of order $1/N$ near the critical point. Now using $D = 3$ for the particular RPS example, the noise kernel (6.40) reduces to the simple expression

$$
-\overline{\frac{\partial^2 \hat{\mathcal{L}}}{\partial \eta_i \partial \eta_j}} = \left[\bar{a} - \frac{a-b}{3} + \frac{1}{N}\right]\left(\delta_{ij} - \frac{1}{D}\right). \tag{6.41}
$$

Incorporating the source (6.41) and the time-ordered exponentials (6.38) in the expression (6.28) for the correlation function, and performing the elementary integrals over $\tau - \tau'$, gives

$$
\begin{aligned}
\left\langle \begin{bmatrix} r' \\ \bar{r}\theta' \end{bmatrix} \begin{bmatrix} r' & \bar{r}\theta' \end{bmatrix} \right\rangle_\tau &= \begin{bmatrix} e^{-2A_1\omega(\tau-\tau')} & \\ & 1 \end{bmatrix} \left\langle \begin{bmatrix} r' \\ \bar{r}\theta' \end{bmatrix} \begin{bmatrix} r' & \bar{r}\theta' \end{bmatrix} \right\rangle_{\tau'} \begin{bmatrix} e^{-2A_1\omega(\tau-\tau')} & \\ & 1 \end{bmatrix} \\
&+ \left(\bar{a} - \frac{a-b}{3} + \frac{1}{N}\right) \begin{bmatrix} \dfrac{1 - e^{-4A_1\omega(\tau-\tau')}}{4A_1\omega N} & \\ & \dfrac{\tau - \tau'}{N} \end{bmatrix} + O(\bar{R}).
\end{aligned}
\tag{6.42}
$$

Restoring the combination of aggregated constants to their expression in terms of the parameters of the problem,

$$
4A_1\omega N = \frac{2N}{D}\left(a - b - \frac{2D^2}{N}\right), \tag{6.43}
$$

and using large $(\tau - \tau')$ to ignore exponentials in the non-zero damping radial eigenvalue, we obtain (6.29).

Bibliography

[1] Bertini L, De Sole A, Gabrielli D, Jona-Lasinio G and Landim C 2002 Macroscopic fluctuation theory for stationary non equilibrium states *J. Stat. Phys.* **107** 635–75

[2] Bertini L, De Sole A, Gabrielli D, Jona-Lasinio G and Landim C 2009 Towards a nonequilibrium thermodynamics: a self-contained macroscopic description of driven diffusive systems *J. Stat. Phys.* **135** 857–72

[3] Smith E 2011 Large-deviation principles, stochastic effective actions, path entropies, and the structure and meaning of thermodynamic descriptions *Rep. Prog. Phys.* **74** 046601
[4] Hofbauer J and Sigmund K 1998 *Evolutionary Games and Population Dynamics* (New York: Cambridge University Press)
[5] Boland R P, Galla T and McKane A J 2008 How limit cycles and quasi-cycles are related in systems with intrinsic noise *J. Phys. A: Math. Theor.* **41** 435003
[6] Boland R P, Galla T and McKane A J 2009 Limit cycles, complex Floquet multipliers, and intrinsic noise *Phys. Rev.* E **79** 051131
[7] Nowak M A 2006 *Evolutionary Dynamics: Exploring the Equations of Life* (New York: Belknap)
[8] Coleman S 1985 *Aspects of Symmetry* (New York: Cambridge)
[9] Smith E 1998 Carnot's theorem as Noether's theorem for thermoacoustic engines *Phys. Rev.* E **58** 2818–32
[10] Smith E 1999 Statistical mechanics of self-driven Carnot cycles *Phys. Rev.* E **60** 3633–5, PMID: 11970197
[11] Tinkham M 2004 *Introduction to Superconductivity* 2nd edn (New York: Dover)

IOP Publishing

Symmetry and Collective Fluctuations in Evolutionary Games

Eric Smith and Supriya Krishnamurthy

Chapter 7

Neutral directions and evolutionary entropy

Externally imposed symmetries are responsible for neutrality in evolutionary dynamics. They produce qualitatively different phenomena than degeneracies among multiple stable population states that arise out of symmetry breaking. Both cases lead to persistent dynamics at the population level, but in neutral systems large fluctuations in population composition can be dense in time. They therefore feed back to change population dynamics and to give steady states different from the those predicted by the replicator dynamic, by amounts that may not vanish in large-population limits. The repeated prisoners' dilemma (RPD) game is the most well-studied example, in which neutrality between a naïve cooperative strategy and a cooperative strategy that protects itself is essential to the survivability of the naïve strategy. We show, first, how fluctuations lead to a game-theoretic parametrization of the average population state that differs from the game describing the individual interactions by terms that are the correct evolutionary dynamic construction for entropy corrections. We then show that neutrality, which is essential to make the repeated game's solution very different from that of the prisoners' dilemma stage game, is a *fragile* property to attain. The symmetry decomposition of repeated games from chapter 3 is used to study recursive strategies that produce neutrality by different means and we use symmetry to understand why different strategies fail to do so in different evolutionary contexts. Finally, we consider crossover and the fitness of particular moves in multistage strategies and show why both linear regressions and interaction effects are required to yield models that correctly account for the way neutrality contributes to protecting naïve strategies.

7.1 Collective fluctuations that affect model estimation and model interpretation

In this chapter we consider a game in which no symmetries are spontaneously broken but a symmetry is imposed as a defining feature of the model, which makes

doi:10.1088/978-0-7503-1137-3ch7

both the play and the fitness of two types of agents indistinguishable when only they occur in a population. Exogenous symmetries of this kind are less widely modeled with evolutionary game theory than the bifurcations of the previous two chapters, because they result in mean-field approximations (MFAs) that show *no fitness difference* between agent types in certain population contexts. Therefore the mean-field equations are satisfied by all population compositions in an extended interval, which is the equivalent either in game theory, or in evolutionary dynamics, of failure to identify a solution. The problems created by such *neutrality* in the fitness map are widely understood, but to our knowledge, a systematic solution of these problems based on symmetries and collective fluctuations has not been carried out.

We choose a canonical form of the RPD game [1] as our example because it touches upon many of the key concepts that arise from the use of games to model evolution and development. Repetition redirects the force of selection to favor a large fraction of occurrences of a cooperative move C in successful strategies. The move C is strictly dominated in the normal form for any single ply, treated as a game on its own, so the prevalence of C in solutions to the repeated game is a large departure of qualitative behavior. However, neutrality in the normal form of the repeated game is essential to attaining this result. The neutrality is between a naïve C strategy and a *defensive* strategy that can protect both naïve C and itself.

At the same time as neutrality is essential, it is *fragile*. The formally indeterminate solution of a strictly neutral evolutionary game can be regulated with a small rate of uniform mutation. Starting from any population that plays only C moves but has an unknown ratio of naïve to defensive cooperative types (who do not need to use their defensive moves), mutation introduces naïve 'defector' types who act as a kind of *police*. The defectors create an advantage for defensive types over naïve cooperators, ensuring that the defensive types are not lost through genetic drift and remain sufficiently numerous to stabilize the population.

However, in the large-population limit, naïve defectors created by rare mutations are too few to contribute significant fitness differentials to other types and the mean-field equilibrium based on their function as police is not, in fact, the equilibrium about which the population settles. We compute the correct equilibrium, which is stabilized not by the vanishingly small mean fraction of police, but by large fluctuations among all types in the population. This role of fluctuations leads to a normal-form expansion for fitness with the properties of RPD either at the individual or the population-average level, but with different coefficients at the two levels.

The other fragilities stemming from a solution that requires neutrality come from non-deterministic moves (called 'trembles') in the repeated game and from crossover leading toward linkage equilibrium if we model the game moves as genes. We explore the consequences of trembles and develop a small combinatorial example to suggest that neutrality of this form will become typical in models with large extensive-form games. Finally, we consider the dynamical maintenance of linkage disequilibrium in RPD and show why fourth-order associations in the regression model for fitness are essential to explain linkage and hence the maintenance of neutrality.

7.2 Externally imposed symmetries: neutrality and the fragility of the mean-field replicator equation

The normal-form prisoners' dilemma is a single-stage game (so moves and strategies are the same thing) between two players, whose types may be *C*ooperate (*C*) or *D*efect (*D*). It is the simplest non-cooperative game in which a Pareto-superior strategy profile[1] (C, C) is strictly dominated by the Pareto-inferior profile (D, D) [2]. Therefore it has become the most widely used stage game against which to study what can be accomplished with repetition. The finitely repeated prisoners' dilemma in evolutionary game theory uses a combination of recursively defined strategies in an extensive-form game with a replicator dynamic to achieve high frequency of (C, C) move profiles in the strategies that determine population-averaged fitness. It also achieves a high penetrance[2] of the naïve-*C* genotype in populations where, alone, it would be excluded for the same reason the *C* move is dominated in the stage game. The RPD model may be understood as showing how repetition together with evolutionary stability can select Pareto superiority over stage-game dominance, while remaining within the space of non-cooperative solution concepts.

The version of minimal RPD that we will use here to illustrate the consequences of neutrality is played by two memory-zero strategies labeled ALLC and ALLD, identical to *C* and *D* in the stage game, and a single memory-one strategy. The zero-memory strategies are defined on the maximally coarse-grained tree of figure 3.2 and the memory-one strategy uses only opponent's move in the previous round, and so is one of the six additional strategies on the coarse-grained game tree of figure 3.5. The memory-one strategy, known as tit-for-tat (TFT) [3], consists of an opening move *C*, followed by a recurrent-stage move that copies the opponent's move in the preceding round. The sub-set of transitions between move profiles from the finite-state machine of figure 3.7—with an initial state added—that are visited by *deterministic* RPD are shown in figure 7.1.[3]

The population state in RPD is denoted by a vector

$$\mathrm{n} \equiv \begin{bmatrix} \mathrm{n_C} \\ \mathrm{n_D} \\ \mathrm{n_T} \end{bmatrix}. \tag{7.1}$$

[1] A profile such as (C, C) is called *Pareto superior* to another such as (D, D) if, in the former, at least one of the two players receives a higher payoff and neither player receives a lower payoff. In this case, both players receive higher payoffs.

[2] In genetics, the term *penetrance* is defined to be the fraction of individuals in a population carrying a particular allele. Since the execution of *C* moves is a property both of the move assignments in successful strategies and of player pairings that cause the *C* move to be played, the frequency of *C* moves requires penetrance of the appropriate 'allele', but also stabilization of the appropriate population states.

[3] We consider the effect of 'trembling-hand' moves in section 7.3.1.

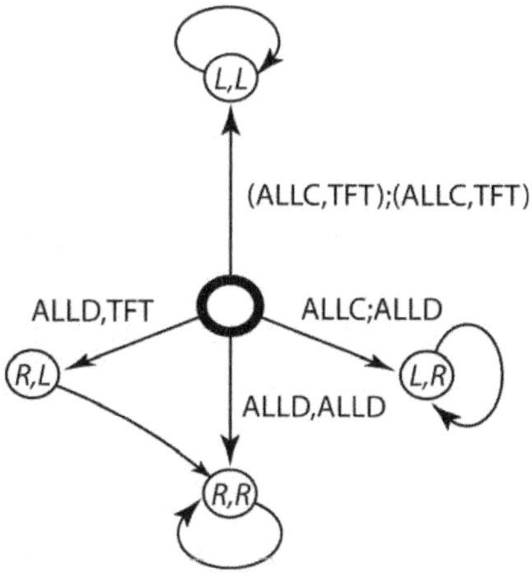

Figure 7.1. Sub-set of transitions from the finite-state automaton (FSA) of figure 3.7 produced by the deterministic RPD. Each possible *pair* of information sets in figure 3.7 corresponds to a state of the FSA. An initial state (heavy circle) represents the state of play before either player has moved. The moves out of this state correspond to the moves from information sets 0 and 1 in figure 3.7. After any stage has been played, since only memory-one is possible, pairs of information sets are uniquely labeled by the pairs of moves that lead to them, in the same manner as sufficient statistics are associated with symbolic histories in [4, 5]. (Player roles are symmetric and only ordered matches are shown to simplify the diagram while emphasizing strategy asymmetries. The transition labeled (ALLC, TFT); (ALLC, TFT) stands for any pairings of these two strategies. We retain the designation L and R for moves from chapter 3, which reflect the *mechanics* of these strategies in the game tree, whether or not the payoffs give them the interpretation of C and D moves.)

The normal-form payoff, following Rapoport's notation [2], is given by

$$[a] = \bar{a}\begin{bmatrix}1\\1\\1\end{bmatrix}\begin{bmatrix}1 & 1 & 1\end{bmatrix} + \begin{bmatrix}R & S & R\\ T & P & P+\epsilon_{TP}\\ R & P+\epsilon_{SP} & R\end{bmatrix}. \tag{7.2}$$

The payoffs R, S, T and P, for moves C and D are those of the stage game. The corrections to infinitely-repeated play when TFT plays ALLD, $\epsilon_{TP} \equiv (T - P)/N_{\text{ply}}$ and $\epsilon_{SP} \equiv (S - P)/N_{\text{ply}}$, account for first-round effects.

The equality of the four R entries in 7.2 (any combination of ALLC and TFT) is the important feature that distinguishes deterministic RPD from the prisoners' dilemma stage game. This feature makes a *neutral space* of all population states with $n_D = 0$. The connection of these population states by mutation and replication makes the space into a *neutral network* [6–11]. Neutrality is an externally imposed symmetry of the model. Together with the ordering of payoffs R, S, T and P for the stage game, these define the qualitative features of minimal RPD.

Figure 7.2 shows solutions to the mean-field dynamics (2.15) with payoffs (7.2). For the parameters used, the corrections ϵ_{TP} and ϵ_{SP} are ignorable compared to mutation and the ratio of selection strength to mutation rate is set by N. Neutrality between ALLC and TFT results in a contour of slowest diffusion (the concentrating

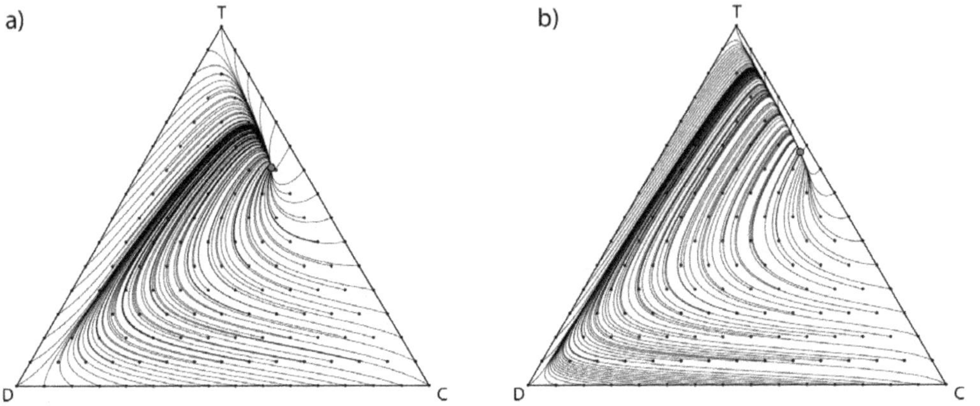

Figure 7.2. Mean-field population trajectories for the RPD defined by payoffs (7.2). Populations of pure types ALLC, ALLD and TFT are labeled C, D and T. Parameters of individual play are $R = 0.6$, $S = 0.1$, $T = 0.8$, $P = 0.3$ and $N_{\text{ply}} = 150$ in both panels. Here a small mutation rate produces a rest point offset a small distance from the interior of the C–T line and slow convergence to the rest point parallel to that line. The relative strength of mutation to payoffs scales as $1/N$. Increasing the number of players—$N = 100$ (*a*) versus $N = 300$ (*b*)—therefore slows convergence and moves the slow-flow line including the mean-field rest point toward the C–T axis.

contour for flow lines) toward the rest point of the classical equations. Comparing the two panels with $N = 100$ versus $N = 300$ in figure 7.2 shows that the speed of approach along this line becomes slower (flowlines bend more sharply into it) and the offset of the line from the pure C–T axis smaller, both effects scaling as $\sim 1/N$.

7.2.1 Regulating the mean field with population size converges, but to the wrong answer

At any finite N, the RPD is no longer a neutral theory when the average effects of mutation are incorporated into the dynamics. Moreover, one can see from figure 7.2 that, while the rest point approaches the C–T boundary as $1/N \rightarrow 0$, the point approached is in the interior of the boundary (it does not converge to either of the vertices) and so defines a regular limit. A sequence of mean-field models at increasing N (see figure 7.3 below) shows that this regularity persists in the large-N limit.

One might have expected that such an approach—using a weak perturbation such as mutation to make the MFA well defined and then taking the limit of the regulator's strength (or the mutation rate's) going to zero to define the neutral theory—would be an appropriate solution when the limit is regular as it is here. However, that conclusion is incorrect both quantitatively and conceptually. The limit of mean-field analyses with weak regulators is not merely a singular perturbation in the sense that it is algebraically well defined at all finite N but undefined at '$N = \infty$'. Rather, the regulator becomes weak *regularly* in $1/N$, allowing fluctuation effects to become increasingly important, which the mean-field analysis omits entirely. The proper point at which to discount the mean-field estimate is the point at which quantitatively the regulating effect of the *average number of* defectors becomes weaker than

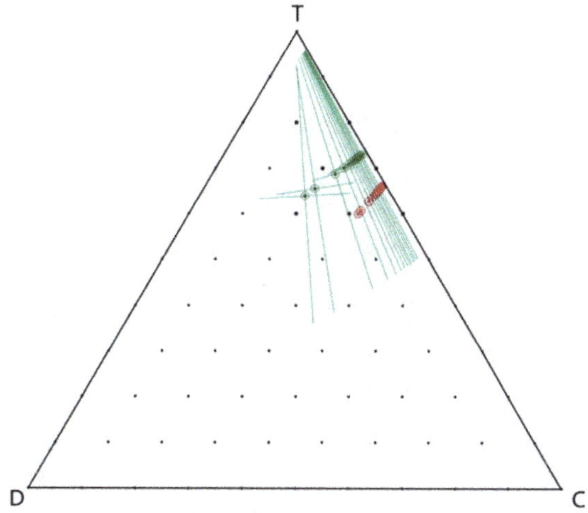

Figure 7.3. Self-consistent solutions for rest points of the mean behavior of the (infinite-population) stochastic replicator with neutral directions (green points with crosses) show systematic deviation from the rest points of the deterministic replicator (red points).

the *average regulating effect of the transient number of defectors* sampled by large population fluctuations. The largeness of collective fluctuations is of course precisely a consequence of the weakness of the average regulator.

Collective fluctuations due to increasing neutrality have a distinct source and distinct scaling from fluctuations due to increasing mutation rate. The latter scale only as $1/\sqrt{N}$, as a fraction of the total population, because they self-average under the central limit theorem. Excursions due to neutrality occur essentially as drift *at any finite N*, reaching magnitudes $\sim N^0$, until transient populations are generated that can trigger significant mean regression.

7.2.1.1 Large collective fluctuations make equations of motion responsive to averages of population states

The speed of approach toward a rest point in the deterministic evolutionary game equation determines the rate of mean regression for fluctuations in the stochastic process. For a fixed rate of random type change, the half-width of fluctuations is inverse to the rate of mean regression. The convergence scales as $1/N$, while the population size scales as N along the direction of slowest approach in figure 7.2. Therefore, these two scaling effects cancel and in the large-population limit fluctuations of a finite *fraction* of agents characterize the stationary distribution at large N.[4] The problem for a computational method is to provide the correct

[4] In contrast, all mean regression in the coordination game of chapter 5 and regression normal to the limit cycle in the rock–paper–scissors game of chaper 6 scale as N^0. This does not cancel the central-limit theorem averaging of fluctuations, so the fluctuation magnitude as a fraction of total population in these previous cases scales as $1/N$, giving convergence to the replicator equation. Recall also that the rate of Brownian motion tangential to the limit cycle in (6.29) is independent of N, similar to the large fluctuations in this model.

weighting with which transiently sampled population states contribute to average behavior.

7.2.2 The correct regulation with collective fluctuations

We begin with the alternative power-series expansion (4.120) of the action S introduced in section 4.4.3, appropriate when fluctuations may not be small and third-order terms may not be suppressed relative to the classical stationary-path equations.

We will not attempt to compute here the corrections to dynamical stationary paths, which require the evaluation of time-ordered exponentials about dynamical backgrounds in the solution (4.106) for $\left\langle \phi_n' \phi_m' \right\rangle$. We will compute corrections to asymptotic steady states, by solving (4.121) for the flow to fixed points, but evaluating both derivatives of $\bar{\mathcal{L}}$ and $\left\langle \phi_n' \phi_m' \right\rangle$ as if the backgrounds were constant. The resulting flow will deviate from the true fluctuation-corrected equations of motion, but will converge to it in neighborhoods of the fixed points where change of the true backgrounds becomes slow.

7.2.3 Evolutionary entropy corrections and the 'free fitness'

Stationary paths of the form (4.121) are the minima of a function of the background fields known as the *effective action*[5]. Effective actions are computed by first introducing time-dependent sources into the path integral (4.66) to produce a *moment-generating functional*, whose logarithm is called the *cumulant-generating functional.* These are the dynamical generalizations of, respectively, the partition function and the free energy from equilibrium thermodynamics. The functional Legendre transform of the cumulant-generating functional is the effective action, which is the non-equilibrium generalization of the equilibrium entropy.

In quantum field theory, this functional (of classical background fields) is known as the *quantum effective action* [13]. It is the basis for a wide variety of background-field methods and renormalization schemes based on them. We have termed its counterpart for classical stochastic processes the 'stochastic effective action', and developed its role in large-deviations theory and a thermodynamics of histories in [12]. As we noted in section 5.2.1, the stochastic effective action is also known in Freidlin–Wentzell theory as the *quasipotential* [14], because it is a path-space generalization of the equilibrium (Helmholtz or Gibbs) free energy.

Elegant graphical methods exist [13] for defining and computing the cumulant-generating functional and effective action in many contexts. For the classical solutions of the Doi–Peliti dynamical system, in which $\bar{\phi}^{\dagger} \equiv 1$, the fluctuation correction to the

[5] The systematic development of the concept of effective action is extensive enough to fall outside the scope of this treatment and for this we refer the reader to [12]. The opening paragraphs of this section summarize the main elements of the construction. The term 'effective action' was introduced in section 4.4, where some approximation schemes derived from it were summarized. For elementary stationary-point expansions, the approximations are simple enough that the integral S, regarded as a function of classical fields, serves as an approximation to the effective action. The first feedback of fluctuations to expand the effective action beyond the classical approximation, and to alter the stationary point conditions, was introduced in section 4.4.3. That expansion is more fully developed in this chapter.

first variational derivative, given in (4.121), is simply the first derivative with respect to the background $\bar{\phi}^\dagger$ of an action in which $\bar{\mathcal{L}}$ is replaced by the quantity

$$\mathcal{L}^{\text{eff}} = \bar{\mathcal{L}} + \frac{1}{2} \sum_{m,n=1}^{D} \overline{\frac{\partial^2 \mathcal{L}}{\partial \phi_m \partial \phi_n}} \left\langle \phi_n' \, \phi_m' \right\rangle + \cdots . \tag{7.3}$$

$\mathcal{L}^{\text{eff}}$ could be called an 'effective Liouville function'; it is the function that at the *classical* level produces rest points equal to the first moments actually produced by the population dynamics. Because fluctuation corrections are not added to the effective action (they are already included), its parameters have the interpretation of regression coefficients for the average population behavior, as if that behavior reflected the fitness and mutation coefficients of a set of 'effective agents' that are too large to undergo significant fluctuation and whose individual interaction parameters therefore provide a good description of whole-population behavior. With respect to the Doi–Peliti Hamiltonian dynamical system, this effective Liouville function is a direct counterpart to the *effective Hamiltonian* of a one-loop renormalized quantum or stochastic field theory.

The fluctuation corrections in stochastic evolutionary games have the same origin as the entropy corrections in thermodynamics, which distinguish internal energies U from *free energies* $F \equiv U - T\mathcal{S}$ [15]. Averaging a Gibbs function $e^{-H/k_B T}$ (where H is the Hamiltonian), in an ensemble whose fluctuations are governed by H, yields the partition function $e^{-F/k_B T}$. In the Doi–Peliti sum over histories, $\mathcal{L}$ is the Hamiltonian, and its self-consistent average is $\mathcal{L}^{\text{eff}}$.

To make this correspondence more explicit, notice that, in the generating functional (4.66), if all values $\left(\phi^\dagger, \phi\right)$ are integrated except those at a single time, the remaining integral defines a density over a single set of coherent-state fields,

$$1 = \int_0^T \mathcal{D}\phi^\dagger \mathcal{D}\phi \, e^{-S} e^{\left(\phi_0^\dagger - 1\right)\cdot\left(\bar{n} - \phi_0\right)} \equiv \int \frac{d^D\phi_t^\dagger \, d^D\phi_t}{\pi^D} W_{\phi_t^\dagger, \phi_t}. \tag{7.4}$$

$W_{\phi_t^\dagger, \phi_t}$ is called the *Wigner function* corresponding to the probability density ρ_n at time t. It is the distribution over the continuous, but overcomplete, set of coherent-state indices at a single time which gives the equivalent weight function to the one given by ρ_n for number states. At late times in a system whose stationary paths converge to a stable rest point, both ρ_n and $W_{\phi^\dagger, \phi}$ become t-independent. The stationary-path condition computed by varying the stochastic effective action, with time-independent backgrounds, simply equals the average of the classical stationary-path condition in the Wigner distribution for fluctuations about that average:

$$\begin{aligned} \overline{\frac{\partial \mathcal{L}^{\text{eff}}}{\partial \phi_i^\dagger}} &\equiv \overline{\frac{\partial \mathcal{L}}{\partial \phi_i^\dagger}} + \frac{1}{2} \sum_{m,n=1}^{D} \overline{\frac{\partial^3 \mathcal{L}}{\partial \phi_i^\dagger \partial \phi_m \partial \phi_n}} \left\langle \phi_n' \, \phi_m' \right\rangle \\ &= \int \frac{d^D\phi^\dagger d^D\phi}{\pi^D} W_{\phi^\dagger, \phi} \frac{\partial \mathcal{L}}{\partial \phi_i^\dagger}. \end{aligned} \tag{7.5}$$

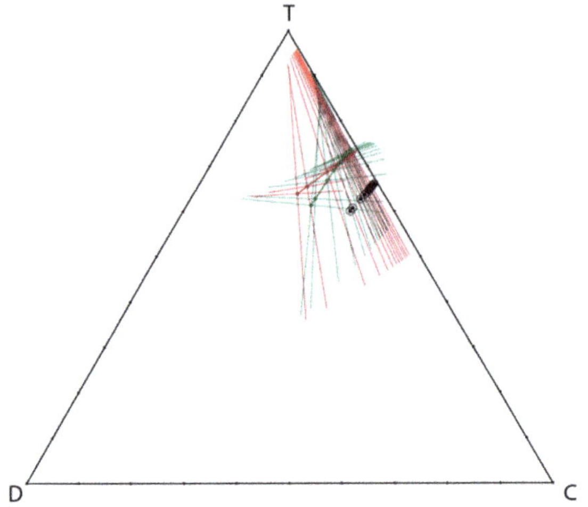

Figure 7.4. Simulation results (green) compared to the leading-order analytic fluctuation corrections (red) and to the naïve mean-field limit (black points), for the same sequence of N values as in figure 7.3.

(All the low-order approximations that have been derived here for coherent-state fields have counterparts in action-angle variables and specific forms are provided in appendix 4.6.2.)

Figure 7.3 and 7.4 show the consequences of fluctuation corrections to the mean large-N rest points. The circles without axes drawn through them in the figures are the rest points corresponding to figure 7.2 for a sequence of increasing N.[6] The mean population compositions, when self-consistently averaged over fluctuations by (7.5) so that $\overline{\partial \mathcal{L}^{\text{eff}} / \partial \phi_i^{\dagger}} = 0, \forall i$, are shown as the circles with crosses. The crosses identify both the major and minor axes of the noise covariance in the population state from the combination of noise accumulation and relaxation. Both sequences have regular limits at large N. However, the naïve mean-field rest points differ from the evaluation of (4.121) by ~10%.

The naïve mean-field and one-loop corrected rest points are compared to simulation results in figure 7.4. Equation (4.121) converges at large N to simulation values within ~1%. Since we lack a formal small parameter such as $1/N$ to control the fluctuation expansion, it was not assured that the first-order correction alone would produce such close agreement.[7] Proofs of convergence of the all-orders expansion remain a topic for future work.

This example illustrates an important general way to understand and resolve singular limits in evolutionary game theory. The zero-policing limit of the mean-field

[6] N values used were, from left to right in the figure: 90, 100, 150, 200, 250, 300, 350, 400, 450, 500, 600, 700, 900.

[7] Note that the significance of convergence between the Gaussian fluctuation model and simulations at large N is not due to expansion of the orders of fluctuation, but rather to the fact that the model is only nearly neutral within a domain of distance $1/N$ from the boundary. It is necessary that the mean-field non-neutrality in the interior give way to control by fluctuations, before the simplified Gaussian fluctuation expansion is expected to become a good approximation.

analysis is an apparently removable singularity at $1/N \rightarrow 0$. The appearance is deceptive: the singularity signals a real effect, though not the one that it appears to signal within mean-field theory. The only reason a singularity can exist is that, in a finite range of small $1/N$, policing is too weak to exercise any significant influence on the population. Some other regime must take over to control dynamics, even if mean-field theory cannot suggest what that regime is or estimate its properties.

As a category of mechanisms, the policing function is *fragile* at large N. In the stochastic theory, fragile policing never reaches the singular limit, because within the interior of the simplex, the stabilizing function shifts from the mean-field parameters to the distribution over fluctuations.

7.3 On neutrality and fragility

Neutrality between a defensive strategy (in the worked example, this has been TFT), and the unconditionally cooperative strategy ALLC, is essential for the normal-form payoffs to direct the force of selection to overcome the dominance of ALLC by ALLD in the stage game. In addition to the fragility with which policing by ALLD stabilizes populations of TFT and ALLC against drift along the neutral boundary at large N, the property of neutrality itself is fragile in many limits. A symmetry analysis of the game trees and strategies that are used to work around various dimensions of fragility can be very helpful in understanding when different games even address the same coordination problem and when they do not.

7.3.1 Trembles in finitely RPD and other sources of fragility in overcoming stage-game dominance

The most widely studied fragility of RPD replaces the deterministic repeated game by one with 'trembling-hand' moves [16], which have a small independent probability in each round to be opposite to the move in the deterministic strategy. Trembles, in very long repeated games, unravel the ability of TFT to coordinate and hence to achieve neutrality with ALLC and assure either high frequency of (C, C) move profiles or high penetrance of the ALLC genotype, as shown in figure 7.5. In the limit of infinite repetition, the coordination of deterministic RPD is a (non-removable!) singular limit with respect to any probability of trembles. This fragility, like the fragility of policing, can be regulated by choosing the number of repetitions in relation to the rate at which trembles cause coordination in TFT to decay to the equivalent of random play. However, sufficiently long repeated games are required to reduce first-move effects in the payoffs (7.2) below the randomizing strength of mutation, so the fragility of trembles is meaningful and is tied to the fragility of policing at large N.

One way to address the fragility of coordination that is displayed by TFT with trembles is to replace TFT in the role of the defensive strategy with a parity strategy known as 'Pavlov' (or win–stay/lose–shift; WSLS) [17–19]. The parity

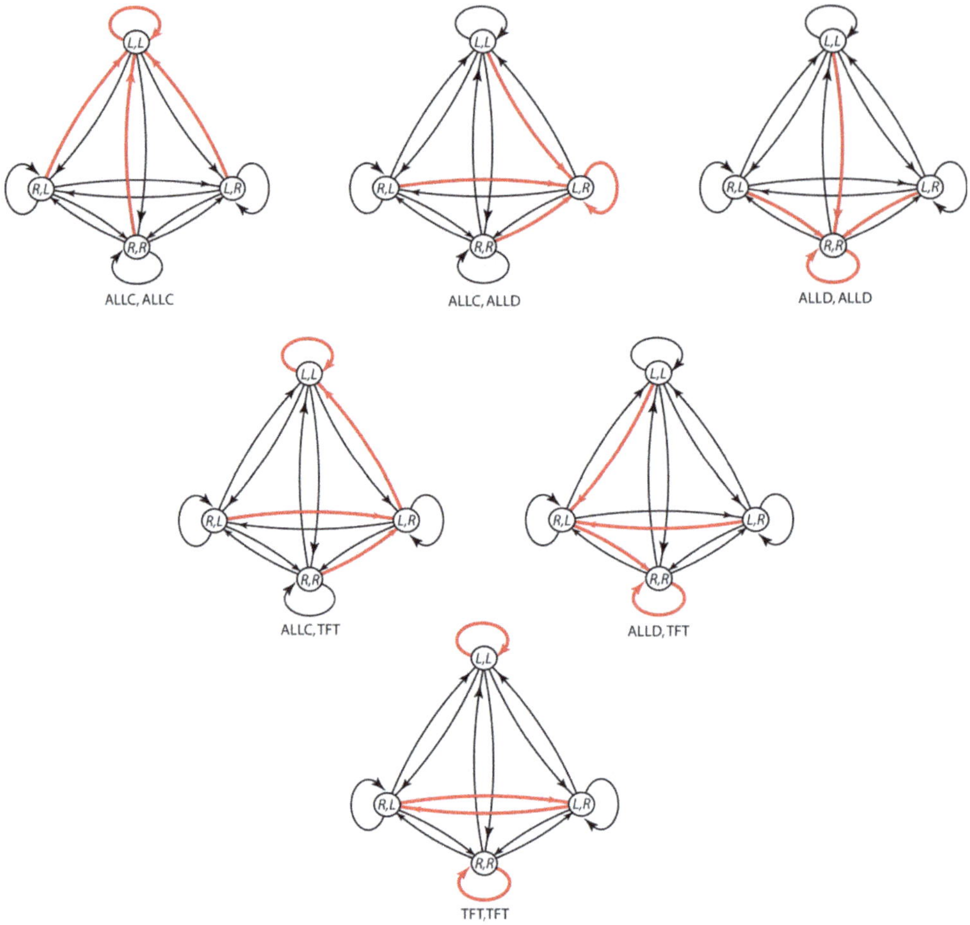

Figure 7.5. Decomposition of the possible recurrent moves (red) from figure 3.7 according to the matched strategies. (Only ordered matches are shown, as in figure 7.1, to aid viewing.) All strategy pairs formed from ALLC, ALLD and TFT—except TFT,TFT—are robust against both recombination and trembles (trembles result in moves in black), returning players within no more than two moves to the absorbing states. TFT,TFT, however, will sustain three recurrent patterns, one of which involves a two-state alternation. If the opening move is mixed as in section 7.3.3, each sub-graph is stable. If instead non-canonical starting states result from unbiased trembles, the three sub-graphs (*L*,*L*, *L*,*R*/*R*,*L*, *R*,*R*) are visited in ratios 1:2:1. Mixing of sub-graphs in these ratios leads to the same mixture of payoffs as pairing of equal-weight mixed strategies (purely random play) for both players. In repeated games with very many plies, convergence to this distribution over the repeated move profiles removes the neutrality of TFT with ALLC, requiring replacement of TFT by WSLS [17] to evolve cooperation. The rate of trembles in relation to the number of repetitions may be used to tune the transition between the deterministic and random limits.

strategy is defined on the game tree of figure 3.6 and moves left (C) for even parity and right (D) for odd parity. A pair of WSLS players, if one of them trembles, enters a three-move cycle that returns them to repeated (C, C) move profiles until the next tremble.

WSLS is sometimes presented as a more robust solution than TFT to overcoming the ‘paradox of altruism’, but this reading of a limited range of evolutionary contexts is incomplete as emphasized by the poor performance of the strategy in other contexts. Pavlov strategies were known already to Rapoport [2], who characterized them with the name ‘The Simpleton’ [3]. In the first tournament of Axelrod that was won by TFT, WSLS placed in the lowest quartile [20]. That tournament used deterministic play, for which TFT’s fragility to trembles was not exposed, but its flexibility against an open-ended slate of opponents was a strength. In contrast, WSLS fared poorly because the same predictability that makes it robust against trembles makes it exploitable by many innovations.

The symmetry analysis of chapter 3 provides a principled way to see that selection of the (C, C) move profile is inherently fragile and may be undermined in a variety of ways. Strategies that are immune to some sources of fragility are *by the same traits* more susceptible to others. Thus while some strategies may be intrinsically poor, the best that one may be able to do with a good strategy is match its strength to the evolutionary setting.

The comparison of the game trees in figure 3.5 and figure 3.6 shows that the main difference between TFT and WSLS is that one retains left/right symmetry and the other violates it. Except in the opening stage, the assignment of TFT moves on the ‘other’s-move’ game tree of figure 3.5, is symmetric under reflection about the vertical in the figure, replacing left for right branches in both the tree and the assignments (where left is called C and right is called D). The first stage is not symmetric, because its assignment of C to a unique information set for each player changes under reflection to an assignment of D. However, the first move contributes a fraction $1/N_{\text{ply}}$ to payoff, while the recurrent move accounts for the remainder. This reflection symmetry makes TFT equally flexible in matching C or D moves by opponents in contexts of any duration. However, in order for TFT to succeed when matched with itself, it must preserve the coordination from an initial (C, C) move profile to arbitrarily late times; otherwise its symmetry makes it no better than random.

In contrast, the WSLS move assignments on the parity tree of figure 3.6 are asymmetric under reflection: the information sets are preserved but the moves are switched in both the opening and the recurrent stages. The map from both the stage-dominant (D, D) and the goal (C, C) profiles towards a C move makes WSLS less flexible than TFT. It does not need to preserve any intra-game coordination with WSLS opponents from the opening move, making it nearly immune to trembles that degrade intra-game coordination, because the preference for C is inherited when the agent is born. However, because it is asymmetric it is readily exploited.

Respecting or violating symmetry thus selects a *timescale* to control fitness. Symmetric TFT is responsive within generations through flexibility at the ‘development’ level, while WSLS can respond only between generations as selection judges whether its inherent asymmetry is well or poorly matched to the payoff structure and any restrictions imposed in the arena of selection.

7.3.2 Neutrality will be a typical feature of extensive-form games

In deterministic RPD, the neutrality between ALLC and TFT is a form of *payoff neutrality*, externally imposed but not a feature required by the structure of the game itself. A simple counting argument suggests that, in extensive-form games with more plies, a different *structural* neutrality may become very typical. Structural neutrality is distinguished from mere payoff neutrality, in that structural neutrality is enforced by the combinatorics of the extensive-form game, no matter what payoffs are assigned at the leaves.

The counting argument goes as follows; we will use the total information simultaneous-move tree of figure 3.1 as an example. Equation (3.2) gives the total number of strategies for one player on a tree of N_{ply} plies. The total number of strategy profiles (pairs of strategies for both players) is then

$$|\{i\}|^2 = 4^{\frac{2^{2N_{\text{ply}}}-1}{2^2-1}}. \tag{7.6}$$

However, the number of leaves on the game tree is only $4^{N_{\text{ply}}}$. Therefore in a large game tree many strategy pairs must lead to the same leaf and, no matter how payoffs are assigned, to the same payoff. Starting from any given strategy pair, most nodes in the tree will not be visited. For many changes of move in the strategy of either player, the nodes also will not be visited by the revised strategy pair, so the change must be neutral to the payoff of both players. If the mutation process defines a notion of adjacency between strategies, then neutral sets are given the topology of neutral networks.

To demonstrate structural neutrality, we continue within the context of trembling-hand play of a repeated game with a balanced collection of memory-one strategies built from a two-player normal-form stage game. Rapoport's notation will continue to denote payoffs in the stage game, but in this example we are not concerned with their numerical values or the particular equilibria of the stage game. For simplicity, we continue to restrict attention to games with symmetric player roles.

We consider the three memory-one trees in figures 3.4, 3.5 and 3.6. These trees jointly admit 18 new strategies with equal numbers of L and R moves, in addition to the two strategies from figure 3.2. From these four trees, we select $2^3 = 8$ strategies (two binary complements from each tree) that contain TFT and its opposite (from figure 3.5), WSLS and its opposite (from figure 3.6), and strategies REP and ALT (from figure 3.4) that repeat or alternate the player's own previous move.

The recurrent finite-state diagrams for all pairs of these strategies are shown in table 7.1. Ignoring first-move effects, the payoffs from a long-time trembling hand recurrent move for all pairings of these strategies against one another are shown in table 7.2.

Table 7.1. Recurrent finite-state automata visited by the eight strategies in trembling-hand prisoners' dilemma. Each node indicates a recurrent pattern of move profiles and the sequence of row- and column-player moves in that pattern is indicated. Arcs between the nodes are created by trembles with rare frequency per move. Transients and opening-move effects are not shown. In cases where nodes have cycles of different lengths, the probabilities on arcs are equal per term in the cycle. The practical consequence is that all cycles containing all four move profiles lead to equal visits when averaged over the late-time distribution over the recurrent states. Entries in the lower triangle (omitted for clarity) are the mirror images of those in the upper triangle.

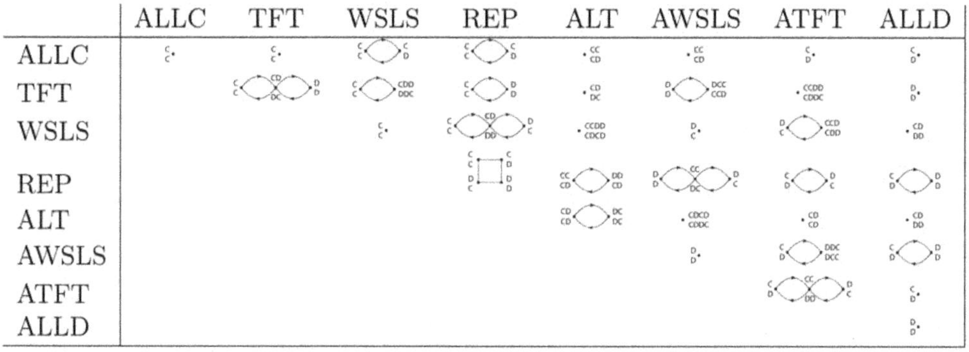

The eight strategies may be assigned a three-digit binary 'genotype'[8], given by

$$\begin{aligned} 000 &\leftrightarrow \text{ALLC} \\ 001 &\leftrightarrow \text{TFT} \\ 010 &\leftrightarrow \text{WSLS} \\ 011 &\leftrightarrow \text{REP} \\ 100 &\leftrightarrow \text{ALT} \\ 101 &\leftrightarrow \text{AWSLS} \\ 110 &\leftrightarrow \text{ATFT} \\ 111 &\leftrightarrow \text{ALLD}, \end{aligned} \tag{7.7}$$

and illustrated in figure 7.6. We may introduce a 'Hamming' mutation process that allows single bit-flips, moving player types one step over edges in the cube. In this mutation metric, each strategy is antipodal to its binary complement and the two zero-memory strategies ALLC and ALLD are each connected to their three refinements in the memory-one trees.

The neutral sets of an evolutionary game with these strategy profiles are readily seen from table 7.2. The mutation process requires a six-dimensional hypercube to illustrate, but the neutral networks formed by the Hamming mutation process can be illustrated with ties in the original matrix, as shown in figure 7.7.

[8] This assignment is partly arbitrary. The non-arbitrary criteria are that mirror-image strategies are placed at antipodes of the hypercube and the two biased-cooperative recurrent moves ALLC and WSLS are neighbors. TFT was chosen as a neighbor of ALLC because it is commonly paired with an opening *C* move, so the neutrality of these two strategies forms a network under mutation of only the recurrent move between the two. Other assignments would yield neutral networks that differ in detail, but show similar distributions of links.

Table 7.2. Payoffs for the player whose strategies are listed across the top of the table, when paired with the player listed on the left. Strategies have memory zero or one, and even numbers of L and R moves. X stands for the equally mixed payoff $(R + T + P + S)/4$, which would be achieved also by random play. Note that, like the payoffs (7.2) of RPD, the table goes into its transpose under $T \leftrightarrow S$.

	ALLC	TFT	WSLW	REP	ALT	AWSLS	ATFT	ALLD
ALLC	R	R	$\frac{R+T}{2}$	$\frac{R+T}{2}$	$\frac{R+T}{2}$	$\frac{R+T}{2}$	T	T
TFT	R	X	X	$\frac{R+P}{2}$	$\frac{T+S}{2}$	X	X	P
WSLW	$\frac{R+S}{2}$	X	R	X	X	S	X	$\frac{P+T}{2}$
REP	$\frac{R+S}{2}$	$\frac{R+P}{2}$	X	X	X	X	$\frac{T+S}{2}$	$\frac{P+T}{2}$
ALT	$\frac{R+S}{2}$	$\frac{T+S}{2}$	X	X	X	X	$\frac{R+P}{2}$	$\frac{P+T}{2}$
AWSLS	$\frac{R+S}{2}$	X	T	X	X	P	X	$\frac{P+T}{2}$
ATFT	S	X	X	$\frac{T+S}{2}$	$\frac{R+P}{2}$	X	X	T
ALLD	S	P	$\frac{P+S}{2}$	$\frac{P+S}{2}$	$\frac{P+S}{2}$	$\frac{P+S}{2}$	S	P

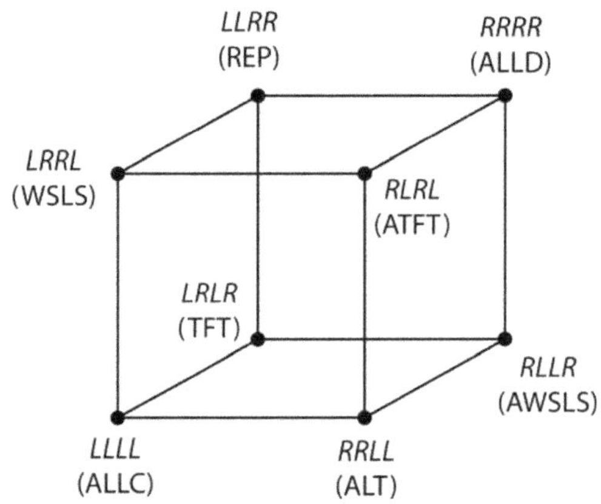

Figure 7.6. An assignment of the eight strategies of table 7.2 to the vertices of a cube. Mutation moves player types between vertices connected by an edge. The left or right move at each node in the information sets (2–3) is indicated above the strategy name at each vertex.

Even for this small strategy set, the neutral space consists of two large, interwoven cycles of 12 pairs (the two long loops in the figure), one chain of three pairs, ten sets with two pairs and 17 singletons. The adjacency structure of these neutral networks to one another under the Hamming mutation can be seen from the graph of the hypercube in figure 7.8.[9] In the figure, horizontal moves represent strategy changes by single bit flips on the hypercube of figure 7.6 for the first player; vertical moves

[9] In the figure, payoff values are indicated by the color scheme. (The three intermediates $(R + P)/2$, $(T + P)/2$, $(S + P)/2$ are not labeled for clarity.) Neutral networks in the hypercube are drawn with heavy links in the same colors as the nodes they connect.

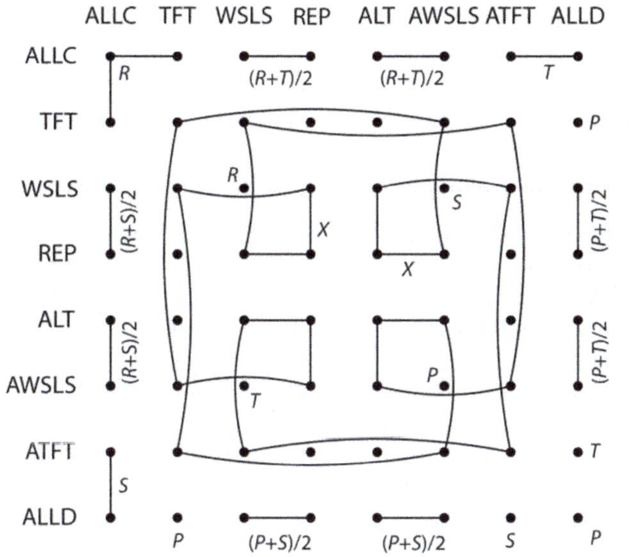

Figure 7.7. The neutral networks of figure 7.6 drawn on the payoff table 7.2. Neutral networks are labeled with the values of the nodes to which they correspond and as in the table X stands for $(R + T + P + S)/4$. There are two neutral networks for each of $(R + T)/2$, $(R + S)/2$, $(P + T)/2$, $(P + S)/2$ and X. Eight internal nodes, which have no non-trivial networks, remain unlabeled for clarity.

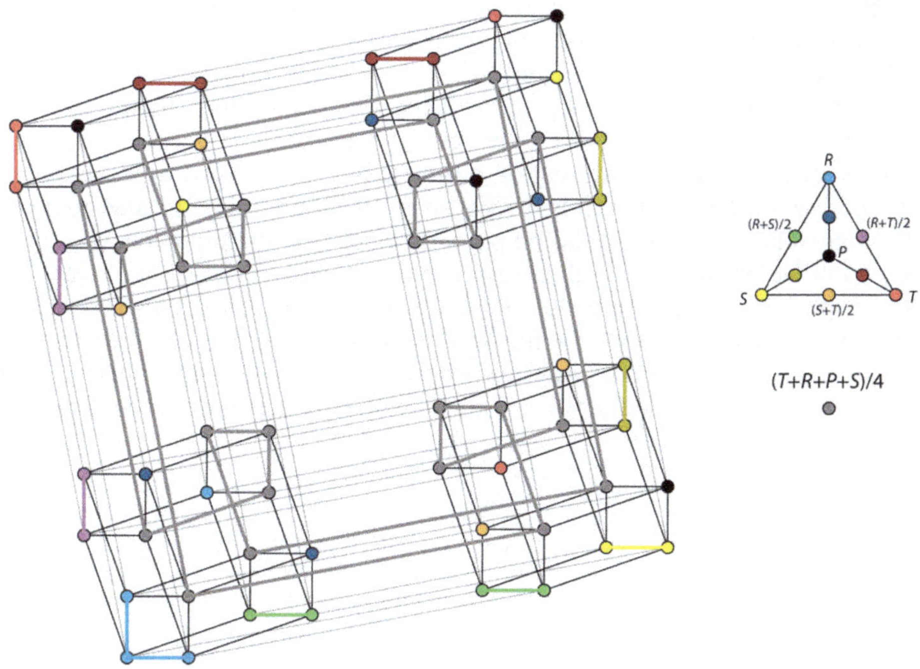

Figure 7.8. Hypercube of strategy pairs from table 7.2, showing the adjacency induced from the mutation metric of figure 7.6.

represent strategy changes by single bit flips for the second player. The smallest squares represent flips in the direction from ALLC to TFT. The squares formed by the next level of nesting represent flips in the direction from ALLC to WSLS. The outermost level of nesting represents flips in the direction from ALLC to ALT.

7.3.3 Multilevel selection and explanatory sufficiency in the repeated-game setting

Many approaches to overcoming paradoxes of stage-game dominance require the maintenance of *linkage*[10] between moves that enable coordination and moves that confer fitness. Examples include costly signaling or the recognition of kinship or identity[11]. The requirement to actively maintain linkage is another source of fragility in such coordinated solutions.

In this section we expand minimal RPD into a gene-based population process with genes described from the coarse-grained extensive-form tree of figure 3.5. We regard the information sets on the tree as loci and permit crossover between the first move and the recurrent move of the game tree. The moves C or D are allowed as alleles on every information set. The possible opening moves will therefore be C and D, and four recurrent-move alleles will be possible. Three are C, D, and the recurrent move from TFT, which we denote by T. (The fourth is the complement to recurrent T, shown as ATFT in the previous sub-section. We will suppress this possibility from the analysis in order to simplify the presentation and because it never plays a significant role in late-time stable populations.)

The resulting process demonstrates the following general properties involving genes based on the extensive form: (i) the asymmetry in fitness contribution between different loci, here opening-move and recurrent-move on the coarse-grained tree; (ii) the non-equivalence of different recurring moves for providing contextual importance to the opening move; and (iii) the way an additive model can justify treating both opening and recurrent moves as genes, because linear regressions on them capture part of the determination of fitness but, at the same time, selection on covariance between them which is driven by interaction terms is needed to account for the main feature of the population process, which is to maintain a high frequency of (C, C) move profiles and high penetrance of the ALLC recurrent move.

In the spirit of refining regression models to identify mechanisms statistically, we will construct best estimators for model fitness in a sequence of increasing polynomial order, starting from additive models and terminating when a stable, dynamically sufficient model has been identified. As higher-order models are used,

[10] In population genetics, *linkage* refers to excess co-occurrence of certain pairs of alleles within genotypes, relative to the frequency that would result from random shuffling through crossover at the same overall allele frequencies. Since crossover during reproduction drives populations towards a randomly shuffled distribution of genotypes (a condition termed *linkage equilibrium*), excess co-occurrence (or 'linkage dis-equilibrium') must be actively maintained by selection on covariance if it is to persist. We consider active maintenance of linkage disequilibrium at length in chapter 8.

[11] Since many of the moves described occur at a very high level of development (or are institutional in origin), we refer heavily to the 'statistical' nature of the gene and to 'crossover' in models of linkage as a coarse-grained reflection of any process that disrupts the integration of behaviors while preserving the identity of the individual behaviors in some adequate approximation.

parts of the regression coefficients, which depended on population state in lower-order regressions, are successively transferred to higher-order interactions where they become constants. Nothing less than a a fourth-order interaction term in the genetic model for fitness leads to regression coefficients that can successfully account for emergence of (C, C) as the most frequent move profile, without explicit reference to the population state as an external variable[12].

7.3.3.1 Genotypes and the normal form resulting from crossover in the memory-one extensive-form game

When crossover and mutation at all loci are introduced into evolutionary RPD, the genotype space expands to include eight types (of which we consider six). Using (′) to denote a genotype with the same recurrent move as the types in canonical RPD and opposite opening move, we denote the composition vector by

$$\mathrm{n} \equiv \begin{bmatrix} \mathrm{n_C} \\ \mathrm{n_{C'}} \\ \mathrm{n_D} \\ \mathrm{n_{D'}} \\ \mathrm{n_T} \\ \mathrm{n_{T'}} \end{bmatrix}. \tag{7.8}$$

The repeated-play payoffs to agents in this larger population a'_{ij} equal the values of a_{ij} for types with the same move in the recurrent stage to $O(1/N_{\mathrm{ply}})$, for all pairings except T and T′. For these latter four pairings, the payoffs become

$$[a']_{\mathrm{TT'}} = \bar{a}\begin{bmatrix} 1 \\ 1 \end{bmatrix}\begin{bmatrix} 1 & 1 \end{bmatrix} + \begin{bmatrix} R & \frac{1}{2}(S+T) \\ \frac{1}{2}(S+T) & P \end{bmatrix}. \tag{7.9}$$

The values in (7.9) follow from the states visited in the bottom diagram in figure 7.5.

7.3.3.2 Dependence of fitness on opening and recurrent moves

The concept of the 'contribution to fitness of a gene' comes from regression of fitness against equivalence classes of genotypes, which are the alleles of the gene. Therefore, in order to make models of genes and gene interactions, we first require a notation for these equivalence classes. We will use superscript ^{0} to label moves in the opening position on the game tree and $^\infty$ to label moves in the recurrent position[13].

We denote the possible opening moves by C and D, and the possible recurrent moves by C, D, and T. As in the examples of the figures, we will suppose that the

---

[12] For comparison, the linked-heterosis model developed next will require only second-order interactions.

[13] Changing first moves in ALLC and ALLD has no effect in the limit of large games. Substituting TFT′ for TFT, however, has the effect of producing a 'dual' game. Duality here means that the roles of C and D are exchanged, and the sign of the payoffs is then reversed. Thus, as C is a repelling point in the game with TFT, D is the corresponding attracting point in the game with TFT′, which is thus eliminated apart from mutation and crossover effects.

payoffs satisfy $R + P - S - T = 0$. We will not write expressions for mutation and crossover explicitly, but we suppose that such terms are kept non-zero to ensure that memory of initial population states is always lost eventually and replaced by consequences of selection and that the population does not become trapped in absorbing boundary states. We also assume that rates of mutation and crossover are large enough that we may ignore terms of order $1/N_{\mathrm{ply}}$. Approximation $\approx$ instead of equality = will indicate corrections at $O\left(1/N_{\mathrm{ply}}\right)$.

The population numbers for each of the recurrent moves are then the sums

$$\begin{aligned} \mathrm{n}_C^{\infty} &\equiv \mathrm{n}_{\mathrm{C}} + \mathrm{n}_{\mathrm{C}'} \\ \mathrm{n}_D^{\infty} &\equiv \mathrm{n}_{\mathrm{D}} + \mathrm{n}_{\mathrm{D}'} \\ \mathrm{n}_T^{\infty} &\equiv \mathrm{n}_{\mathrm{T}} + \mathrm{n}_{\mathrm{T}'}. \end{aligned} \tag{7.10}$$

The corresponding population numbers for each of the first moves are given by

$$\begin{aligned} \mathrm{n}_C^{0} &\equiv \mathrm{n}_{\mathrm{C}} + \mathrm{n}_{\mathrm{D}'} + \mathrm{n}_{\mathrm{T}} \\ \mathrm{n}_D^{0} &\equiv \mathrm{n}_{\mathrm{C}'} + \mathrm{n}_{\mathrm{D}} + \mathrm{n}_{\mathrm{T}'}. \end{aligned} \tag{7.11}$$

The fitness coefficients associated with the recurrent-move alleles in (7.10) are readily obtained by taking the appropriate sums of the replicator equation on the corresponding set-unions of complete genotypes and are given by

$$\begin{bmatrix} f_C^{\infty} \\ f_D^{\infty} \\ f_T^{\infty} \end{bmatrix} \approx \begin{bmatrix} R & S & R \\ T & P & P \\ R & P & X \end{bmatrix} \begin{bmatrix} \mathrm{n}_C^{\infty} \\ \mathrm{n}_D^{\infty} \\ \mathrm{n}_T^{\infty} \end{bmatrix}. \tag{7.12}$$

The regression coefficients R, S, T and P, are the same constants as for canonical RPD, but X now depends on population composition. With the assumption $R + P - S - T = 0$, it is given by

$$X\mathrm{n}_T^{\infty} \approx R\mathrm{n}_{\mathrm{T}} + P\mathrm{n}_{\mathrm{T}'}. \tag{7.13}$$

The fitnesses of types incorporating recurrent C and D are approximately independent of the opening move,

$$\begin{bmatrix} f_{\mathrm{C}} - f_C^{\infty} \\ f_{\mathrm{C}'} - f_C^{\infty} \end{bmatrix} \approx \begin{bmatrix} f_{\mathrm{D}} - f_D^{\infty} \\ f_{\mathrm{D}'} - f_D^{\infty} \end{bmatrix} \approx \begin{bmatrix} 0 \\ 0 \end{bmatrix}, \tag{7.14}$$

while those incorporating recurrent T depend on the first move to a degree that again depends on the population composition, as

$$\begin{bmatrix} f_{\mathrm{T}} - f_T^{\infty} \\ f_{\mathrm{T}'} - f_T^{\infty} \end{bmatrix} \approx \frac{R - P}{2} \begin{bmatrix} \mathrm{n}_{\mathrm{T}'} \\ -\mathrm{n}_{\mathrm{T}} \end{bmatrix}. \tag{7.15}$$

Because $R - P > 0$, mutant/recombinant TFT′ will be suppressed by selection relative to standard TFT, justifying our omission of TFT′ to produce the foregoing minimal example of evolutionary entropy effects.

If we denote by f_C^0 and f_D^0 the first-move fitnesses, and note that the mean fitness for alleles is the same quantity as the mean fitness defined for the population as a whole, which we have denoted by ϕ, then the fitness differences for opening-move alleles are

$$\begin{bmatrix} f_C^0 - \phi \\ f_D^0 - \phi \end{bmatrix} \approx \frac{1}{N}\begin{bmatrix} \mathrm{n}_D^0 \\ -\mathrm{n}_C^0 \end{bmatrix}\left\{\left(\frac{\mathrm{n}_{\mathrm{T}}}{\mathrm{n}_C^0} - \frac{\mathrm{n}_{\mathrm{T}'}}{\mathrm{n}_D^0}\right)\begin{bmatrix} R & P & \frac{R+P}{2} \end{bmatrix}\begin{bmatrix} \mathrm{n}_C^\infty \\ \mathrm{n}_D^\infty \\ \mathrm{n}_T^\infty \end{bmatrix} + \frac{R-P}{2}\left(\frac{(\mathrm{n}_{\mathrm{T}})^2}{\mathrm{n}_C^0} + \frac{(\mathrm{n}_{\mathrm{T}'})^2}{\mathrm{n}_D^0}\right)\right\}$$

$$\to \left(\frac{\mathrm{n}_T^\infty}{N}\right)^2 \begin{bmatrix} \mathrm{n}_D^0 \\ -\mathrm{n}_C^0 \end{bmatrix} \frac{R-P}{2}. \tag{7.16}$$

The first line is the general form and the second line indicates the form if the population is found in a linkage equilibrium where $\mathrm{n}_{\mathrm{T}}/\mathrm{n}_C^0 = \mathrm{n}_{\mathrm{T}'}/\mathrm{n}_D^0 = \mathrm{n}_T^\infty/N$.

The first term on the first line of (7.16) is a constant background part of the fitness of recurrent T, which simply carries along whichever first move is linked to it in excess, while the second term is a linkage-dependent, but always positive, preference for TFT over TFT′. If $\mathrm{n}_{\mathrm{T}'} \to 0$, the correlation with n_C^0 is total and the fitness in the first line is governed by the regression coefficients $\begin{bmatrix} R & P & R \end{bmatrix}$, with positive sign. If $\mathrm{n}_{\mathrm{T}} \to 0$, the correlation with n_D^0 is total and the fitness is governed by regression coefficients $\begin{bmatrix} R & P & P \end{bmatrix}$, with negative sign. Unless the population is sufficiently far from a linkage equilibrium in favor of $\mathrm{n}_{\mathrm{T}'}$, we will have $f_C^0 \geqslant \phi \geqslant f_D^0$, and n_D^0 will be suppressed in proportion to its correlation with recurrent T.[14]

In the linkage-equilibrium case, we have

$$\begin{bmatrix} f_{\mathrm{T}} - f_T^\infty \\ f_{\mathrm{T}'} - f_T^\infty \end{bmatrix} \to \frac{N}{\mathrm{n}_T^\infty}\begin{bmatrix} f_C^0 - \phi \\ f_D^0 - \phi \end{bmatrix}. \tag{7.17}$$

Therefore, in the MFA, the direction of change from an initial state in linkage equilibrium is given by

$$\frac{\mathrm{d}}{\mathrm{d}t}\begin{bmatrix} \log\left(\dfrac{\bar{n}_{\mathrm{T}} N}{\bar{n}_C^0 \bar{n}_T^\infty}\right) \\ \log\left(\dfrac{\bar{n}_{\mathrm{T}'} N}{\bar{n}_D^0 \bar{n}_T^\infty}\right) \end{bmatrix} \equiv \begin{bmatrix} f_{\mathrm{T}} - f_T^\infty - \left(f_C^0 - \phi\right) \\ f_{\mathrm{T}'} - f_T^\infty - \left(f_D^0 - \phi\right) \end{bmatrix}$$

$$\to \left(\frac{N}{\mathrm{n}_T^\infty} - 1\right)\begin{bmatrix} f_C^0 - \phi \\ f_D^0 - \phi \end{bmatrix}$$

$$\to \frac{\mathrm{n}_T^\infty}{N}\left(1 - \frac{\mathrm{n}_T^\infty}{N}\right)\begin{bmatrix} \mathrm{n}_D^0 \\ -\mathrm{n}_C^0 \end{bmatrix}\frac{R-P}{2}. \tag{7.18}$$

[14] Note that opening D may still be maintained by mutation in types where it is paired with recurrent C or D, where it is neutral, and then re-mixed through either recombination or mutations of the recurrent move.

7.3.3.3 Indicator functions and regressions of fitness

We now illustrate the genetic decomposition of fitness. Instead of writing fitness as a function with coefficients that must be specified for each of 36 possible pairs of fully specified types, we attempt to find more compact representations in which contributions to fitness may be assigned to different pairs of genes, or in rare cases, to interactions that involve combinations of genes for one or more of the players in a pair. If this procedure can account for the observed fitness values of the previous section, without reference to an average population state, the construction to that order is dynamically sufficient. If it cannot, higher-order terms must be added until no more explicit reference to an average over population states is required.

The lowest-order models, although they involve pairs of agents, refer to only one gene within each agent. We will term this order of description 'additive'. We will show how, in cases where the additive model is not sufficient, the addition of interaction terms both reduces the size of the coefficients that refer to averaged population states and in some cases makes small changes to the best-estimate coefficients for the additive terms.

When interaction terms are required to produce a dynamically sufficient fitness model, if we assume that crossover is possible, we implicitly bring in a potential need for multilevel selection. Not only must gene frequencies be preserved by replication and selection; the linkage of multiple genes in a parent type must be preserved for the phenotype to be reproduced. Since we do not make explicit models of reproduction from pairs of parents with crossover, we will not model the preservation of linkage disequilibrium in this chapter (that phenomenon is introduced in chapter 8), but we note in passing the association between selectively significant interaction terms and multiple levels at which selection must preserve characters.

Let $i \in \{C, C', D, D', T, T'\}$ indicate a fully specified agent type. Continue to let $\left\{C^0, D^0\right\}$ label opening moves and $\left\{C^\infty, D^\infty, T^\infty\right\}$ label recurrent moves. Use lowercase Greek indices for opening moves $\alpha, \beta \in \left\{C^0, D^0\right\}$ and capital Roman indices for recurrent moves $I, J \in \left\{C^\infty, D^\infty, T^\infty\right\}$. A fully specified agent type index corresponds to a concatenated pair of indices for opening and recurrent moves, as $i \leftrightarrow \alpha I$. With these notations we may introduce indicator functions for the sets corresponding to alleles, defined in the preceding section.

The Kronecker delta on fully specified types is denoted $\delta_{\alpha I,i}$. In terms of these elementary indicator functions, the indicator function for an opening move may be written

$$\sigma^0_{\alpha i} \equiv 1 - \prod_I \left(1 - \delta_{\alpha I,i}\right). \tag{7.19}$$

The corresponding indicator for a recurrent move is

$$\sigma^\infty_{Ii} \equiv 1 - \prod_\alpha \left(1 - \delta_{\alpha I,i}\right). \tag{7.20}$$

It is straightforward to check that the counting rules for set unions are satisfied,

$$\begin{aligned}\sum_i \sigma^0_{\alpha i}\mathrm{n}_i &= \mathrm{n}^0_\alpha \\ \sum_i \sigma^\infty_{Ii}\mathrm{n}_i &= \mathrm{n}^\infty_I \\ \sum_i \sigma^0_{\alpha i}\sigma^\infty_{Ii}\mathrm{n}_i &= \mathrm{n}_{\alpha I}.\end{aligned} \tag{7.21}$$

Because the recurrent move dominates for most types, the simplest regression is on the recurrent-move 'gene' in (3.1),

$$\begin{aligned}f^\infty_i &\approx \sum_j \sum_{IJ} \sigma^\infty_{Ii}\sigma^\infty_{Jj} a_{IJ}\mathrm{n}_j \\ &= \sum_{IJ} \sigma^\infty_{Ii} a_{IJ} n^\infty_J.\end{aligned} \tag{7.22}$$

For this approximation the regression coefficients on recurrent moves are the same as those in the CDT game of section 7.2—abusing notation, $a_{IJ} = a_{ij}$, except for the recurrent-T regression. The least-squares regression coefficient may readily be computed to be

$$\begin{aligned}a_{T^\infty T^\infty} &= \left\langle \frac{R\mathrm{n}_\mathrm{T} + P\mathrm{n}_{\mathrm{T}'}}{n^\infty_T} \right\rangle \\ &= \frac{R+P}{2} + \frac{R-P}{2}\left\langle \frac{\mathrm{n}_\mathrm{T} - \mathrm{n}_{\mathrm{T}'}}{n^\infty_T} \right\rangle,\end{aligned} \tag{7.23}$$

where we use $\langle\ \rangle$ as a reminder that these regression coefficients are now functions of the (hidden) dynamical population variables n_T, $\mathrm{n}_{\mathrm{T}'}$. The regression (7.22) recovers the forms for all f^∞_I of the preceding sub-section, which were computed from the replicator dynamic acting on set unions. It both conflates types T and T′, and produces a regression coefficient which may depend sensitively on the samples from the population process used to compute it.

We could try to improve the approximation (7.22) with a regression that assumes additive genetic variance, by including an indicator function for the first move (out of many such possibilities, an interaction with opponent first moves is shown here),

$$\begin{aligned}f^{\mathrm{lin}}_i &\approx \sum_j \left(\sum_{IJ} \sigma^\infty_{Ii}\sigma^\infty_{Jj} a^{\mathrm{lin}}_{IJ} + \sum_{\alpha\beta} \sigma^0_{\alpha i}\sigma^0_{\beta j} b^{\mathrm{lin}}_{\alpha\beta} \right)\mathrm{n}_j \\ &= \sum_{IJ} \sigma^\infty_{Ii} a^{\mathrm{lin}}_{IJ}\mathrm{n}^\infty_J + \sum_{\alpha\beta} \sigma^0_{\alpha i} b^{\mathrm{lin}}_{\alpha\beta}\mathrm{n}^0_\beta.\end{aligned} \tag{7.24}$$

No regression of this form can generate coefficients $b_{\alpha\beta}$ large enough to be useful in the general case, because all but the f_{T} and $f_{\mathrm{T'}}$ fitness functions are well approximated by their f^{∞} estimators. Thus even in this simple game, additivity is violated at the single-gene level.

Therefore we consider interactions. The simplest interaction is between an agent's first move and recurrent move, still depending on the environment only through the opponent's recurrent move. We write this regression as

$$\begin{aligned} f_i^{(1)} &\approx \sum_j \sum_{IJ} \sigma_{Ii}^{\infty} \sigma_{Jj}^{\infty} \left(a'_{IJ} + \sum_{\alpha} \sigma_{\alpha i}^{0} a_{\alpha IJ}^{(1,0)} \right) \mathrm{n}_j \\ &= \sum_{IJ} \sigma_{Ii}^{\infty} \left(a'_{IJ} + \sum_{\alpha} \sigma_{\alpha i}^{0} a_{\alpha IJ}^{(1,0)} \right) \mathrm{n}_J^{\infty}. \end{aligned} \tag{7.25}$$

Again the recurrent-move regression coefficients almost always equal those in canonical RPD—$(a'_{IJ} = a_{ij})$—but now the constant part of the T–T interaction has changed from the value in (7.23) because an interaction term predicts a separate component of the variance. The constant coefficient

$$\begin{aligned} a'_{T^{\infty}T^{\infty}} &= a_{T^{\infty}T^{\infty}} - \left(\frac{R-P}{4} \right) \left\langle \frac{\mathrm{n}_{\mathrm{T}} - \mathrm{n}_{\mathrm{T'}}}{n_T^{\infty}} \right\rangle \\ &= \frac{R+P}{2} + \frac{R-P}{4} \left\langle \frac{\mathrm{n}_{\mathrm{T}} - \mathrm{n}_{\mathrm{T'}}}{n_T^{\infty}} \right\rangle, \end{aligned} \tag{7.26}$$

and the first-order interaction term is now stable,

$$a_{C^0 T^{\infty} T^{\infty}}^{(1,0)} = -a_{D^0 T^{\infty} T^{\infty}}^{(1,0)} = \frac{R-P}{4}. \tag{7.27}$$

Equation (7.25) now produces the correct averaged forms for f_{T} and $f_{\mathrm{T'}}$, and half of the correct (constant) difference from (7.15).

To obtain stable regression coefficients, we must introduce a second-order interaction. The relevant regression is written

$$\begin{aligned} f_i^{(2)} &\approx \sum_j \sum_{IJ} \sigma_{Ii}^{\infty} \sigma_{Jj}^{\infty} \left(a''_{IJ} + \sum_{\alpha\beta} \sigma_{\alpha i}^{0} \sigma_{\beta j}^{0} a_{\alpha I\beta J}^{(1,1)} \right) \mathrm{n}_j \\ &= \sum_{IJ} \sigma_{Ii}^{\infty} \left(a''_{IJ} \mathrm{n}_J^{\infty} + \sum_{\alpha\beta} \sigma_{\alpha i}^{0} a_{\alpha I\beta J}^{(1,1)} \mathrm{n}_{\beta J} \right). \end{aligned} \tag{7.28}$$

Now the constant coefficient is stable at value

$$a''_{T^{\infty}T^{\infty}} = \left(\frac{R+P}{2} \right). \tag{7.29}$$

For the simple example we have used to illustrate, with $R + P - S - T = 0$, only an antisymmetric interaction between self- and opponent-first-moves is required,

$$a^{(1,1)}_{C^0T^\infty C^0T^\infty} = -a^{(1,1)}_{D^0T^\infty D^0T^\infty} = \frac{R-P}{2}. \tag{7.30}$$

The regression coefficients in (7.28) with second-order interactions in both opening and recurrent moves are thus needed to reconstruct the extensive-form game in a 'genetic' description.

Bibliography

[1] Nowak M A 2006 *Evolutionary Dynamics: Exploring the Equations of Life* (New York: Belknap)

[2] Rapoport A 1976 *The 2 × 2 Game* (Ann Arbor, MI: University of Michigan Press)

[3] Axelrod R 2006 *The Evolution of Cooperation* (New York: Perseus)

[4] Crutchfield J P and Young K 1989 Inferring statistical complexity *Phys. Rev. Lett.* **63** 105–8

[5] Shalizi C R, Shalizi K L and Haslinger R 2004 Quantifying self-organization with optimal predictors *Phys. Rev. Lett.* **93** 118701

[6] Fontana W, Wagner G and Buss L W 1994 Beyond digital naturalism *Artificial Life* **1** 211–27

[7] Gruener W, Giegerich R, Strothmann D, Reidys C, Weber J, Hofacker I L, Stadler P F and Schuster P 1996 Analysis of RNA sequence structure maps by exhaustive enumeration: I. Neutral networks *Monatsh. Chem.* **127** 355–74

[8] Gruener W, Giegerich R, Strothmann D, Reidys C, Weber J, Hofacker I L, Stadler P F and Schuster P 1996 Analysis of RNA sequence structure maps by exhaustive enumeration: II. Structures of neutral networks and shape space covering *Monatsh. Chem.* **127** 375–89

[9] Reidys C, Stadler P F and Schuster P 1997 Generic properties of combinatory maps: neutral networks of RNA secondary structures *Bull. Math. Biol.* **59** 339–97

[10] Ancel L W and Fontana W 2000 Plasticity, evolvability and modularity in RNA *J. Exp. Zool. (Mol. Dev. Evol.)* **288** 242–83

[11] Fontana W 2002 Modeling 'evo-devo' with RNA *Bioessays* **24** 1164–77

[12] Smith E 2011 Large-deviation principles, stochastic effective actions, path entropies, and the structure and meaning of thermodynamic descriptions *Rep. Prog. Phys.* **74** 046601

[13] Weinberg S 1996 *The Quantum Theory of Fields* vol 2 (New York: Cambridge)

[14] Freidlin M I and Wentzell A D 1998 *Random Perturbations in Dynamical Systems* 2nd edn (New York: Springer)

[15] Huang K 1987 *Statistical Mechanics* (New York: Wiley)

[16] Harsanyi J C and Selten R 1988 *A General Theory of Equilibrium Selection in Games* (Cambridge, MA: MIT Press)

[17] Nowak M A 1993 A strategy of win-stay, lose-shift that outperforms tit-for-tat in prisoner's dilemma *Nature* **364** 56–58

[18] Kraines D and Kraines V 1995 Evolution of learning among Pavlov strategies in a competitive environment with noise *J. Conflict Resolution* **39** 439–66

[19] Imhof L A, Fudenberg D and Nowak M A 2007 Tit-for-tat or win-stay, lose-shift *J. Theor. Biol.* **247** 574–80

[20] Axelrod R 1997 *The Complexity of Cooperation* (Princeton, NJ: Princeton University Press)

Chapter 8

Complex neutral spaces and 'dressed' genes

If the presence of symmetries at the macroscale is to be the relevant source for robust multilevel dynamics in evolving systems, it will be important to show how symmetry can lead to unlimited numbers of macrostates. Only then can it create a possibility for open-ended evolutionary dynamics. Here we describe a model due to Richard Lewontin with this property, albeit in a very simple form. For the first time in this chapter we consider a diploid 'organism' as an essential level of organization, represented by the play of a large extensive-form game by two chromosomes. Pairs of chromosomes replicate with crossover and the number of genes in the chromosomes, represented as moves in the game, is the scaling variable between microdynamics and macrodynamics. Fitness in this model favors parallel anti-coordination between pairs of counterpart-genes in the two chromosomes, a relation known as heterosis. Fluctuations in fitness drive populations toward states consisting of two mirror-image sub-populations, the members of each having high covariance of alleles at different loci, known in population genetics as 'linkage'. The tension between selection which maintains linkage and crossover which disrupts linkage leads to covariance among alleles that decays exponentially with their separation. Although the genes at one level are fundamental, with weak crossover it is the correlated regions within chromosomes that become the relevant units of transmission and selection; individual genes without covariance contribute a vanishingly small part of fitness. This model illustrates a form of multilevel selection within chromosomes, with a mathematical dependence on covariance similar to that among population members in Hamilton's rule for multilevel selection based on relatedness.

8.1 Scaling from cooperative effects among components in development

The examples of scaling of macroscopic properties we have demonstrated in the previous chapters all resulted from fluctuations in the relative numbers of different strategies in the population. Changes in population size affect the scale of these fluctuations relative to mean-field parameters and hence their importance for

doi:10.1088/978-0-7503-1137-3ch8

macroscopic dynamics. In the next example we introduce a new source of large-number scaling and associated fluctuation phenomena.

The following model invokes three levels of organization: genes, chromosomes and diploid organisms. Within each generation chromosomes are drawn in pairs, each pair forming a diploid organism, which corresponds to a single play of a 'game of development'. The outcome of the game determines the fitness of both partners, which may then replicate with crossover. A property familiar from genic-selection arguments [1, 2] applies: only genes are assured to be replicating units under all transmission events. However, chromosomes are not simply bags of genes. They are tuples of distinct loci in a specific linear order, which we represent as sequential plies in an extensive-form game. As introduced in chapter 3, crossover between two 'parent' strategies consists of swapping their moves on all stages below some randomly chosen stage k of the game (which we will call the *crossover point*) and then replicating one of the chromosomes that results from the crossover rather than the original parent's type. Sequence in the game tree determines which moves in a strategy remain together because they fall on the same side of a crossover point. The co-occurrence pattern of alleles within chromosomes may be propagated over many generations before it is interrupted by crossover. Since both allele numbers and allele covariances are propagated and are targets of selection, both levels are essential. They are also independent: neither in transmission nor in selection are gene identities interchangeable with the values of gene covariances.

In the model of this chapter, fluctuations in the population composition of alleles for each gene will still depend on the population size. Therefore increasing the population size will confer some stability on the 'additive' contribution to fitness from individual genes. However, the fluctuations in fitness of a whole chromosome are no longer ameliorated by large population numbers, because each chromosome is paired with only one partner within a generation. Fluctuations in chromosome fitness depend on the range of variation in genotypes present in the population. In this model they will depend on a very large number of 'interaction' terms, whose aggregate contribution can be made much larger than the contribution of any single gene.

Multilevel selection and the concept of the dressed gene

The contribution of a very large number of covariance terms between alleles at different loci to the fitness of any particular allele, forces upon us a fundamental reconceptualization of what constitutes a unit of selection. The naïve gene built into the definition of the model, which formally enjoys a privileged status as the only unit of replication that is always respected, is nevertheless *not* the practically relevant evolutionary unit or the unit that defines particular population states. The latter two roles are filled by a much more amorphous entity, whose definition must be arrived at statistically. The relevant unit of selection at a locus is not only the allele at that locus, but the distribution of alleles at neighboring loci that are co-present in typical chromosomes. The tightness of covariance between alleles at pairs of loci falls off with the distance between them, reflecting the balance between crossover and selection. At low crossover rates, groups of closely neighboring alleles may be transmitted together in almost every generation, making them little different in practice from single alleles. We will call the local allele together with its distribution of covariant alleles the 'dressed gene', and the microscopic allele at a single locus the corresponding 'bare gene'.

The emergence of dressed genes from dynamics, and their multilevel character, will be familiar to statistical physicists from experience with dressed particles in condensed matter and elementary particle physics. The fact that they arise from the basic assumptions of population genetics and development should have put to rest naïve arguments for genic selection when this model was first introduced in the 1970s [3] and replaced it with a more nuanced and empirically grounded statistical understanding of units of selection. We hope that, in emphasizing the mathematical equivalence of dressed genes to dressed particles, whose importance as a fundamental reconceptualization of the nature of objects is recognized in two well-vetted fields, this presentation will contribute to bringing about an eventual conceptual revision in genetics as well.

Scaling relations with the number of components of genomes and complex symmetry breaking

The scaling variable that determines the range of variants in the population, and hence the variance in fitness, will be the *number of genes* that make up a chromosome. Microscopically, the number of possible genotypes grows exponentially with the number of genes, since the genotype is simply a product of its genes. The more interesting result is that the number of possible distinct, quasi-stable *population states*, each indexed by a consensus genotype, also grows with the number of genes. The distinct population states are not uniform distributions over all genotypes, so the model exhibits symmetry breaking. However, unlike the coordination game, in which symmetry breaking always results in a pair of states, in this game the number of possible distinct states is unlimited. The model is therefore an example of what we termed 'complex' symmetry breaking in chapter 1. We will show that the process of forming stable population distributions is one of solving a large number of mutually incompatible constraints. It therefore is characterized by *frustration* and produces a large number of about equally good solutions, which is the origin of its complexity. Because of the similarity of this form of frustration and macrostate degeneracy to similar phenomena in glasses, we also refer to this as a population model of 'glassy' symmetry breaking.

8.2 Complex discrete symmetry breaking through selection on both genes and covariance

The example we will develop originates in a model of linked heterosis[1] introduced by Richard Lewontin [3]. The linked heterosis model was presented as one mechanism

[1] In genetics, *linkage* refers to a larger than random probability that pairs of alleles on a chromosome will be maintained together against the mixing effects of crossover during replication. An initially linked pair of alleles will be preserved with exponentially decreasing probability over time, with the rate of mixing depending on the likelihood that crossover will occur between them and on the frequency of alternative alleles in the population. Linkage can be actively maintained if selection favors the co-occurrence of certain pairs of alleles and thus tends to cull chromosomes that have experienced crossover between them. *Heterosis* refers to the relative fitness of non-haploid (diploid or higher polyploid) organisms with mixed alleles on different chromosomes (termed heterozygous) versus organisms with the same alleles (termed homozygous). Heterotic organisms are those in which the heterozygous organisms are more fit than either of the homozygous forms.

to explain the high frequency of observed heterozygosity in natural populations, without recourse to the assumptions of neutral theory, but with plausible (meaning, small) levels of heterosis at each locus. Lewontin demonstrated that selection can act cooperatively at many loci, amplifying the effects of heterosis, by effectively selecting for linkage between nearby loci. The model isolates selection on linkage by removing all other sources of fitness differentiation among alleles through an imposed symmetry[2].

This example illustrates a new role for collective fluctuations, as transient sample excesses of linkage in the population become the only source of fitness differences among genotypes. Through competition among diploid pairs, linkage becomes self-amplifying. The model also illustrates how plies in an extensive form game, under appropriate information conditions, can be made equivalent to loci on a chromosome with respect to epistasis and crossover, removing the notion of a 'sequence of play' that is a default property of game trees. Finally, it illustrates the parallel role of repetition within and between generations, by showing the homology between linkage in a repeated game and the expression for relatedness in Hamilton's rule derived in section 5.4.

8.2.1 An anti-coordination game that isolates the dynamics of linkage from a neutral manifold of allele frequencies

An extensive-form game that captures both heterosis and linkage is defined on the tree of figure 3.3, with single crossovers per chromosome per generation eligible to occur between any pair of successive plies[3]. The game has two players. Each player can move either left (L) or right (R) at each of a set of N_{ply} information sets and the structure of information sets precludes defining strategies in which any move is contingent on any other move either by the player or by the opponent. A genotype is an N_{ply}-letter sequence and any of the $2^{N_{\text{ply}}}$ sequences is permitted.

The labels for alleles are $I \in \{L^0, R^0, \ldots, L^{N_{\text{ply}}-1}, R^{N_{\text{ply}}-1}\}$, where the superscript indexes the locus. Indicator functions σ_{Ii} take value one if strategy i has move I at the

[2] Random matching together with recombination that is not subsequently filtered by selection, lead eventually to the condition of Hardy–Weinberg linkage equilibrium [4] (HWLE; also known as the *Wright manifold*). This thoroughly studied limit in population genetics leads to simplifications for the identification and analysis of extensive-form games, considered in [5]. We will assume random matching and thus Hardy–Weinberg equilibrium, but in the limit of strong selection, the absorbing states of the Lewontin model will lie maximally far from linkage equilibrium.

Because the strength of selection determines whether heterogeneity or drift to fixation will be the persistent state of the population, we regard the contribution of linkage to the strength of selection as a kind of Hill–Robertson effect. Like other Hill–Robertson effects [6], the cause of population structure will not be found in the direct selection coefficients, but will depend on fluctuations in fitness.

[3] The competing effects of crossover and selection are essential to the maintenance of ongoing dynamics in this model. Under selection alone the model would 'freeze out' to one of its absorbing states, but crossover frustrates this selective pressure toward convergence. In the calculations of this chapter we compute only the effects of selection explicitly to show how these depend on fluctuations. We return to consider the effects of crossover in section 8.3.1, for which we draw on results of [3].

appropriate locus. Since a fully specified strategy must have $\sigma_{L^{k}i} + \sigma_{R^{k}i} = 1$ at every k, we may represent strategies compactly in terms of a 'spin' variable

$$\sigma_i^k \equiv \sigma_{L^k i} - \sigma_{R^k i}, \tag{8.1}$$

which takes value ± 1 according to the move at locus k.

Plies serve as loci and a model that treats all loci identically defines a kind of repeated game. The definition of *heterosis* (also called 'hybrid vigor') in a diploid organism is that the heterozygous genotype is fitter than either of the homozygous genotypes. The simplest possible payoff function showing heterosis treats both homozygotes equally and treats the contribution to fitness from every locus equally. Translating this description into a normal-form stage game is equivalent to keeping only the L and R types from the coordination model of chapter 5 and exchanging a and $-a$ in the upper-left 2×2 sub-matrix in (5.2) to produce an *anticoordination* game instead. Setting $a = 1$ without loss of generality, the payoff when two genotypes i and j are matched is symmetric in i and j, given by

$$a_{ij} = -\frac{1}{2N_{\text{ply}}} \sum_{k=0}^{N_{\text{ply}}-1} \sigma_i^k \sigma_j^k. \tag{8.2}$$

These two equations define the genotype space and the payoffs per match for a pair of individual chromosomes.

8.2.1.1 Fluctuations are responsible for population covariance and linkage disequilibrium

The models from earlier chapters included fluctuation due to sampling noise in the events of reproduction and death, but made fitness a deterministic function of the frequencies of agents in the population. That fitness model is exact only if each agent is matched with all other agents exactly once within a generation and it is an appropriate approximation if each agent has a large number of randomly matched encounters within a generation.

Here for the first time we make the opposite assumption and model sampling variance in fitness itself by making the fitness of a chromosome the outcome of a single pairing to form a diploid organism. We suppose that each chromosome participates in only one random pairing with another chromosome between the event when it is born and the event when either both partners are culled, or the pair together are chosen to produce a pair of offspring[4]. The large sample fluctuations that come from pair matching will be essential to maintaining covariance among

[4] Note also that the offspring genotypes will generally no longer be the same as those of the parents because crossover is part of the transmission process. Both parents may remain in the population, so reproduction with crossover generates novelty at the chromosome level though not at the allele level. Only mutation can potentially produce novelty in an allele and it is not our concern in this chapter, since selection tends to preserve both alleles at every locus.

alleles at different loci in populations filtered by selection, a condition known as *linkage disequilibrium*[5]. Therefore we begin by contrasting the mean fitness of alleles, which approaches zero for balanced populations, with the fluctuations in fitness due to sampling of partners within generations.

The population-averaged frequency of the indicator function $\sigma^k = \sigma_{L^k} - \sigma_{R^k}$ at any locus k on which we focus attention[6], in a population state n, is given by

$$\langle \sigma^k \rangle_{\mathrm{n}} \equiv \sum_j \frac{\mathrm{n}_j}{N} \sigma_j^k. \tag{8.3}$$

(The index j runs over all $2^{N_{\mathrm{ply}}}$ genotypes, and n_j is their number in the population vector n.) The fitness of any complete genotype i in population n, as it would appear in expectation under random pairing against all members of the population, is then

$$f_i = -\frac{1}{2N_{\mathrm{ply}}} \sum_{k=0}^{N_{\mathrm{ply}}-1} \sigma_i^k \langle \sigma^k \rangle_{\mathrm{n}}. \tag{8.4}$$

The corresponding population-averaged fitness is

$$\phi = \sum_i \frac{\mathrm{n}_i}{N} f_i = -\frac{1}{2N_{\mathrm{ply}}} \sum_{k=0}^{N_{\mathrm{ply}}-1} \langle \sigma^k \rangle_{\mathrm{n}}^2. \tag{8.5}$$

These are reference values, which provide estimates of the cumulative selection strength on alleles over many generations. In any population with balanced alleles (all $\langle \sigma^k \rangle \approx 0$), both f_i and ϕ will be near zero. Therefore it can only be fluctuation about these values resulting from *particular* pairings that leads to non-random association at the population level.

The average correlation of alleles at two loci k and k' in the population is given by

$$C_{kk'} \equiv \sum_i \frac{\mathrm{n}_i}{N} \sigma_i^k \sigma_i^{k'}. \tag{8.6}$$

Subtracting the expected values of these alleles in the population state n leaves the covariance between loci,

$$\mathrm{Cov}\left(\sigma^k, \sigma^{k'}\right) = C_{kk'} - \langle \sigma^k \rangle_{\mathrm{n}} \langle \sigma^{k'} \rangle_{\mathrm{n}}. \tag{8.7}$$

The covariance at a single locus k is the variance of σ^k, which evaluates to $\mathrm{Cov}(\sigma^k, \sigma^k) = \mathrm{Var}(\sigma^k) = 1 - \langle \sigma^k \rangle_{\mathrm{n}}^2$.

[5] For a discussion of terminology related to linkage, recall footnote 10 in chapter 7.

[6] We will refer to σ^k as the 'focal' allele when considering its fitness, which will generally depend on the covariance of its frequency with that of all other alleles among genotypes present in the evolving population.

Referring back to the form of the Price equation (2.12) and letting σ^k for an allele be the indicator function whose frequency in the population we wish to follow, shows that allele frequency is controlled by the covariance (8.7) in state n, as

$$\begin{aligned}\frac{\mathrm{d}}{\mathrm{d}t}\langle\sigma^k\rangle_{\mathrm{n}} &= -\frac{1}{2N_{\mathrm{ply}}}\sum_{k'=0}^{N_{\mathrm{ply}}-1}\mathrm{Cov}\left(\sigma^k, \sigma^{k'}\right)\langle\sigma^{k'}\rangle_{\mathrm{n}} \\ &= -\frac{1}{2N_{\mathrm{ply}}}\left(1 - \langle\sigma^k\rangle_{\mathrm{n}}^2\right)\langle\sigma^k\rangle_{\mathrm{n}} - \frac{1}{2N_{\mathrm{ply}}}\sum_{k'\neq k}\mathrm{Cov}\left(\sigma^k, \sigma^{k'}\right)\langle\sigma^{k'}\rangle_{\mathrm{n}}.\end{aligned} \tag{8.8}$$

Comparing (8.8) to the expression (5.36) for Hamilton's rule, we see that linkage disequilibrium plays the same role with respect to moves in the extensive-form game that relatedness plays among populations that fluctuate on the generation scale.

If covariance in the population remains near zero, the rate of mean regression for the focal allele σ^k will fall as $1/2N_{\mathrm{ply}}$ with increasing N_{ply}, because of the normalization assumed for fitness in (8.2). However, if covariance is large over a range of adjacent plies whose number grows with N_{ply}, the mean regression for $\langle\sigma^k\rangle$ may remain stronger than $1/2N_{\mathrm{ply}}$, ensuring the preservation of heterozygosity against drift in the population even with very weak heterosis per locus. This mechanism for the maintenance of heterozygosity was the motivation behind this model in [3].

A similar construction may be performed for whole genotypes. We now suppose that a single pairing of zygotes determines the fitness of both partners within each generation. The mean fitness of type i relative to the population average remains the same as the expectation over multiple pairings, given by

$$f_i - \phi = -\frac{1}{2N_{\mathrm{ply}}}\sum_{k=0}^{N_{\mathrm{ply}}-1}\left(\sigma_i^k - \langle\sigma^k\rangle_{\mathrm{n}}\right)\langle\sigma^k\rangle_{\mathrm{n}}. \tag{8.9}$$

The variance in the expected number of offspring left by any individual of type i, due to random pairing of players, is much larger for single pairings than for samples of the population average. The sample fluctuation now depends on details of the probabilities of reproduction, which we absorb into a characteristic time increment Δt. The time increment per typical reproduction scales the overall variance in reproductive success, which is then given by

$$(\Delta t)^2\sum_j\frac{\mathrm{n}_j}{N}\left(a_{ij} - f_i\right)^2 = \frac{(\Delta t)^2}{4N_{\mathrm{ply}}^2}\sum_{k,k'}\sigma_i^k\sigma_i^{k'}\mathrm{Cov}\left(\sigma^k, \sigma^{k'}\right). \tag{8.10}$$

The expected number of offspring from *all* parents of type i in the population is then

$$\langle\mathrm{n}_i'\rangle = \mathrm{n}_i\left(1 + \Delta t\left(f_i - \phi\right)\right). \tag{8.11}$$

(Equation (8.11) is just the replicator equation restored from continuous-time to a discrete time interval characteristic of the generation time, on which (8.10) will give the appropriate expression for the variance in chromosome fitness.)

By the central limit theorem for independently drawn partners for each of n_i individuals, the corresponding variance in offspring due to the distribution over pairings is then

$$\left\langle \left(n_i' - \langle n_i' \rangle\right)^2 \right\rangle = (\Delta t)^2 n_i \sum_j \frac{n_j}{N} \left(a_{ij} - f_i\right)^2$$

$$= \frac{(\Delta t)^2 n_i}{4N_{\text{ply}}^2} \sum_{k,k'} \sigma_i^k \sigma_i^{k'} \mathrm{Cov}\left(\sigma^k, \sigma^{k'}\right). \tag{8.12}$$

These expressions for reproductive variance due to sampling may be used to estimate dynamics for the covariance itself, which we do next.

8.2.2 Variation in the linked heterosis model and the heritable component of fitness

The important property of the linked heterosis model is that *covariance* of alleles is the sole eligible target for selection, because all other properties have been made equivalent by imposed symmetries[7]. A stochastic differential equation or Langevin equation provides a simple framework for computations that is sufficient to capture covariance effects. From a Langevin equation we may derive the effective fitness that results from the positive selection for population covariance. From the derivations in section 4.4.2.2 we also understand how the Langevin formulation relates to more complete treatments of the master equation.

Consider the role of fluctuations due to random pairing, about a background in which $\langle \sigma^k \rangle_{\text{n}} = 0, \forall k$, so that the population is evenly mixed on average for L and R alleles at each locus and the only characteristic remaining to be selected is the pattern of co-occurrence among loci. Then $f_i = 0, \forall i$ and $\phi = 0$ as well. We will make the further simplifying assumption concerning whole genotypes i, that $n_i = n_{-i}$, for all pairs $(i, -i)$ which differ by sign change of each σ_i^m. Such symmetrized pairs share all fluctuation properties and our goal is to understand how fluctuations affect heterotic loci as a result of the correlation structures of the genomes that contain them.

8.2.2.1 The Langevin equation to capture the fitness consequences of random pairing of players

Let the consequences of random pairing be represented by a Langevin field, so that in the notation of the Price equation, the number of descendents from individuals of type i is a random variable written

$$n_i' = \left(1 + \xi_i\right) n_i. \tag{8.13}$$

[7] In an anticoordination game, unlike a coordination game, the most fit allele is whichever is in the minority, so that the only stable fixed points of the replicator dynamic are those in which all loci are maximally heterozygous.

By (8.11), the expectation $\langle \xi_i \rangle_t = \Delta t \left(f_i - \phi \right) = 0$, where $\langle \ \rangle_t$ denotes the average over pairings in some interval Δt starting from time t. Then, by (8.12) at $f_i = 0$, the variance in ξ_i is given by

$$\langle \xi_i^2 \rangle_t = \frac{(\Delta t)^2}{\mathrm{n}_i} \sum_j a_{ij}^2 \frac{\mathrm{n}_j}{N}. \tag{8.14}$$

To conserve total number, it is necessary that $\sum_j \mathrm{n}_j \xi_j \equiv 0$ in each realization of the variables ξ, leading to the requirement that in correlations

$$\sum_j \mathrm{n}_j \left\langle \xi_i \xi_j \right\rangle_t \equiv 0, \ \forall i. \tag{8.15}$$

Now we consider the fluctuations in a succeeding interval Δt, starting at a time $t' \equiv t + \Delta t$. Putting aside mutation or crossover effects, which we will assume occur at a low rate, the number of second-generation descendants is

$$\mathrm{n}_i'' = \left(1 + \xi_i'\right)\left(1 + \xi_i\right)\mathrm{n}_i. \tag{8.16}$$

In the context of the fluctuation at t, the Langevin field ξ_i' now has a non-zero expectation at t' due to fitness change in the population,

$$\langle \xi_i' \rangle_{t'} = \Delta t \left(f_i' - \phi' \right), \tag{8.17}$$

in which

$$f_i' = f_i + \sum_j a_{ij} \frac{\mathrm{n}_j}{N} \xi_j, \tag{8.18}$$

and $\phi' = \sum_i \mathrm{n}_i f_i' / N$ as usual. We assume that apart from the mean value (8.17), fluctuations ξ_i' are uncorrelated with fluctuations ξ_i.

8.2.2.2 Relaxation response to fluctuations and effective fitness

The apparent fitness of any individual in a type i over two time lags, taking into account the relaxation of the population in response to previous fluctuations, is

$$\begin{aligned} \frac{\langle\langle \mathrm{n}_i'' - \mathrm{n}_i \rangle\rangle_{t',t}}{\mathrm{n}_i} &= \left\langle \langle \xi_i \rangle_{t'} \xi_i \right\rangle_t \\ &= \Delta t \left\langle \left(f_i' - \phi' \right) \xi_i \right\rangle_t \\ &\equiv 2 \Delta t \left(f_i^{\mathrm{eff}} - \phi^{\mathrm{eff}} \right). \end{aligned} \tag{8.19}$$

If we choose Δt to represent approximately the time over which fluctuation correlations have decayed, the *effective fitness* f_i^{eff} will correspond to the coefficient obtained by regression of the mean population change on genotypes that differ only by their correlations. The quantity determining the effective fitness in (8.19) is the correlation

$$\left\langle f_i' \xi_i \right\rangle_t = \sum_j a_{ij} \frac{\mathrm{n}_j}{N} \left\langle \xi_j \xi_i \right\rangle_t$$

$$= \sum_{j \neq i} \left(a_{ij} + 1 \right) \frac{\mathrm{n}_j}{N} \left\langle \xi_j \xi_i \right\rangle_t . \tag{8.20}$$

In the second line we have used the cancellation of the covariance terms in (8.15), together with the identity $a_{ii} \equiv -1$, $\forall i$.

In order to evaluate (8.20), it is necessary to obtain an estimate for the off-diagonal covariance terms $\left\langle \xi_j \xi_i \right\rangle_t$. To obtain this, we consider the off-diagonal counterpart to (8.19)

$$\left\langle \langle \xi_i' \rangle_{t'} \xi_k \right\rangle_t = \Delta t \left\langle \left(f_i' - \phi' \right) \xi_k \right\rangle_t$$

$$= \frac{\Delta t}{N} \left\{ a_{ik} \mathrm{n}_k \langle \xi_k^2 \rangle_t - \mathrm{n}_i \langle \xi_i \xi_k \rangle_t + \sum_{j \neq i,k} \mathrm{n}_j a_{ij} \left\langle \xi_j \xi_k \right\rangle_t + O\left(\frac{1}{N} \right) \right\}, \tag{8.21}$$

in which $O(1/N)$ subsumes terms from $\langle \phi' \xi_k \rangle_t$ which we do not write down.

We now make the self-consistency ansatz that for a suitable choice of Δt, the two terms

$$\left\langle \langle \xi_i' \rangle_{t'} \xi_k \right\rangle_t + \left\langle \langle \xi_k' \rangle_{t'} \xi_i \right\rangle_t \approx \langle \xi_i \xi_k \rangle_t \tag{8.22}$$

in steady state. These two terms are the relaxation responses to primary fluctuations caused by random pairing, during the time interval before the primary correlations have regressed to the mean.

From the self-consistency condition (8.22) we obtain a *Dyson equation* for the off-diagonal correlation terms,

$$\left(1 + \Delta t \frac{\mathrm{n}_i + \mathrm{n}_k}{N} \right) \langle \xi_i \xi_k \rangle_t - \frac{\Delta t}{N} \sum_{j \neq i,k} \mathrm{n}_j \left(a_{ij} \left\langle \xi_j \xi_k \right\rangle_t + \left\langle \xi_i \xi_j \right\rangle_t a_{jk} \right)$$

$$= \frac{\Delta t}{N} a_{ik} \left(\mathrm{n}_i \langle \xi_i^2 \rangle_t + \mathrm{n}_k \langle \xi_k^2 \rangle_t \right). \tag{8.23}$$

We now make use of the assumption that sufficiently many types are populated; that number conservation does not enforce stronger correlations between types than those induced by fluctuating fitness. Under this assumption $\mathrm{n}_i/N \ll 1, \forall i$. Then the leading term in the solution to the Dyson equation (8.23) is

$$\langle \xi_i \xi_k \rangle_t \approx \frac{\Delta t}{N} a_{ik} \left(\mathrm{n}_i \langle \xi_i^2 \rangle_t + \mathrm{n}_k \langle \xi_k^2 \rangle_t \right)$$

$$= \frac{(\Delta t)^3}{N} a_{ik} \sum_j \left(a_{ij}^2 + a_{kj}^2 \right) \frac{\mathrm{n}_j}{N}. \tag{8.24}$$

The second line in (8.24) makes use of (8.14). The important conclusion of (8.24) is that among non-identical types, the equal-time covariance of fluctuations $\langle \xi_i \xi_k \rangle_t$ has the sign of the pair-fitness function a_{ik}, weighted by the rate of primary fluctuation that is generated for the two types.

Plugging (8.24) back into the expressions (8.19, 8.20) for the two-timelag effective fitness yields

$$f_i^{\mathrm{eff}} = \frac{1}{2} \left\langle f_i' \xi_i \right\rangle_t$$

$$\approx \frac{(\Delta t)^3}{2N} \sum_{k \neq i} \left(a_{ik}^2 + a_{ik} \right) \frac{\mathrm{n}_k}{N} \sum_j \left(a_{ij}^2 + a_{kj}^2 \right) \frac{\mathrm{n}_j}{N}$$

$$\approx \frac{(\Delta t)^3}{2N} \sum_{k \neq i} a_{ik}^2 \frac{\mathrm{n}_k}{N} \sum_j \left(a_{ij}^2 + a_{kj}^2 \right) \frac{\mathrm{n}_j}{N}. \tag{8.25}$$

The approximation in the second line of (8.25) is that terms at $O(1/N)$ as well as higher-order terms in the Dyson equation (8.24) have been dropped from the expression for $\langle \xi_i \xi_k \rangle_t$. A further approximation at the same order is made in the third line, in which the linear term in a_{ik} sums to $O(1/N)$. The reason is that, for any two indices k and $-k$, defined so that $\sigma_k^m = -\sigma_{-k}^m, \forall m$, the pair fitness functions are defined to have the property $a_{ki} = -a_{-ki}, \forall i$. Meanwhile, $a_{kj}^2 = a_{-kj}^2$, so the argument in the sum on k is antisymmetric under $k \to -k$. If the sum were complete and the approximation (8.24) were applicable to $k = i$, then in any background n symmetric under $i \leftrightarrow -i$, the antisymmetric term would vanish entirely. The residual from complete vanishing is $-2\mathrm{n}_i \sum_j a_{ij}^2 \mathrm{n}_j / N^2$, which is $O(n_i/N)$ relative to the sum on symmetric terms in a_{ik}^2.

8.2.2.3 The heritable component of fitness

We may estimate the result of running a regression on the complicated sum (8.25), to obtain the heritable component of the effective fitness. Such estimates are qualitatively coarser than the $O(1/N)$ approximations that have been made up to now, in that dependencies on the fine structure of the population will be absorbed into overall prefactors.

The overlap function $a_{ik}^2 \sim 1$ for $k \sim i$ and $k \sim -i$, and zero for types with small correlation with i. Meanwhile, $a_{kj}^2 \sim a_{ij}^2, \forall j$, for all such k with large overlap. Therefore, an estimate for the effective fitness becomes

$$\begin{aligned} f_i^{\mathrm{eff}} &\sim \frac{(\Delta t)^3}{N} \sum_j \frac{\mathrm{n}_i}{N} \frac{\mathrm{n}_j}{N} a_{ij}^2 \\ &= \frac{(\Delta t)^3 \mathrm{n}_i}{N^2} \frac{1}{4N_{\mathrm{ply}}^2} \sum_{k,k'} \sigma_i^k \sigma_i^{k'} \mathrm{Cov}\left(\sigma^k, \sigma^{k'}\right). \end{aligned} \tag{8.26}$$

In the second line we have re-introduced appropriate factors f_i, which were set to zero for the balanced-allelic populations assumed in order to simplify the Langevin analysis, and we have made use of (8.10) to re-express the dependence on population composition $\{\mathrm{n}_j\}$ in terms of the allele covariance. We see that the heritable component of fitness is expected to be $\sim(\Delta t)\mathrm{n}_i/N^2$ times the variance in individual reproductive success over pair assignments as given in (8.10).

Although many approximations were invoked to remove higher-order correlations than the leading one, the resulting dynamic is straightforward to express and is what would be expected: fluctuations in genotype numbers n_i transiently polarize the population to favor the growth of genotypes that are complementary (heterotic at many loci) to the transiently increased type i. The strength of polarization is proportional to the reproductive variance given in (8.10). Positive covariance between fluctuations in the number n_i, and fluctuations in the complements that enhance the fitness of i through heterosis, lead to an *effective fitness*.

The effective fitness (8.26) is largest for genotypes that overlap with the covariance in the population, and their own reproduction then reinforces this covariance, leading to growth in the linkage of alleles in the population. The variance in total reproductive success, averaged over the population, is then proportional to

$$\sum_{i,j} \frac{\mathrm{n}_i}{N} \frac{\mathrm{n}_j}{N} \left(a_{ij} - \phi\right)^2 = \frac{1}{4N_{\mathrm{ply}}^2} \sum_{k,k'} \left(C_{k,k'}^2 - \langle \sigma^k \rangle_{\mathrm{n}} \langle \sigma^{k'} \rangle_{\mathrm{n}} \right). \tag{8.27}$$

The population-level variance (8.27) is maximized when $C_{k,k'} \to 1, \forall k, k'$ and $\langle \sigma^k \rangle_{\mathrm{n}} \to 0, \forall k$. This condition is met on any of $2^{(N_{\mathrm{ply}}-1)}$ independent solutions, each having exactly two population types that oppose each other at each locus. Which type carries which allele at a locus is otherwise arbitrary, reflecting the independence and L–R exchange symmetry of payoffs at each move. Populations made from two equal-size sub-populations of genotypes heterotic at all loci would be absorbing states of the replicator dynamic, in the absence of mutation or crossover.

8.3 Dressed genes and cooperativity within and between chromosomes

The analysis given in this chapter has been notably less complete than the analyses of simple games in the preceding three chapters. We have not attempted to write down a closed-form expression for the large-deviations behavior of distributions and where we have computed estimates for fitness or fluctuation properties we have relied on a number of simplifying conditions and assumptions that symmetric genotypes will have approximately mirror-image behavior. A more thorough analysis would be desirable, but in extensive-form games where the number of distinct strategies—and potentially the number of distinct stable or quasi-stable population states—grows with N_{ply}, analysis may become much more laborious than it is for simple games. Up to this point, we have treated the distribution over sequences implicitly and computed its consequences for fitness. We have not considered crossover and the implications for dynamics, which determine the stable distribution.

In the remaining sub-sections, we will estimate the phenomenological properties of distributions that result from the interaction of crossover with selection. As the previous section showed, selection acting alone would concentrate populations on distributions of two mirror-image genotypes; which two could be any sequence of N_{ply} letters L and R and its binary complement. Mutation and crossover, acting without selection, would randomize genomes and drive the population toward a uniform distribution over all genotypes. The tension between these two effects acting in opposition leads to a new dimensional property of ordered states, which is a characteristic correlation length for alleles within genomes. The correlated regions take over from the elementary loci to become the fundamental dynamical units transmitted under replication and preserved by selection. Their extended, statistical nature not only forces a conceptual departure from naïve genic selection interpretations, but also leads to a form of population dynamics by creep, which is a third category distinct from either escapes or simple Brownian motion in the previous chapters.

8.3.1 Crossover and the correlation length

[3] (chapter 6) shows that, if crossover is introduced with equal probability between any two adjacent plies k, k', the resulting stable distributions have correlation coefficients falling off as $C^2_{k,k'} \approx e^{-2|k-k'|/N_{\mathrm{corr}}}$. $C^2 = 1$ when $k = k'$ from the definition of variance and the fact that the population is equally balanced in the frequency of L and R alleles. For loci k' sufficiently near k that the two are likely to undergo many rounds of selection together before they are split by crossover, the allele at k' can come to predict the allele at k almost as well. Only among alleles that are well separated is crossover so likely to occur that selection can do little to maintain a pattern of co-occurrence between them. The correlation length N_{corr} is proportional to the ratio between the rate of crossover and the magnitude of the payoff per allele, which is $1/N_{\mathrm{ply}}$ in (8.2). Therefore a fixed probability of crossover between any two adjacent loci leads to a correlation length scaling as a fixed fraction of the whole-genome length, which defines the continuum-limit description of the chromosome as a unit of selection [3].

A finite fraction of a chromosome in a correlated state refers to a distribution over what may be an indefinite number of allelic states, becoming an infinite number of alleles as $N_{\mathrm{ply}} \to \infty$. In this asymptote, the contribution to fitness from the distribution goes to a fixed limit, even as the contribution from the bare gene goes to zero. The limiting distribution and limiting fitness are best regarded as properties of an *equivalence class* of chromosome models at large N_{ply}. The combination of symmetry and the scaling (8.2) in Lewontin's model allows us to construct the strongest possible contrast to genic selection within a model for which genes are still assumed as building blocks. Neither the bare gene, nor even any definite value of N_{ply}, matter in the limit, but only the stable properties of an equivalence class of distributions over alleles. The distribution, as the fundamental entity, defines the concept of the 'dressed gene' that we noted in the introduction to the chapter. Like dressed particles, dressed genes are fuzzy structures that extend over distances $\sim N_{\mathrm{corr}}$ on the chromosome. A single bare gene may contribute to the properties of a range of dressed genes, depending on its covariance with the centroid of the region of interest.

The dressed gene concept originates, in all important mathematical respects, from the same scaling behavior that defines dressed particles in renormalized condensed matter physics [7, 8]. Dressed genes are notably *not granular* even though the elementary bare genes are. Lewontin's construction is the more remarkable in that it was not conceived to imitate condensed matter physics. It defined the 'effective theory' of the chromosome from elementary assumptions and the problem of accounting for observed levels of heterozygosity with plausible levels of heterosis and other known mechanisms of heredity and selection.

8.3.2 The representations of the product symmetry group formed by ordered population states

The number of possible genotypes occupied with significant frequency in any such correlated equilibrium is $\sim 2^{(N_{\mathrm{ply}}-N_{\mathrm{corr}})}$. Therefore the number of independent and mutually exclusive population-state equilibria must scale as $\sim 2^{N_{\mathrm{corr}}}$. Any population equilibrium will be characterized by a pair of mirror images of a 'consensus' sequence which produce the greatest heterosis when the two complements are matched. A scaling in which the number of distinct macrostates grows exponentially with system size is characteristic of glassy systems. The growth rate is slower than that for the number of microstates; therefore the number of microstates per macrostate also grows exponentially and the macrostate stability can increase with scale, although not as quickly as it does for systems with simple symmetry breaking.

8.3.3 Cooperativity and creep

In the presence of both mutation and sampling fluctuations from matching, population states polarized around any given consensus pair are not asymptotically stable, because symmetry among all loci guarantees an exact degeneracy of all $2^{(N_{\mathrm{ply}}-1)}$ mirror-image genotypes as possible consensus pairs. Therefore polarized population states will be subject to drift on a timescale which is slower than the generation timescale and increasing with N_{corr}. Drift of this form in systems with large numbers of mutually exclusive equilibria and with very slow time constants, is sometimes given the name *creep*.

The number of microstates within an ordered population state should affect its mobility under creep. More concentrated distributions, resulting when mutants are more strongly suppressed by normalizing selection toward the consensus sequence, should have lower genetic variance and hence lower rates of drift. The reduction in population variance will be expressed directly as an increase in N_{corr}. The change in drift speed with the selection/crossover ratio should simply be a reflection of the greater standing variance seen in typical populations at any time. Strong crossover leads back to populations with little covariance within chromosomes, little variation in the members that can cooperatively support fitness and fast drift: in other words, to the loss of the linkage disequilibrium and stable heterozygosity.

Bibliography

[1] Williams G C 1966 *Adaptation and Natural Selection* (Princeton, NJ: Princeton University Press)

[2] Dawkins R C 1976 *The Selfish Gene* (New York: Oxford University Press)

[3] Lewontin R C 1974 *The Genetic Basis of Evolutionary Change* (New York: Columbia University Press)

[4] Gillespie J H 2004 *Population Genetics: A Concise Introduction* (Baltimore, MD: Johns Hopkins University Press)

[5] Cressman R 2003 *Evolutionary Dynamics and Extensive Form Games* (Cambridge, MA: MIT Press)

[6] Lehmann L and Rousset F 2009 Perturbation expansions of multilocus fixation probabilities for frequency-dependent selection with applications to the Hill–Robertson effect and to the joint evolution of helping and punishment *Theor. Population Biol.* **76** 35–51

[7] Anderson P W 1972 More is different *Sci., New Ser.* **177** 393–6

[8] Goldenfeld N 1992 *Lectures on Phase Transitions and the Renormalization Group* (Boulder, CO: Westview)

IOP Publishing

Eric Smith and Supriya Krishnamurthy

Chapter 9

One game theory

The preceding chapters set up a functional view of game theory that is defined from the mathematics of interaction, transmission and information. The solution concepts used are principally adaptive (or equivalently, Bayesian). Our development of the extensive form and recursive strategies, however, presents a very *material* point of entry into strategic solution concepts. Strategic solutions anticipate the future, but in real systems they rely on models to do so and those models must be created and maintained by reinforcement dynamics. In this final chapter we combine adaptive and strategic dynamics by studying recursive strategies on extensive-form games as an inductive counterpart to selection's deductive use of history. We review the fundamental theorem of evolutionary game theory and the folk theorems of repeated games as the two classes of results that, respectively, connect adaptive and strategic equilibria of the normal form and the extensive form with its generator. Using these we show how not only simple fixed points, but even complex mixed equilibria resulting from mutation, map to the correlated equilibria of repeated games in the strategic interpretation. Our aim is a formulation of game theory in which the basic abstractions correspond to robust, statistically and functionally defined entities, and in which convergence in population dynamics may draw on both adaptive and strategic mechanisms.

9.1 How do games arise in nature?

Sometimes purely technical analyses aimed at quantifying a class of phenomena can lead to a reconsideration of basic concepts. The previous chapters presented a framework to derive the large-deviations behavior of stochastic evolutionary games and exhibited a few small examples of the way the qualitative character of dynamics relates to symmetry. Our four simple game examples show how, in addition to creating several kinds of multiscale dynamics, stochasticity can serve as a filter, which separates out the aspects of dynamics that are robust against empirical mis-estimation, analytic

doi:10.1088/978-0-7503-1137-3ch9

approximation and some imprecision in scale, from an infinite collection of details that are fragile, difficult to determine and in most cases irrelevant. We have emphasized robustness and the role of symmetry as a classifier when we introduced basic concepts of either evolutionary dynamics or games, because most of the applications of evolutionary game theory will be to natural systems for which these abstractions are not given.

We have avoided narrative motivation or analogy as the basis for introducing the main abstractions of the formalism of evolutionary games in order to arrive at a system of concepts that are derived from robust functions. The resulting functional characterization of evolutionary games splits into three distinct and essential sub-systems. Both games and evolution presume a context of populations in which time is organized into generations. Games subsume the structure of interactions within generations and all the dimensions of function that are constructed in this interval but not directly persistent in time. Transmission determines what can persist by explicit replication across generation boundaries. Selection subsumes the processes that impart and maintain information in population states (or distributions over population states), within the constraints imposed by the game model of interaction and the transmission model of heredity. These three categories of function are extremely general and may be applied to a very broad range of natural phenomena. The categories are sensible without appeal to metaphor and this is essential since deriving the case-specific applications of concepts such as unit of selection, as chapters 7 and 8 showed in particular, can be a technically non-trivial exercise.

Beyond solution concepts It has been traditional to regard evolutionary stability as merely one among many solution concepts for games. The notion of imparting information to populations through games is somewhat less restrictive than a formal notion of solution concept, however, and we might wish to regard the three-part decomposition of population dynamics into interaction/transmission/selection as defining the emergence of games in natural phenomena, within which strategic as well as adaptive behavior can be described. This use of games becomes particularly interesting in cases where we do not wish to take concepts such as strategic intentionality as primitives, but rather to understand how they emerge from the causal dynamics of natural processes that are not by their own constitution strategic or intentional. Rather than splitting game theory into two branches defined by evolutionary versus strategic solution concepts, we seek an integrated concept of game in which the outcome of robust dynamics draws on both kinds of action.

In this final chapter, we will consider the relations of strategic to evolutionary convergence on solutions. We use repeated games (specifically the repeated prisoner's dilemma (RPD)) as an example, in order to again use a constructive, mathematically explicit mapping to show how a single architecture of games serves to organize both strategic and adaptive dynamics. The common element in repeated extensive-form games and in evolutionary updating is recursion.

We show that these two uses of recursion are, respectively, inductive and deductive. Strategic action in a (non-finite) repeated game entails an implicit or explicit model of an indefinitely extended future, yet such models at best can only be anchored in sample estimators from finite experience. Hence strategic action in real systems is unavoidably inductive. It contrasts with adaptation, under which a population state constitutes a model of uncertainty about the actual selective criteria, which is updated incrementally from each new sample of evidence. Arguing that any commitments to solutions of inductive problems involve trade-offs of efficacy, robustness and cost which are functions of model complexity, we explain what we think evolutionary solutions can contribute to regulating some of the difficult uses of induction in strategic game theory.

9.2 Repetition in games

Repeated games are a small sub-set of the extensive forms[1], generated by a (depth-independent) algorithm that recursively appends the same ply to each leaf in an existing game tree. The ply, in this context, is referred to as a *stage game*. It serves as the generator of structure for the whole extensive form and also exists as a finite game in its own right which may be used as a basis for comparative analysis of solution properties. Repeated games have been the most common platform to frame a variety of 'paradoxes' in game theory, most prominently the 'paradox of cooperation'.

The notion of paradox is possible because some solution concepts, such as the Nash equilibrium or evolutionary stable state (ESS), are sufficiently general that they can be applied to both the stage game and the repeated game. Solutions to the repeated game, which are histories, may be projected onto moves or distributions of moves for the stage game. Sometimes these projections are similar or identical to the equilibria of the stage game. The cases in which they explain a 'paradox' are those in which projected solutions to the repeated game can produce distributions of move profiles significantly different from the solutions to the stage game.

9.2.1 Three views of repetition in games

The role of repetition in games can be viewed three ways: as a partly arbitrary way to frame the notion of paradox; as a general framework that relates an algebra of complex solutions to the simple solutions of the generator of the algebra, or as a solution to the problem of induction. Although they seem very different, each is true from a particular perspective, and each emphasizes a different interpretation of which aspect of the real world repetition is meant to capture.

In what follows, when we refer to 'repeated games', we intend games with many plies, including cases with an indefinite or even an infinite number, which are thus quantitatively and even qualitatively distinct from the stage game.

[1] Recall that the extensive forms provide a representation of the unfolding of process for *all* fully defined games, including such structurally complex and heterogeneous cases as chess, Go and poker. In these combinatorially complex games, the different moves at each node lead the game down sub-trees with quite different topologies, different spaces of possible subsequent moves and different maps from paths of play to payoffs.

9.2.1.1 It is a straw man

From a strictly formal point of view that regards only the normal form of a game as the definition of the corresponding problem, the posing of 'paradoxes' by means of repeated games is a straw man. That is: the modeler introduces the stage game and its solutions as a reference for what is 'expected', but then declares that the real system is described not by the stage game but by the corresponding repeated game and that this is why the observed solutions 'paradoxically' differ from the expected ones.

The timescale and correlation length of an interaction is as much a defining property of the game that represents it as any of the moves and payoffs. In the normal form all of these properties are treated on an equal footing. If the empirically valid description of a natural system is the repeated game then it is not the stage game and vice versa. Since the normal forms of the repeated game and the stage game are drastically different in dimension and sometimes in structure, the projection of moves from the Nash equilibria of the repeated game onto moves (or distributions of moves) in the stage game need not correspond to the Nash equilibria of the stage game. In other words, if the repeated game is the valid description, the stage game is introduced into the discussion only as a mis-identification, and its equilibria are irrelevant.

9.2.1.2 It is a general-purpose framework

The formalist criticism that the stage game is unnecessary if the extensive-form game has been given is naïve but useful because it brings in other aspects of the normal form besides its equilibria, such as compressibility and information. The framework of repetition in extensive-form games is general. The information that distinguishes cases comes from the stage game that defines the generator of the extensive form. Another way to say this is that the stage game has an independent status as a *compression* of the extensive form to its generating algorithm. It does not immediately follow from the use of an equilibrium concept such as Nash equilibrium for the repeated game that the *same* equilibrium concept reflects the most informative comparison to the stage game. However, for simple stage games such as the prisoners' dilemma, the stage game has so little structure that it is hard to argue for much ambiguity in what information it carries.

The technical problem of repeated game analysis then becomes the construction of a kind of commutation diagram, as illustrated in figure 9.1. The stage game generates the extensive form by recursion. Nash equilibrium selects a sub-set of strategies in the extensive form, which may be mapped by projection onto the stage game. The comparison of these to the equilibria selected directly from the stage game then defines something like an algebra produced by recursion.

9.2.1.3 It is a form of induction

Both of the preceding points of view emphasize the normal form and selective criteria such as Nash equilibrium based on the normal form. The notion of 'paradox' comes from regarding equilibrium as a comparable notion when it is applied to the stage game and to the repeated game.

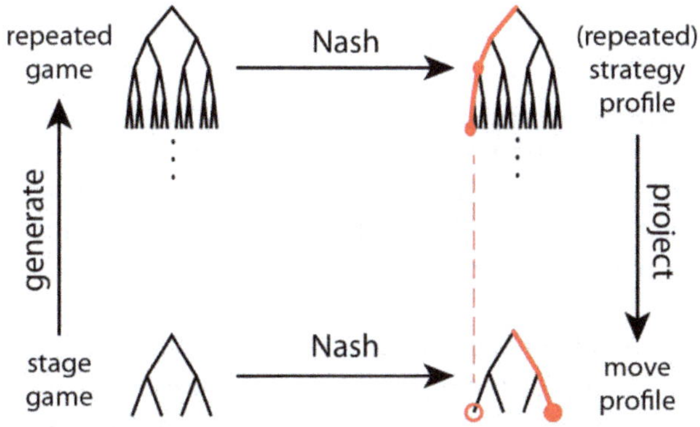

Figure 9.1. The commutation diagram relating stage games and repeated games through generic functions of recursive generation (left-hand side), Nash equilibrium (horizontal) and projection of move profiles (right-hand side). Nash equilibrium selects paths down either game tree (red lines). Move profiles at the terminal leaves of each stage (red dots) can be directly compared. The projection (red dashed line) of a distribution of outcomes from stages in the repeated game onto outcomes of the stage game (the unfilled, red-outlined leaf node) may or may not coincide with the Nash outcome for the stage game in isolation.

Rather than viewing the Nash equilibrium in the stage game and in the repeated game as forms of the 'same' solution concept, we could instead regard the stage game as fundamental in both cases and the repeated game as a way to shift from the Nash solution concept to a qualitatively different inductive solution concept. The stage game in this interpretation becomes a more complicated entity: one of its functions is to contribute a part of the total payoff from the outcome of each round; the other is to change the state of the complex of players by adding a step to their history, which they may be able to remember and which may therefore have consequences for the player. The *value*[2] of the stage within the larger game likewise has two components. One has the same denomination as the payoff that the game delivers during one round of play. The other derives from the state in which it leaves the players.

The inductive character of repeated game solutions is formalized if we limit the memory or complexity of strategies to be less than the multiplicity of states on the game tree, as illustrated in chapter 3. The adoption of a particular recursive move is a commitment to the set of input configurations that move can discriminate and the distribution over states of play that it will generate from any distribution of partners. Since the strategy is drawn from a smaller set of possibilities than the states of play of the game, it is an induction that must respond to more than one state of the world with the same action.

[2] In economic optimization problems, a (possibly multicomponent) function known as the *value* is introduced to decompose the solution of the complete problem into a series of recursive solutions to intermediate problems. The recursion relations are known as *Bellman equations* [1] and the process of solving them goes under the heading 'dynamic programming'.

9.2.2 Induction and deduction: an integrative view of repeated play in an evolutionary context

Our functional characterization of games at the level of interaction, transmission and information-incorporating processes (including but not limited to *ex post* selection) neither entails nor precludes high-level cognitive concepts such as strategic intentionality. These are higher-level abstractions, which must be defined operationally through further assumptions.

9.2.2.1 A player of a repeated game is a control system

One route to formalizing strategic complexity comes from control theory. The strategic agent is, *inter alia*, a control system that attempts to keep its environment within a certain sub-set of the environment's possible states. Note that this may not imply changing the environment—it is enough to control the agent's situation and the responses to its environment that determine how the environment acts on it—though sometimes change of the environment also occurs.

9.2.2.2 A strategy is a model of the world

James Conant and Ross Ashby argued [2, 3] that, tautologically, any control system contains an implicit or explicit *model* of the environment to be controlled, so we may identify strategic intentionality with the existence of such a model.

For an agent to be intentional, it is not necessary that a *separate* model of a goal state directs the selection of a strategy. The strategy itself, selected by whatever means, entails a prediction about what the environment may be and what sequence of responses to that environment will lead to a particular suite of outcomes.

9.2.2.3 Inductive models paid for through selection

Even very sophisticated cognitive systems are not infinitely foreward-seeing. They rely on internal models of finite complexity, which are built and maintained by what are ultimately reinforcement methods (perhaps acting very indirectly through many levels of structure). Where limits of complexity arise in real systems, they must result from the trade-off among the range of cases an inductive rule can discriminate, the marginal payoff value of more discrimination[3] and the ability of selection to support the costs of strategic complexity.

The problem of induction from a finite sample to an indefinite future is regarded in logic as having no general solution [6]. Yet every material system has no choice but to commit to inductive solutions, as it meets an unlimited world with a finite resource of internal complexity. An interesting class of problems for an integrated game theory is how the complexity of induction and selection is balanced and by which mechanisms each is served when both are present in a system. We will not

[3] This marginal value need not be positive. In the Axelrod tournaments [4], where greater complexity was permitted, the simple strategy tit-for-tat (TFT) remained among the most successful. Similar results are known for minority games [5] in which agents may be capable of resolving histories of limited depth as the inputs to selecting their next moves. Optimal depths exist for coordination in the group which leads to expected payoff.

address this question in any generality, but we can show under what mappings selection and induction construct equivalent solutions to a coordination problem.

9.2.3 Mirror relations in time and new complexities

Pure or mixed populations that are ESSs in the zero-mutation limit, corresponding to pure or mixed-strategy equilibria in the strategic formulation, are the universal scaffolding on which adaptive, Bayesian and Nash solution concepts converge. Because they are fixed points, they are independent of the direction or even the existence of time. Actual evolutionary-stable population states at non-zero mutation are deformations from these elementary fixed points in the presence of noise, which mixes optimization with degradation of any form of information. Identifying the relation between evolutionary solutions in the presence of mutation and finite-complexity inductive solutions, takes place in two stages. The first is a mapping from evolutionary to Nash fixed points of the repeated game. The second is interpreting the evolutionary solution as a particular inductive solution and referring the criteria for selection to the properties of the stage game.

9.2.3.1 The fundamental theorem of evolutionary game theory: selection and rationality arrive at the same solutions

Given a normal form, whether of a stage game or a repeated game, the fundamental theorem of evolutionary game theory [7, 8] establishes relations between the population states produced by replicator dynamics and the Nash equilibria of the corresponding non-cooperative strategic game[4] under models of rational choice. The following three correspondences hold between the evolutionary formulation and the rational-choice formulation:

1. The evolving population that embodies the diversity of its strategies in the set of possible types for agents corresponds to the single optimizing strategic player for whom diverse strategies are options of play. Mixed populations (of randomly matched players) correspond to mixed strategies; pure populations correspond to pure strategies.
2. Distributions over population states that become concentrated on a single (mixed or pure) state correspond to commitments by the rational agent to a single (mixed or pure) strategy.
3. The rest points of the evolutionary game equation correspond to non-cooperative (Nash) equilibria of the normal form; stable rest points correspond to strict Nash equilibria.

Therefore, given a re-formulation from a stage game to a repeated game, the replicator equation and the non-cooperative rational-choice equilibrium will extract fixed points from it in closely related ways. If we think of rational choice as a result of convergence

[4] More formally, *Nash equilibrium* refers to mutual-best response in the setting of a non-cooperative game. Here 'non-cooperative' signifies the lack of knowledge or control of any agent over any other agent's strategic choices, in contrast to 'cooperative' solution concepts such as those based on the coalitional form representation of payoffs [9]. We will often use the term *non-cooperative equilibrium* to emphasize this distinction.

under some form of Bayesian optimization, it may even be possible to formulate the rational-choice equilibrium as a kind of time reverse of the evolutionary equilibrium.

The fundamental theorem is a strong and general correspondence and it is part of the understanding of the way a common fixed-point structure underlies evolutionary and strategic solutions to games. However, in the case of repeated games, it relates evolutionary stability (a fixed point toward which populations converge under repeated selection) to stability of *normal-form* Nash equilibria, which are defined on the extensive-form game as an aggregate, one-shot interaction. To compare repetition in selection and in strategies, we must make the further step from the aggregate extensive form to the stage game, which introduces conceptually new problems.

9.2.3.2 The folk theorems of repeated game theory: when the normal form fails to provide a satisfying solution concept

The trouble that remains with repeated game theory, which has made the field an ongoing area of active work, is that the kind of *minimal* assumption of repetition required to make the projection of the repeated-game moves to the stage game qualitatively different from the stage-game Nash equilibria, generally admits 'too many' solutions. In order to avoid unraveling of repeated-game strategies by backward induction, a minimal assumption generally requires infinite (or at least indefinite) repetition. However, the set of inter-ply correlations made available by infinite repetition is so rich that an enumerably infinite set of strategy profiles, producing payoffs anywhere within a convex hull[5] from a sub-set of the stage-game payoffs, then become non-cooperative equilibria. The content of the *folk theorems* of rational-choice repeated game theory [10–14] is that any strategy profile within the *feasible, strictly individually rational* set, can be made a Nash equilibrium. If one wants to select a particular strategy profile, or a set with particular move profiles, further assumptions must be made and these may become as ad hoc as the outcome they select. The requirement for infinite repetition may be weakened by placing limits on players' abilities to perform induction, but these limits must then be justified. Alternatively, infinite repetition may be kept, but the selection of a particular strategy profile then requires introducing a complex and largely unformalized notion of 'prior agreement' about how adherence to the strategy is to be enforced. Because much of game theory is dedicated to the precise formalization of agency, interaction and strategy, notions of prior agreement that can be quite complex and that are not themselves grounded in mechanism seem unsatisfying[6].

[5] Payoffs falling within a convex hull arise in infinitely repeated games with discounted, additive payoff functions. The possibility to cycle through different stage-game moves at an infinite number of times within the course of play produces summed payoffs that may be infinitely many different convex combinations of stage-game payoffs.

[6] Here we endorse a point of view toward game theory expressed well by both Martin Shubik [9] and Herbert Gintis [15]. The main value added by a formal theory of games comes not from their capacity to re-iterate pre-existing narratives about incentive, but from the requirement they impose for a precise description of the structure of interaction and its consequences. Any theory of the source of incentives or the actions they induce should grow out of fully formulated assumptions about the states and processes available in the system. By this criterion, the concept of 'prior agreement' is much less well developed than either the game or the non-cooperative equilibrium concept that it is meant to supplement.

9.2.3.3 Why map evolution to rational choice?

Our example of RPD in chapter 7 yielded a simple fixed point (although with large fluctuations). The average population state produced by evolution (which is the best counterpart to an ESS in the presence of fluctuations) corresponds to a *mixed* strategy in the rational-choice interpretation. The fixed point extracted by evolution is simple partly because the space of population compositions is low-dimensional: only three recursive strategies are in competition, whereas in conventional rational-choice repeated game theory an indefinite number of recursive strategies would be admitted to consideration. At the same time, the mixed-population fixed point is complex enough that it has no elementary relation to non-cooperative equilibria of the stage game: not only are mixtures of C and D moves present; they are also used differently by agents with no memory and by agents with recursive strategies and their use by the same agent varies against different opponents. It is desirable to know what problem of induction that population solves optimally in the equivalent strategic construction. We therefore show that such a mapping can be constructed and consider its interpretations next.

9.2.4 The correspondence between evolutionary RPD and a particular rational-choice solution

To map an evolutionary solution for a mixed population onto a *particular* non-cooperative equilibrium of a repeated game, the folk theorems say we need two elements. The first is a definition of the feasible set of move profiles by the players. The second is something that functions as a 'trigger' strategy, which is compatible with individual rationality and can be adopted by the players in a way that precludes rational deviation from any pre-arranged move profile within that set. Variations in versions of the folk theorem often turn on the complexity of the trigger strategies that they employ [14].

An example is provided by the canonical RPD game developed in section 7.2, with types ALLC, ALLD and TFT, and deterministic moves. The move profiles produced, by any interactions among these, map directly to either persistent cooperation, or to a very simple form of trigger strategy used in proofs of folk theorems for Nash equilibria of the normal form for the repeated game [11, 13]. These folk theorems are therefore the appropriate versions for comparison.

For many games including this canonical RPD, the set of non-cooperative, mixed strategies of the stage game is too restrictive to define a feasible set that captures long-range correlation in the repeated game. The rational choice folk theorems therefore introduce additional models of 'public signals' that extend the notion of non-cooperative equilibrium to include specific forms of coordination, known as *correlated equilibria.*

To show that the evolutionary dynamic includes counterparts to both requirements of the corresponding folk theorem, we first show how neutrality involving ALLC, TFT and ALLD makes repeated move profiles in evolutionary RPD *strictly individually rational* by the strategic definition. We then show that the non-trivial coordination of recurrent moves within TFT does not require prior agreement to any temporally complex punishment strategies, but is equivalent to a particular use of a public signal in the rational-choice repeated game.

9.2.4.1 Strict individual rationality maps without difficulty

The neutrality structure of RPD ensures that strict individual rationality poses no difficulty in the map from evolutionary to strategic solution concepts. ALLC and TFT are neutral with each other in any combination of play between the two, while ALLD and TFT are neutral when played against ALLD. Thus 'punishment' is simply denying ALLD the gains of exploitation and is an instantaneous best response even in the stage game. It is therefore not necessary to invoke more complex finite-term punishment strategies or indefinite play as are required by sub-game perfect folk theorems.

9.2.4.2 The map from heterogeneous populations to mixed strategies with either cost-free signaling or public signaling

We may convert the probabilities of the four recurrent-stage outcomes, given in terms of population numbers in the replicator dynamic, into two forms which separate purely mixed from coordinated strategies:

$$\begin{bmatrix} P_{C/D} \\ P_{D/C} \\ P_{D/D} \\ P_{C/C} \end{bmatrix} = \frac{1}{N^2}\begin{bmatrix} n_C n_D \\ n_D n_C \\ n_D^2 + 2n_D n_T \\ (n_C + n_T)^2 \end{bmatrix}$$
$$= (1-\varphi)^2 \begin{bmatrix} \tilde{p}(1-\tilde{p}) \\ (1-\tilde{p})\tilde{p} \\ (1-\tilde{p})^2 \\ \tilde{p}^2 \end{bmatrix} + 2\varphi(1-\varphi)\begin{bmatrix} 0 \\ 0 \\ 1-\tilde{p} \\ \tilde{p} \end{bmatrix} + \varphi^2 \begin{bmatrix} 0 \\ 0 \\ 0 \\ 1 \end{bmatrix}$$
$$= (1-\Phi)\begin{bmatrix} p(1-p) \\ (1-p)p \\ (1-p)^2 \\ p^2 \end{bmatrix} + \Phi \begin{bmatrix} 0 \\ 0 \\ 1-p^2 \\ p^2 \end{bmatrix}. \tag{9.1}$$

The parameters in the second line of (9.1) are defined as

$$\varphi \equiv \frac{n_T}{N}$$
$$\tilde{p} \equiv \frac{n_C}{n_C + n_D}, \tag{9.2}$$

while those in the third line are defined as

$$\Phi \equiv \frac{n_T}{n_C + n_T}$$
$$p \equiv \frac{n_C + n_T}{N}. \tag{9.3}$$

We may re-formulate the interpretation of mixed play from a population-dynamic model into an equivalent model of a single type of optimizing player, able to use mixed strategies and also endowed with knowledge of either the opponent's state or of a global signal.

9.2.4.3 The interpretation in terms of opponent type identification

The interpretation of the second line in (9.1) is a literal translation of the effect that memory-one strategies have on the long-term payoff of the repeated game. $\varphi = n_T/N$ is a probability that a player has 'sight', and $1 - \varphi$ is the probability that he is 'blind'. Sight partitions types in the original population between n_T and anything else, and the residual $\{n_C, n_D\}$ identify a single type of blind player who plays a mixed strategy C/D with weights $(\tilde{p}, 1 - \tilde{p})$. Sight is local in that it may be treated as a binary variable sampled with probabilities $(\varphi, 1 - \varphi)$ for each player at the time of play. Thus, the idea that an individual may freely vary φ, in the same way as the population can vary $\tilde{p}$, maps the mutation–selection dynamic in a natural way onto the strategic concept of individual optimization. Mechanistically, (9.1) says that blind players, when paired, mix independently over C/D; a sighted player paired with a blind player coordinates on the blind player's strategy; two sighted players paired with each other not only coordinate but optimize on the Pareto-superior coordinated strategy.

This partition of the population gives the opening move in the three composite strategies the interpretation of a cost-free signal, which may be used to coordinate the subsequent recurrent behavior that determines fitness. The signal is not part of the 2×2 stage game itself and indeed requires repetition over a sufficient number of plies that the cost $\propto 1/N_{\text{ply}}$ may be made as small as necessary to be regarded as cost-free.

9.2.4.4 Interpretation as a correlated non-cooperative equilibrium

The interpretation of the third line in (9.1) replaces local sight with sight of a global signal, provided exogenously. Φ is the probability that the signal is provided in any single instantiation of the stage game, and $1 - \Phi$ is the probability that it is absent. All players simultaneously know whether the global signal exists or not, and if not, they independently mix over C/D with probabilities $(p, 1 - p)$, corresponding to the frequencies of the opening move in the population. In order to produce outcomes with the probabilities of the population model, we must suppose that if the signal exists, it takes some form such as a random variable uniformly distributed in the unit interval. Players coordinate on move C if the variable is above p^2, and move D otherwise. It is a property of this mapping from the population variables to a rational-choice setting that this 'threshold' use of the global signal is defined in a specific way from the same parameter p as the independent mixed strategies. Thus, again, players have a single parameter to optimize. The dynamics of the probability Φ must be translated from the population model in some other terms, for instance as a property of the game rather than an individual optimization parameter. While this interpretation requires greater complexity of the stage game and assumes a rather arbitrary use of the mixing probabilities, it has the feature of expanding the feasible set to include non-cooperative equilibria that coordinate on a globally visible signal, which are the correlated equilibria.

The feasible set defined by the range of values $\Phi \in [0, 1]$ is shown in figure 9.2. Each value of Φ corresponds to a ray from the vertex D to some point on the C–T line in figure 7.2. The value $\Phi = 0$ is the boundary of the feasible set formed by independent mixed strategies, which maps to the D–C line. The opposite boundary $\Phi = 1$ comprises the fully coordinated mixed strategies and maps to the D–T line.

Optimizers responsible only for p would move along the boundary $\Phi = 0$ away from C/C and toward D/D, as all mixed strategies are dominated by the pure-strategy Nash equilibrium of the stage game. Along the fully coordinated boundary

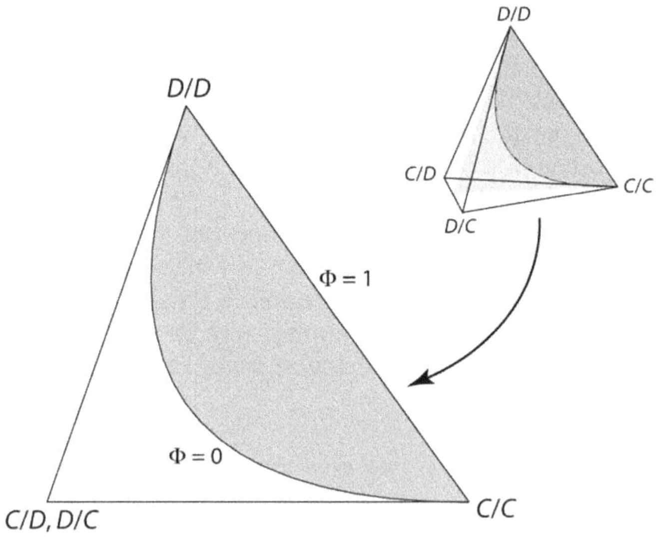

Figure 9.2. The four possible recurrent move profiles in RPD are C/C, D/D, C/D, D/C. All possible matchings of types produce frequencies for these move profiles shown in the three-dimensional probability simplex (inset). The set of role-symmetric feasible strategies is contained within the vertical plane through this simplex at $P_{C/D} = P_{D/C}$, shaded in the inset, and plotted by projecting the full simplex onto this plane in the main panel. Without a public signal to enable coordination, the feasible set is the one-dimensional set of products of mixed strategies, indicated by the heavy curve $\Phi = 0$. When players can use a public signal, the feasible set expands to the shaded lens between $\Phi = 0$ (no signal) and $\Phi = 1$ (signal present in every round). The contour $\Phi = 0$ maps in figure 7.2 onto population configurations on the lower axis between ALLD and ALLC, while the contour $\Phi = 1$ maps onto the upper-left axis between ALLD and TFT.

$\Phi = 1$ motion is away from *D*/*D* and toward *C*/*C*, because these are effectively coalitional strategies ordered by Pareto dominance. The payoff parameters (R, S, T, P), in a population with these three strategies, define a limit point of zero mutation rate along the C–T axis, which we associate with a value $\Phi_{\text{pop-dyn}}$.

The map to the strategic-game formulation requires us to regard Φ as an externally controlled signal frequency and p as a variable optimized by the players. Along each interior contour of fixed $\Phi \in \left(0, \Phi_{\text{pop-dyn}}\right)$, one or more stable interior optima exist for p in the repeated game[7]. This optimum converges to $p \to 1$ for all $\Phi \geqslant \Phi_{\text{pop-dyn}}$.

9.3 Games and the evolution of individuality

In evolutionary dynamics as it emerged from the modern synthesis, the abstraction of the autonomous model of the future—the algorithm for anticipating and responding to an environment, which exists in parallel with other slightly differing models, all of which are compared and then filtered according to fitness—is the *individual*. The individual's form dictates the limits of inductive complexity within the confine of a generation: its genotype encodes a set of elementary rules for converting an initial internal state and an ongoing sequence of environmental signals into a sequence of developmental stages and actions that, in aggregate, define the phenotype and its ecological role. The evolution of inductive strategies for responding to the events in the world is the evolution of the forms of individuality.

The framework for defining, analyzing and interpreting games that we have presented in this monograph may be carried much further into the realm of complexity than we have gone. The questions that we have addressed in very simplified models—how can aspects of phenotype function robustly and how can they be classified and estimated using coarse-grained models, across multiple scales of description?—provide a starting point for the larger problem of understanding the nature and evolution of individuality and ultimately moving beyond one concept to a richer and more empirically derived set of abstract models of organization. We close by noting some active areas of research concerning individuality, and suggest future directions for the use of games.

9.3.1 Evolution in the structure of individuality

The past decade in particular has seen rapid parallel advances in the ability to map and model developmental regulatory networks [16–19], to reconstruct patterns of evolutionary diversification from the paleontological record and comparative genomic analysis [20, 21] and to relate development and diversification to each other [22, 23] and to the ecological contexts that make them possible [24–26]. The picture of evolutionary dynamics that emerges from these studies is one in which elaboration in the structure of development leads to systematic change in the mechanisms

[7] Usually this stable optimum is unique. The reader may check from the graphs that along a narrow range of Φ, for the parameters used in the examples, this solution splits into an unstable rest point and two stable rest points at high and low p.

that generate evolutionary variation, which are essential to the structure of both micro- and macro-evolution.

Low-level regulatory elements, similar in character to elementary moves in an extensive-form game, are often introduced through duplication and divergence of existing components. As further downstream elaborations are appended to the developmental sequence, regulatory sub-systems become increasingly integrated into modules separated by standardized interfaces, as a necessary way to combine robustness and evolvability [27, 28]. As developmental programs become very complicated, the modules may partition into stages which have qualitatively different roles and which admit different categories of variation [23]. Very low-level modules may freeze into *kernels*, within which any variation creates non-viable organisms. Thus an entire category of variation may have no evolutionary potential and mechanisms may arise to preclude its occurrence, which has become a pure source of mortality. Elements that once led to novel and variable functions are thereby converted to constraints, comparable in their effect to external constraints from physical laws, though they originated internally. Intermediate-level modules may take on region-specific switching or gating roles and variations within them may lead to large-scale changes in phenotypic traits such as overall morphology (e.g., the famous innovation in 'body plans' of the Cambrian radiation [29]). While many dependences are among genes for regulatory elements within the genome of an individual cell lineage, it must be remembered that much of the functionality of robust regulatory systems derives from their ability to respond to signals from the context in which a cell lineage differentiates, whether that context comes from other cell populations within the organism, or from environmental factors[8].

These dynamical dependencies, which lie between elementary changes in genes and the generation of phenotype, are the structure in developmental programs that extensive-form games—or perhaps more compact representations such as finite-state machines onto which games can be projected, as illustrated in section 3.4.5—should be called upon to model. The complexity of the games needed to model real systems is suggested by the complexity of models that currently exist for morphogenetic regulatory networks [18, 19, 30]. These will not be simple repeated games, but structurally heterogeneous decision trees with semantically heterogeneous moves. Modeling the evolution of individuality will require defining rules not only for mutating moves within trees, or for crossing over sequences of moves within sub-trees, but also for growing and topologically re-arranging game trees and the information sets on them. The eligible topological transformations will depend on the semantic relations among moves, which will need to be chosen to reflect the functional divergences that are possible for different kinds of regulatory elements.

The modeling requirements are complex, and the degree to which they must appeal to the phenomenology of real systems is significant, but this does not preclude

[8] Examples of developmental robustness include the capacity in complex organisms to re-direct cell lineages to replace cells that are artificially removed in early embryogenesis. This kind of plasticity is absent in *eutelic* organisms such as *Caenorhabditis elegans*, which have tightly programmed cell lineage diversification, but it is essential to robust tissue formation in organisms with numbers of cells much larger than $\sim$1000.

a role for abstraction. The current forefront in understanding development in biology is aimed at precisely the kinds of functional abstraction, and the search for principles that generate robustness, that in simpler contexts have guided our use of symmetry and large-deviations limits.

9.3.2 Beyond individuality: building a richer set of foundations for evolutionary theory

The previous section suggests a number of ways we could enrich our representations of development and reproduction within essentially the formulations of population genetics and games we have reviewed in this manuscript. No major conceptual innovations are required, because these extensions remain within a framework where replicators can be defined at some level and where other organizational forms can be captured within game structures. In cases where organisms do not have well-defined fitness because they are not replicated but rather are *constructed*, as is the case for the heterosis model of chapter 8, we can represent the organism phenotype as the product of a game played by genes, while still treating the gene in formal models as a strict replicator for which fitness is well defined[9].

However, the extensions and improvements of models of development and selection within the Darwinian paradigm are ultimately only the tip of an iceberg, which is the conceptual problem of individuality itself. Not only Darwin's work, but also the formal development within the modern synthesis, was built around the abstraction of individuality. Yet for many forms of evolutionarily relevant biological order, the strong simplifying assumptions behind this concept are not valid and a parallel construction built on fundamentally different abstractions is warranted. (This is apparent already when the chromosome is a unit of selection, because a covariance that is not strictly replicated can still be an important determinant of reproductive success and a target of selection, which is difficult to formalize within conventional definitions.) Perhaps an emphasis that begins in more primitive concepts of agent, actor and explicit representations of steps in development and transmission can be a starting point for discovery of the appropriate abstractions.

9.3.2.1 The concept of individuality is inherently problematic

The concept of individuality that was unproblematic for Darwin is not one kind of relation, but several distinct kinds of relation [31] which happen to coincide with the organism, for the kinds of animals Darwin studied[10]. The organism furnished at the same time an integrated and autonomous unit in development, in physiology and to some degree in behavior; a collection of components that shared the same events of birth and death—and thus, for Darwin in a descriptive mode, the locus at which to define fitness—and a model of the form that was duplicated (with small changes) upon reproduction. Interestingly, the most fundamental concept in Darwin's framework—fitness—becomes undefined for sexually reproducing organisms when

[9] Whether or not this abstraction can be well-validated empirically remains another matter, as we noted in appendix 3.5. Whether the fitness assigned to genes is informative raises the other problem of limits of additive models, discussed in section 7.3.3.

[10] Their appropriateness becomes more difficult even for flowering plants and fungi, as reviewed in [31].

formalized as it was by Fisher [32, 33], a difficulty that sparked the work and also the disputes over genic selection. (See [34] for a compact review and sources.)

A century of natural history since Darwin has shown, however, that the three concepts of individual—in physiology and function, as locus of selection and as replicator—may exist in a given living system only at different levels of organization [31, 34]. In fact, in diverse and ancient clades of small multicellular plants and animals, there may be no level of organization or stage in the lifecycle that meets all relevant evolutionary criteria of individuality [35]. Therefore, any approach to evolution that takes the terms 'individual', 'fitness' and 'selection' as mandated, and then argues how the components in living systems should be shoehorned into these categories, is infeasible[11]. An alternative approach is to recognize that evolutionary dynamics requires both a richer and a more disaggregated suite of basic abstractions of order and corresponding statistical formalisms to handle them.

9.3.2.2 Inheriting the right insight from Fisher

We believe the right abstraction to inherit from Fisher's contribution to the formalization of evolution is not the definition or use of fitness, but rather the process of arriving at that definition. Fisher's insight was to show that a fitness concept was well defined by formulating it as a summary statistic from a diachronic population process [33]. Once the definition was shown to be consistent for a certain class of populations (those with strict replicators, in which each offspring has a unique parent), the questions whether

1. this summary statistic was an *informative* sample estimator for the dynamics of traits carried by individuals and whether it was a *sufficient statistic* for some problems of inference; and
2. whether sample estimators from *synchronic* properties of the individual's type and the population context could be found, which provided good estimators for fitness obtained as a summary statistic in the histories of those populations,

could be pursued as statistical problems without conceptual ambiguity.

The specific statistic used by Fisher is only available for systems in which descent takes a particular form. This does not mean that, for other systems, such as diploid organisms or chromosomes with crossover or epigenetic modifications, other summary statistics might not be informative or sufficient for their population dynamics—only that those summary statistics will not be the same as Fisher's formulation of fitness.

It seems to us that an empirically motivated program begins in the description of the forms of integration and autonomy within the life course that determine survival and reproduction, and of the transmission model, at all the levels where these may be found. The lesson to be learned from Fisher is to search within the summary statistics for the diachronic process, for those which are informative about the

[11] It also appears distinctively non-scientific, to begin with the abstraction and try to partition reality to respect it, rather than beginning with the facts of phenomena and then asking which abstractions can be formed that remain faithful to those facts.

regularities in the population dynamics and which leave residuals that are well explained as random sample fluctuations. One then addresses the problem of causal modeling: whether the informative summary statistics can be well approximated with sample estimators computed from the instantaneous state of the system of interest and its environmental context. Much as Lewontin's linked heterosis model derived the 'dressed gene' (our term, from chapter 8) as a new kind of evolutionary entity, we anticipate that correspondingly new summary statistics will become informative for its population dynamics. These may be acceptably modeled by a Fisherian definition of fitness, as they are for the dressed gene if one does not need to compute the detailed effects of crossover within a correlated region, or they may be much different.

We regard these areas as parts of the frontier where mathematical proofs of concept, such as we have illustrated here, make contact with empirically grounded scientific explanation.

Bibliography

[1] Bellman R E 1957 *Dynamic Programming* (Princeton, NJ: Princeton University Press)

[2] Conant R C and Ashby W R 1970 Every good regulator of a system must be a model of that system *Int. J. Syst. Sci.* **1** 89–97

[3] Ashby W R 1956 *An Introduction to Cybernetics* (London: Chapman and Hall)

[4] Axelrod R 1997 *The Complexity of Cooperation* (Princeton, NJ: Princeton University Press)

[5] Challet D, Marsili M and Zhang Y-C 2004 *Minority Games: Interacting Agents in Financial Markets* (London: Oxford University Press)

[6] Russell B 1967 *A History of Western Philosophy* (New York: Simon and Schuster)

[7] Smith J M and Price G R 1973 The logic of animal conflict *Nature* **246** 15–18

[8] Hofbauer J and Sigmund K 1998 *Evolutionary Games and Population Dynamics* (New York: Cambridge University Press)

[9] Shubik M 1984 *Game Theory in the Social Sciences: Concepts and Solutions* (Cambridge, MA: MIT Press)

[10] Aumann R J and Shapley L S 1994 Long-term competition—a game-theoretic analysis in *Essays in Game Theory: In Honor of Michael Maschler* ed N Megiddo (Heidelberg: Springer) pp 1–15

[11] Fudenberg D and Maskin E 1991 The folk theorem in repeated games with discounting or with incomplete information *Econometrica* **54** 533–54

[12] Fudenberg D and Maskin E 1991 On the dispensibility of public randomization in discounted repeated games *J. Econ. Theory* **53** 428438

[13] Rubinstein A and Wolinsky A 1995 Remarks on infinitely repeated extensive-form games *Games Econ. Behavior* **9** 110–5

[14] Ratliff J 2010 A folk theorem sampler pp 1–34 http://www.virtualperfection.com/gametheory/Section5.3.html

[15] Gintis H 2009 *Game Theory Evolving: A Problem-Centered Introduction to Modeling Strategic Interaction* 2nd edn (Princeton, NJ: Princeton University Press)

[16] Davidson E H 2006 *The Regulatory Genome: Gene Regulatory Networks in Development and Evolution* (San Diego, CA: Academic Press)

[17] Oliveri P, Tu Q and Davidson E H 2008 Global regulatory logic for specification of an embryonic cell lineage *Proc. Natl Acad. Sci. USA* **105** 5955–62

[18] Smith J and Davidson E H 2008 Gene regulatory network subcircuit controlling a dynamic spatial pattern of signaling in the sea urchin embryo *Proc. Natl. Acad. Sci. USA* **105** 20089–94

[19] Ben-Tabou de Leon S and Davidson E H 2009 Experimentally based sea urchin gene regulatory network and the causal explanation of developmental phenomenology *Wiley Interdiscip. Rev. Syst. Biol. Med.* **1** 237–46

[20] Erwin D H, Laflamme M, Tweedt S M, Sperling E A, Pisani D and Peterson K J 2011 The Cambrian conundrum: early divergence and later ecological success in the early history of animals *Science* **334** 1091–7

[21] Erwin D H 2012 Macroevolution: dynamics of diversity *Curr. Biol.* **21** R1000–1

[22] Davidson E H and Erwin D H 2006 Gene regulatory networks and the evolution of animal body plans *Science* **311** 796–800

[23] Erwin D H and Davidson E H 2009 The evolution of hierarchical gene regulatory networks *Nat. Rev. Genetics* **10** 141–8

[24] Dunne J A, Williams R J, Martinez N D, Wood R A and Erwin D H 2008 Compilation and network analyses of cambrian food webs *PLoS Biol.* **6** e102

[25] Erwin D H and Tweedt S 2012 Ecological drivers of the Ediacaran-Cambrian diversification of metazoa *Evol. Ecol.* **26** 417–33

[26] Erwin D H and Valentine J W 2013 *The Cambrian Explosion: The Construction of Animal Biodiversity* (Englewood, CO: Roberts and Company)

[27] Gerhart J and Kirschner M 1997 *Cells, Embryos, and Evolution* (New York: Wiley)

[28] Gerhart J and Kirschner M 2007 The theory of facilitated variation *Proc. Natl Acad. Sci. USA* **104** 8582–9

[29] Gould S J 1989 *Wonderful Life* (New York: Norton)

[30] von Dassow G, Meir E, Munro E M and Odell G M 2000 The segment polarity network is a robust developmental module *Nature* **406** 188

[31] Clarke E 2010 The problem of biological individuality *Biol. Theory* **5** 312–25

[32] Fisher R A 2000 *The Genetical Theory of Natural Selection* (London: Oxford University Press)

[33] Frank S A and Slatkin M 1992 Fisher's fundamental theorem of natural selection *Trends Ecol. Evol.* **7** 92–5

[34] Hull D L 1980 Individuality and selection *Annu. Rev. Ecol. Systematics* **11** 311–32

[35] Herron M D, Rashidi A, Shelton D E and Driscoll W W 2013 Cellular differentiation and individuality in the 'minor' multicellular taxa *Biol. Rev.* **88** 844–61

Lightning Source UK Ltd.
Milton Keynes UK
UKHW050620060821
388415UK00003B/79

9 780750 311380